PEETER JOOT

GEOMETRIC ALGEBRA FOR ELECTRICAL ENGINEERS.

GEOMETRIC ALGEBRA FOR ELECTRICAL ENGINEERS.

PEETER JOOT

Multivector Electromagnetism.

August 2025 – version V0.3.6.4

Peeter Joot: *Geometric Algebra for Electrical Engineers.*, Multivector Electromagnetism., © August 2025

COPYRIGHT

Copyright ©2025 Peeter Joot All Rights Reserved

This book may be reproduced and distributed in whole or in part, without fee, subject to the following conditions:

- The copyright notice above and this permission notice must be preserved complete on all complete or partial copies.

- Any translation or derived work must be approved by the author in writing before distribution.

- If you distribute this work in part, instructions for obtaining the complete version of this document must be included, and a means for obtaining a complete version provided.

- Small portions may be reproduced as illustrations for reviews or quotes in other works without this permission notice if proper citation is given.

Exceptions to these rules may be granted for academic purposes: Write to the author and ask.

Disclaimer: I confess to violating somebody's copyright when I copied this copyright statement.

DOCUMENT VERSION

Version V0.3.6.4

Sources for this notes compilation can be found in the github repository https://github.com/peeterjoot/GAelectrodynamics

The last commit (Aug/11/2025), associated with this pdf was 8718c242ed0b070c3fa6232de4146b1832c7d7a6

Should you wish to actively contribute typo fixes (or even more significant changes) to this book, you can do so by contacting me, or by forking your own copy of the associated git repositories and building the book pdf from source, and submitting a subsequent merge request.

```bash
#!/bin/bash

git clone git@github.com:peeterjoot/latex-notes-
    compilations.git peeterjoot
cd peeterjoot

submods="figures/GAelectrodynamics figures/gabook
    mathematica GAelectrodynamics gapauli latex
    frequencydomain"
for i in $submods ; do
    git submodule update --init $i
    (cd $i && git checkout master)
done

export PATH=`pwd`/latex/bin:$PATH

cd GAelectrodynamics
make scrpage2.sty parameters.sty mmacells.sty all
```

I reserve the right to impose dictatorial control over any editing and content decisions, and may not accept merge requests as-is, or at all. That said, I will probably not refuse reasonable suggestions or merge requests.

Dedicated to:
Aurora and Lance, my awesome kids, and
Sofia, who not only tolerates and encourages my studies, but is also awesome enough to think that math is sexy.

PREFACE

Why you want to read this book. When you first learned vector algebra you learned how to add and subtract vectors, and probably asked your instructor if it was possible to multiply vectors. Had you done so, you would have been told either "No", or a qualified "No, but we can do multiplication like operations, the dot and cross products." This book is based on a different answer, "Yes." A set of rules that define a coherent multiplication operation are provided.

Were you ever bothered by the fact that the cross product was only defined in three dimensions, or had a nagging intuition that the dot and cross products were related somehow? The dot product and cross product seem to be complimentary, with the dot product encoding a projection operation (how much of a vector lies in the direction of another), and the magnitude of the cross product providing a rejection operation (how much of a vector is perpendicular to the direction of another). These projection and rejection operations should be perfectly well defined in 2, 4, or N dimensions, not just 3. In this book you will see how to generalize the cross product to N dimensions, and how this more general product (the wedge product) is useful even in the two and three dimensional problems that are of interest for physical problems (like electromagnetism.) You will also see how the dot, cross (and wedge) products are all related to the vector multiplication operation of geometric algebra.

When you studied vector calculus, did the collection of Stokes's, Green's and Divergence theorems available seem too random, like there ought to be a higher level structure that described all these similar operations? It turns out that such structure is available in the both the language of differential forms, and that of tensor calculus. We'd like a toolbox that doesn't require expressing vectors as differentials, or resorting to coordinates. Not only does geometric calculus provides such a toolbox, it also provides the tools required to operate on functions of vector products, which has profound applications for electromagnetism.

Were you offended by the crazy mix of signs, dots and cross products in Maxwell's equations? The geometric algebra form of Maxwell's equation resolves that crazy mix, expressing Maxwell's equations as a single equation. The formalism of tensor algebra and differential forms also provide simpler ways of expressing Maxwell's equations, but are arguably harder to relate to the vector algebra formalism so familiar to electric engineers and physics practitioners. In this book, you will see how to work with the geometric algebra form of Maxwell's equation, and how to relate these new techniques to familiar methods.

Overview. Geometric algebra generalizes vectors, providing algebraic representations of not just directed line segments, but also points, plane segments, volumes, and higher degree geometric objects (hypervolumes.). The geometric algebra representation of planes, volumes and hypervolumes requires a vector dot product, a vector multiplication operation, and a generalized addition operation. The dot product provides the length of a vector and a test for whether or not any two vectors are perpendicular. The vector multiplication operation is used to construct directed plane segments (bivectors), and directed volumes (trivectors), which are built from the respective products of two or three mutually perpendicular vectors. The addition operation allows for sums of scalars, vectors, or any products of vectors. Such a sum is called a multivector.

The power to add scalars, vectors, and products of vectors can be exploited to simplify much of electromagnetism. In particular, Maxwell's equations for isotropic media can be merged into a single multivector equation

$$\left(\nabla + \frac{1}{c}\frac{\partial}{\partial t}\right) F = J, \tag{0.1}$$

where

- ∇ is the gradient,
- $c = 1/\sqrt{\mu\epsilon}$ is the group velocity for waves in the media (i.e. the speed of light),
- $F = \mathbf{E} + Ic\mathbf{B}$ is the multivector electromagnetic field that combines the electric ($\mathbf{E}$) and magnetic field ($\mathbf{B}$) into a single entity,
- $J = \eta\left(c\rho - \mathbf{J}\right)$ is the multivector current, combining the charge density (ρ) and the current density ($\mathbf{J}$) into a single entity,
- $I = \mathbf{e}_1\mathbf{e}_2\mathbf{e}_3$ is the ordered product of the three $\mathbb{R}^3$ basis vectors, and
- $\eta = \sqrt{\mu/\epsilon}$ is the impedance of the media.

Encountering Maxwell's equation in its geometric algebra form leaves the student with more questions than answers. Yes, it is a compact representation, but so are the tensor and differential forms (or even the quaternionic) representations of Maxwell's equations. The student needs to know how to work with the representation if it is to be useful. It should also be clear how to use the existing conventional mathematical tools of applied electromagnetism, or how to generalize those appropriately. Individually, there are answers available to many of the questions that are generated attempting to apply the theory, but they are scattered and in many cases not easily accessible.

Much of the geometric algebra literature for electrodynamics is presented with a relativistic bias, or assumes high levels of mathematical or physics sophistication. The aim of this work was an attempt to make the study of electromagnetism using geometric algebra more accessible, especially to other dumb engineers[1] like myself.

What's in this book. This book introduces the fundamentals of geometric algebra and calculus, and applies those tools to the study of electromagnetism. Geometric algebra extends vector algebra by introducing a vector multiplication operation, the vector product, incorporating aspects of both the dot and cross products. Products or sums of products of vectors are called multivectors, and are capable of representing oriented point, line, plane, and volume segments.

This book is divided into three parts.

Chapter-1. An introduction to geometric algebra (GA). Topics covered include vectors, vector spaces, vector multiplication, bivectors, trivectors, multivectors, multivector spaces, dot and wedge products, multivector representation of complex numbers, rotation, reflection, projection and rejection, and linear system solution.

The focus of this book are geometric algebras generated from 2 or 3 dimensional Euclidean vector spaces. In some cases higher dimensional spaces will be used in examples and theorems. Some, but not all, of the places requiring generalizations for mixed signature (relativistic) spaces will be pointed out.

---

[1] Sheldon: "Engineering. Where the noble semiskilled labourers execute the vision of those who think and dream. Hello, Oompa-Loompas of science."

Chapter-2. Geometric calculus, Green's function solutions of differential equations, and multivector Green's functions. A multivector generalization of vector calculus, the fundamental theorem of geometric calculus, is required to apply geometric algebra to electromagnetism. Special cases of the fundamental theorem of geometric calculus include the fundamental theorem of calculus, Green's (area) theorem, the divergence theorem, and Stokes' theorems. Multivector calculus also provides the opportunity to define a few unique and powerful (multivector) Green's functions of particular relevance to electromagnetism.

Chapter-3. Application of Geometric Algebra to electromagnetism. Instead of working separately with electric and magnetic fields, we will work with a hybrid multivector field, F, that includes both electric and magnetic field contributions, and with a multivector current, J, that includes both charge and current densities.

Starting with the conventional form of Maxwell's equation, the multivector Maxwell's equation (singular) is derived. This is a single multivector equation that is easier to solve and manipulate than the conventional mess of divergence and curl equations that are familiar to the reader. The multivector Maxwell's equation is the starting point for the remainder of the analysis of the book, and from it the wave equation, plane wave solutions, and static and dynamic solutions are derived. The multivector form of energy density, Poynting force, and the Maxwell stress tensor, and all the associated conservation relationships are derived. The transverse and propagation relationships for waveguide solutions are derived in their multivector form. Polarization is discussed in a multivector context, and multivector potentials and gauge transformations are introduced.

No attempt to motivate Maxwell's equations, nor most of the results derived from them is made in this book.

Prerequisites: The target audience for this book is advanced undergraduate or graduate students of electrical engineering or physics. Such an audience is assumed to be intimately familiar with vectors, vector algebra, dot and cross products, determinants, coordinate representation, linear system solution, complex numbers, matrix algebra, and linear transformations. It is also assumed that the reader understands and can apply conventional vector calculus concepts including the divergence and curl operators, the divergence and Stokes' theorems, line, area and volume integrals, Greens' functions, and the Dirac delta function. Finally, it is assumed that the reader is intimately familiar with conventional electromagnetism, including Maxwell's and the Lorentz force equations, scalar and vector potentials, plane wave solutions, energy density and Poynting vectors, and more.

Thanks: Portions of this book were reviewed or corrected by Steven De Keninck, Dr. Wolfgang Lindner, Prof. Mo Mojahedi, Prof. Alan Macdonald, Prof. Quirino Sugon Jr., Miroslav Josipović, Bruce Gould, Tim Put, David Bond, Bill Ignatiuk, Sigmundur, Zhengbang Zhou, Jack Paladin, Nicky, D, Foreest, Peter Eriksen, Christopher, Wrenn Wooten, Prof. Norman Derby, prlw1 (on github), Ryan Mohseni, Nicholas Dwork, Rasmus Enevoldsen, Frank Dininno, and Timo van Veen. . I'd like to thank everybody who provided me with any feedback (or merge-requests!) This feedback has significantly improved the quality of the text.

Peeter Joot   peeterjoot@pm.me

CONTENTS

Preface xi

1 GEOMETRIC ALGEBRA. 1
   1.1 Prerequisites. 1
      1.1.1 Vector. 1
      1.1.2 Vector space. 3
      1.1.3 Basis, span and dimension. 5
      1.1.4 Standard basis, length and normality. 7
   1.2 Multivectors. 9
   1.3 Colinear vectors. 17
   1.4 Orthogonal vectors. 17
   1.5 Some nomenclature. 19
   1.6 Two dimensions. 21
   1.7 Plane rotations. 23
   1.8 Duality. 29
   1.9 Vector product, dot product and wedge product. 31
   1.10 Reverse. 40
   1.11 Complex representations. 41
   1.12 Multivector dot product. 43
      1.12.1 Dot product of a vector and bivector 45
      1.12.2 Bivector dot product. 47
      1.12.3 Problems. 49
   1.13 Permutation within scalar selection. 49
   1.14 Multivector wedge product. 50
      1.14.1 Problems. 53
   1.15 Projection and rejection. 53
   1.16 Normal factorization of the wedge product. 58
   1.17 The wedge product as an oriented area. 59
   1.18 General rotation. 62
   1.19 Symmetric and antisymmetric vector sums. 65
   1.20 Reflection. 67
   1.21 Linear systems. 69
   1.22 Problem solutions. 77

2 MULTIVECTOR CALCULUS. 93
   2.1 Reciprocal frames. 93

- 2.1.1 Motivation and definition. 93
- 2.1.2 $\mathbb{R}^2$ reciprocal frame. 97
- 2.1.3 $\mathbb{R}^3$ reciprocal frame. 100
- 2.1.4 Problems. 101
- 2.2 Curvilinear bases. 102
  - 2.2.1 Two parameters. 102
  - 2.2.2 Three (or more) parameters. 106
  - 2.2.3 Gradient. 107
  - 2.2.4 Vector derivative. 109
  - 2.2.5 Examples. 110
  - 2.2.6 Problems. 120
- 2.3 Integration theory. 121
  - 2.3.1 Line integral. 121
  - 2.3.2 Surface integral. 124
  - 2.3.3 Volume integral. 126
  - 2.3.4 Bidirectional derivative operators. 127
  - 2.3.5 Fundamental theorem. 128
  - 2.3.6 Stokes' theorem. 130
  - 2.3.7 Fundamental theorem for Line integral. 131
  - 2.3.8 Fundamental theorem for Surface integral. 132
  - 2.3.9 Fundamental theorem for Volume integral. 139
- 2.4 Vector calculus identities. 146
  - 2.4.1 Curl. 146
  - 2.4.2 Chain rule identities. 150
  - 2.4.3 Problems. 153
- 2.5 Multivector Fourier transform and phasors. 154
- 2.6 Green's functions. 155
  - 2.6.1 Motivation. 155
  - 2.6.2 Green's function solutions. 159
  - 2.6.3 Helmholtz equation. 162
  - 2.6.4 First order Helmholtz equation. 164
  - 2.6.5 Spacetime gradient. 166
- 2.7 Helmholtz theorem. 168
- 2.8 Problem solutions. 171

3 ELECTROMAGNETISM. 183
- 3.1 Conventional formulation. 183
  - 3.1.1 Problems. 186
- 3.2 Maxwell's equation. 186

- 3.3 Wave equation and continuity. 189
- 3.4 Plane waves. 192
- 3.5 Statics. 196
  - 3.5.1 Inverting the Maxwell statics equation. 196
  - 3.5.2 Enclosed charge. 198
  - 3.5.3 Enclosed current. 199
  - 3.5.4 Example field calculations. 202
- 3.6 Dynamics. 217
  - 3.6.1 Inverting Maxwell's equation. 217
- 3.7 Energy and momentum. 219
  - 3.7.1 Field energy and momentum density and the energy momentum tensor. 219
  - 3.7.2 Poynting's theorem (prerequisites.) 224
  - 3.7.3 Poynting theorem. 229
  - 3.7.4 Examples: Some static fields. 232
  - 3.7.5 Complex energy and power. 236
- 3.8 Lorentz force. 238
  - 3.8.1 Statement. 238
  - 3.8.2 Constant magnetic field. 241
- 3.9 Polarization. 242
  - 3.9.1 Phasor representation. 242
  - 3.9.2 Transverse plane pseudoscalar. 243
  - 3.9.3 Pseudoscalar imaginary. 250
- 3.10 Transverse fields in a waveguide. 252
  - 3.10.1 Problems. 255
- 3.11 Multivector potential. 256
  - 3.11.1 Definitions. 256
  - 3.11.2 Gauge transformations. 259
  - 3.11.3 Far field. 262
  - 3.11.4 Problems. 266
- 3.12 Dielectric and magnetic media. 267
  - 3.12.1 Statement. 267
  - 3.12.2 Alternative form. 269
  - 3.12.3 Gauge like transformations. 271
  - 3.12.4 Boundary value conditions. 273
- 3.13 Problem solutions 276

A DISTRIBUTION THEOREMS. 289

B	PROOF SKETCH FOR THE FUNDAMENTAL THEOREM OF GEOMETRIC CALCULUS. 293
C	GREEN'S FUNCTIONS. 297
	c.1 Helmholtz operator. 297
	c.2 Delta function derivatives. 300
D	ENERGY MOMENTUM TENSOR (VECTOR.) 303
E	DIFFERENTIAL FORMS COMPARISON. 305
F	HELPFUL FORMULAS. 309
	F.1 Vector relations. 309
	F.2 Blades. 310
	F.3 Multivectors. 311
	F.4 Vector calculus identities. 311

INDEX 313

BIBLIOGRAPHY 317

LIST OF FIGURES

Figure 1.1	Scalar multiples of vectors.	1
Figure 1.2	Addition of vectors.	2
Figure 1.3	Coordinate representation of vectors.	2
Figure 1.4	Graphical representation of bivector addition in the plane.	12
Figure 1.5	Bivector addition.	12
Figure 1.6	Unit bivectors for $\mathbb{R}^3$	13
Figure 1.7	Sum of orthogonal vectors.	18
Figure 1.8	Multiplication by $\mathbf{e}_1\mathbf{e}_2$.	23
Figure 1.9	$\pi/2$ rotation in the plane using pseudoscalar multiplication.	24
Figure 1.10	Rotation in a plane.	25
Figure 1.11	Radial vector in polar coordinates.	26
Figure 1.12	$\mathbb{R}^3$ duality illustration.	30
Figure 1.13	Two vectors in a plane.	32
Figure 1.14	Equilateral triangle in $\mathbb{R}^3$.	34
Figure 1.15	Projection and rejection illustrated.	54
Figure 1.16	Parallelogram representations of wedge products.	61
Figure 1.17	Different shape representations of a wedge product.	61
Figure 1.18	Parallelogram area.	62
Figure 1.19	Rotation with respect to the plane of a pseudoscalar.	66
Figure 1.20	Reflection.	68
Figure 1.21	Intersection of two lines.	73
Figure 1.22	Static load with two members.	74
Figure 1.23	Perpendicular bivectors.	76
Figure 2.1	Oblique and reciprocal bases.	96
Figure 2.2	Contours for an elliptical region.	102
Figure 2.3	Differentials for an elliptical parameterization.	103
Figure 2.4	Two parameter manifold.	106
Figure 2.5	Polar coordinates.	110
Figure 2.6	Spherical polar conventions.	115
Figure 2.7	Spherical polar unit vectors.	116
Figure 2.8	Toroidal parameterization.	119

Figure 2.9	One parameter manifold.	122
Figure 2.10	Two parameter manifold differentials.	125
Figure 2.11	Contour for two parameter surface boundary.	135
Figure 2.12	Sum of infinitesimal loops.	136
Figure 2.13	Three parameter volume element.	141
Figure 2.14	Differential surface of a volume.	143
Figure 2.15	Curl example 1.	147
Figure 2.16	Curl example 2.	147
Figure 2.17	Curl of 3D vector field.	148
Figure 2.18	Contours for harmonic oscillator Green's function.	158
Figure 3.1	Line charge density.	203
Figure 3.2	Rectangular Gaussian pillbox.	208
Figure 3.3	Circular line charge.	209
Figure 3.4	Field due to a circular distribution.	211
Figure 3.5	(a) $A(\tilde{z}, \tilde{\rho})$. (b) $B(\tilde{z}, \tilde{\rho})$.	213
Figure 3.6	Electric field direction for circular charge density distribution near $z = 0$.	214
Figure 3.7	Magnetic field direction for circular current density distribution near $z = 0$.	215
Figure 3.8	Magnetic field for larger z.	215
Figure 3.9	Magnetic field between two current sources.	216
Figure 3.10	Linear polarization.	246
Figure 3.11	Electric field with elliptical polarization.	247
Figure 3.12	Vertical infinitesimal dipole and selected propagation direction.	264
Figure 3.13	Pillbox integration volume.	274
Figure C.1	Neighborhood $\|\mathbf{x} - \mathbf{x}'\| < \epsilon$.	298

1

GEOMETRIC ALGEBRA.

1.1 PREREQUISITES.

Geometric algebra (GA for short), generalizes and extends vector algebra. The following section contains a lightning review of some foundational concepts, including scalar, vector, vector space, basis, orthonormality, and metric.

1.1.1 Vector.

A vector is a directed line segment, with a length, direction, and an orientation. A number of different representations of vectors are possible.

Graphical representation. A vector may be represented graphically as an arrow, with the head indicating the direction of the vector. Multiplication of vectors by positive numbers changes the length of the vector, whereas multiplication by negative numbers changes the direction of the vector and the length, as illustrated in fig. 1.1. Addition of vectors is performed by connecting the arrows heads to tails as illustrated in fig. 1.2. In this book a scalar is a number, usually real, but occasionally complex valued. The set of real numbers will be designated $\mathbb{R}$.

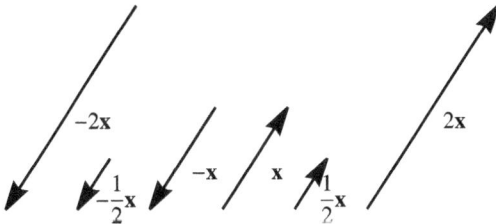

Figure 1.1: Scalar multiples of vectors.

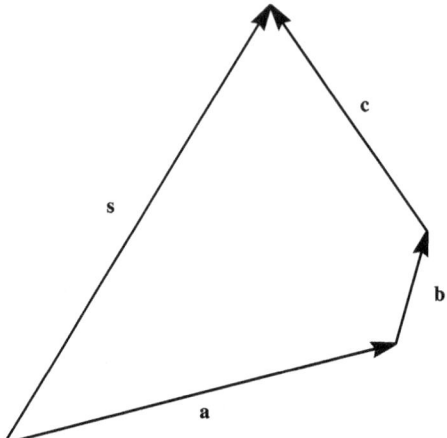

Figure 1.2: Addition of vectors.

Coordinate representation. The length and orientation of a vector, relative to a chosen fixed point (the origin) may be specified algebraically as the coordinates of the head of the vector, as illustrated in fig. 1.3.

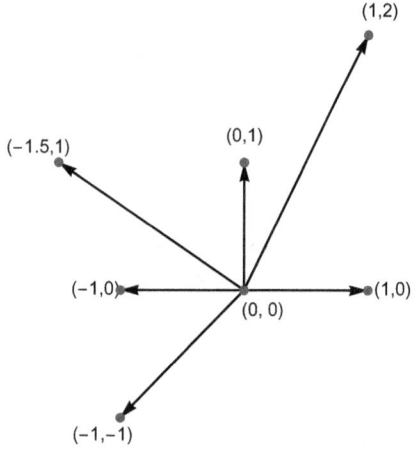

Figure 1.3: Coordinate representation of vectors.

Two dimensional vectors may be represented as pairs of coordinates (x, y), three dimensional vectors as triples of coordinates (x, y, z), and more generally, N dimensional vectors may be represented as coordinate tuples $(x_1, x_2, \cdots, x_N)$. Given two vectors, say $\mathbf{x} = (x, y)$, $\mathbf{y} = (a, b)$, the sum of the vectors is just the sum of the coordinates $\mathbf{x} + \mathbf{y} = (x + a, y + b)$.

Numeric multiplication of a vector rescales each of the coordinates, for example with $\mathbf{x} = (x, y, z)$, $\alpha \mathbf{x} = (\alpha x, \alpha y, \alpha z)$.

It is often convienient to assemble such lists of coordinates in matrix form as rows or columns, providing a few equivalent vector representations as shown in table 1.1.

Table 1.1: Equivalent vector coordinate representations.

Tuple	Row	Column
$(x_1, x_2, \cdots, x_N)$	$\begin{bmatrix} x_1 & x_2 & \cdots & x_N \end{bmatrix}$	$\begin{bmatrix} x_1 \\ x_2 \\ \vdots \\ x_N \end{bmatrix}$

In this book, the length one (unit) vector in the i'th direction will be given the symbol $\mathbf{e}_i$. For example, in three dimensional space with a column vector representation, the respective unit vectors along each of the x, y, and z directions are designated

$$\mathbf{e}_1 = \begin{bmatrix} 1 \\ 0 \\ 0 \end{bmatrix}, \quad \mathbf{e}_2 = \begin{bmatrix} 0 \\ 1 \\ 0 \end{bmatrix}, \quad \mathbf{e}_3 = \begin{bmatrix} 0 \\ 0 \\ 1 \end{bmatrix}. \tag{1.1}$$

Such symbolic designation allows any vector to be encoded in a representation agnostic fashion. For example a vector $\mathbf{x}$ with coordinates x, y, z is

$$\mathbf{x} = x\mathbf{e}_1 + y\mathbf{e}_2 + z\mathbf{e}_3, \tag{1.2}$$

independent of a tuple, row, column, or any other representation.

1.1.2 Vector space.

Two representation specific methods of vector addition and multiplication have been described. Addition can be performed graphically, connecting vectors heads to tails, or by adding the respective coordinates. Multiplication can be performed by changing the length of a vector represented by an arrow, or by multiplying each coordinate algebraically. These rules can be formalized and abstracted by introducing the concept of vector space, which describes both vector addition and multiplication in a representation agnostic fashion.

> **Definition 1.1: Vector space.**
>
> A vector space is a set $V = \{\mathbf{x}, \mathbf{y}, \mathbf{z}, \cdots\}$, the elements of which are called vectors, which has an addition operation designated + and a scalar multiplication operation designated by juxtaposition, where the following axioms are satisfied for all vectors $\mathbf{x}, \mathbf{y}, \mathbf{z} \in V$ and scalars $a, b \in \mathbb{R}$.
>
> | | |
> |---|---|
> | V is closed under addition | $\mathbf{x} + \mathbf{y} \in V$ |
> | V is closed under scalar multiplication | $a\mathbf{x} \in V$ |
> | Addition is associative | $(\mathbf{x} + \mathbf{y}) + \mathbf{z} = \mathbf{x} + (\mathbf{y} + \mathbf{z})$ |
> | Addition is commutative | $\mathbf{y} + \mathbf{x} = \mathbf{x} + \mathbf{y}$ |
> | There exists a zero element $\mathbf{0} \in V$ | $\mathbf{x} + \mathbf{0} = \mathbf{x}$ |
> | For any $\mathbf{x} \in V$ there exists a negative additive inverse $-\mathbf{x} \in V$ | $\mathbf{x} + (-\mathbf{x}) = \mathbf{0}$ |
> | Scalar multiplication is distributive | $a(\mathbf{x} + \mathbf{y}) = a\mathbf{x} + a\mathbf{y}$, $(a + b)\mathbf{x} = a\mathbf{x} + b\mathbf{x}$ |
> | Scalar multiplication is associative | $(ab)\mathbf{x} = a(b\mathbf{x})$ |
> | There exists a multiplicative identity | $1\mathbf{x} = \mathbf{x}$ |

One may define finite or infinite dimensional vector spaces with matrix, polynomial, complex tuple, or many other types of elements. Some examples of general vector spaces are given in the problems below, and many more can be found in any introductory book on linear algebra. The applications of geometric algebra to electromagnetism found in this book require only real vector spaces with dimension no greater than three. Definition 1.1 serves as a reminder, as the concept of vector space will be built upon and generalized shortly.

Exercise 1.1 $\mathbb{R}^N$

Define $\mathbb{R}^N$ as the set of tuples $\{(x_1, x_2, \cdots) \mid x_i \in \mathbb{R}\}$. Show that $\mathbb{R}^N$ is a vector space when the addition operation is defined as $\mathbf{x} + \mathbf{y} \equiv (x_1 + y_1, x_2 + y_2, \cdots)$, and scalar multiplication is defined as $a\mathbf{x} \equiv (ax_1, ax_2, \cdots)$ for any $\mathbf{x} = (x_1, x_2, \cdots) \in \mathbb{R}^N$, $\mathbf{y} = (y_1, y_2, \cdots) \in \mathbb{R}^N$, and $a \in \mathbb{R}$.

Exercise 1.2   Polynomial vector space.

Show that the set of N'th degree polynomials $V = \left\{ \sum_{k=0}^{N} a_k x^k \mid a_k \in \mathbb{R} \right\}$ is a vector space.

Exercise 1.3   Pauli matrices.

The Pauli matrices are defined as

$$\sigma_1 = \begin{bmatrix} 0 & 1 \\ 1 & 0 \end{bmatrix}, \quad \sigma_2 = \begin{bmatrix} 0 & -i \\ i & 0 \end{bmatrix}, \quad \sigma_3 = \begin{bmatrix} 1 & 0 \\ 0 & -1 \end{bmatrix}. \quad (1.3)$$

Given any scalars $a, b, c \in \mathbb{R}$, show that the set $V = \{a\sigma_1 + b\sigma_2 + c\sigma_3\}$ is a vector space with respect to matrix addition. Determine the form of the zero and identity elements. Given a vector $\mathbf{x} = x_1\sigma_1 + x_2\sigma_2 + x_3\sigma_3$, show that the coordinates x_i can be extracted by evaluating the matrix trace of the matrix product $\sigma_i \mathbf{x}$.

1.1.3   Basis, span and dimension.

Definition 1.2: Linear combination

Let $S = \{\mathbf{x}_1, \mathbf{x}_2, \cdots, \mathbf{x}_k\}$ be a subset of a vector space V. A linear combination of vectors in S is any sum

$$a_1\mathbf{x}_1 + a_2\mathbf{x}_2 + \cdots + a_k\mathbf{x}_k.$$

For example, if $\mathbf{x}_1 = \mathbf{e}_1 + \mathbf{e}_2, \mathbf{x}_2 = \mathbf{e}_1 - \mathbf{e}_2$, then $2\mathbf{x}_1 + 3\mathbf{x}_2 = 5\mathbf{e}_1 - \mathbf{e}_2$ is a linear combination.

Definition 1.3: Linear dependence.

Let $S = \{\mathbf{x}_1, \mathbf{x}_2, \cdots, \mathbf{x}_k\}$ be a subset of a vector space V. This set S is linearly dependent if one can construct any equation

$$\mathbf{0} = a_1\mathbf{x}_1 + a_2\mathbf{x}_2 + \cdots + a_k\mathbf{x}_k,$$

for which not all of the coefficients a_i, $1 \leq i \leq k$ are zero.

For example, the vectors $\mathbf{x}_1 = \mathbf{e}_1 + \mathbf{e}_2, \mathbf{x}_2 = \mathbf{e}_1 - \mathbf{e}_2, \mathbf{x}_3 = \mathbf{e}_1 + 3\mathbf{e}_2$ are linearly dependent since $2\mathbf{x}_1 - \mathbf{x}_2 - \mathbf{x}_3 = 0$, and the vectors $\mathbf{y}_1 = \mathbf{e}_1 + \mathbf{e}_2 + \mathbf{e}_3, \mathbf{y}_2 = \mathbf{e}_1 + \mathbf{e}_3, \mathbf{y}_3 = 3\mathbf{e}_1 + \mathbf{e}_2 + 3\mathbf{e}_3$ are linearly dependent since $\mathbf{y}_1 + 2\mathbf{y}_2 - \mathbf{y}_3 = 0$.

Definition 1.4: Linear independence.

Let $S = \{\mathbf{x}_1, \mathbf{x}_2, \cdots, \mathbf{x}_k\}$ be a subset of a vector space V. This set is linearly independent if there are no equations with $a_i \neq 0$, $1 \leq i \leq k$ such that

$$\mathbf{0} = a_1\mathbf{x}_1 + a_2\mathbf{x}_2 + \cdots + a_k\mathbf{x}_k.$$

For example, the vectors $\mathbf{x}_1 = \mathbf{e}_1 + \mathbf{e}_2, \mathbf{x}_2 = \mathbf{e}_1 - \mathbf{e}_2, \mathbf{x}_3 = 2\mathbf{e}_1 + \mathbf{e}_3$, are linearly independent, as are the vectors $\mathbf{y}_1 = \mathbf{e}_1 + \mathbf{e}_2 + \mathbf{e}_3, \mathbf{y}_2 = \mathbf{e}_1 + \mathbf{e}_3, \mathbf{y}_3 = \mathbf{e}_2 + \mathbf{e}_3$.

Definition 1.5: Span.

Let $S = \{\mathbf{x}_1, \mathbf{x}_2, \cdots, \mathbf{x}_k\}$ be a subset of a vector space V. The span of this set is the set of all linear combinations of these vectors, denoted

$$\text{span}(S) = \{a_1\mathbf{x}_1 + a_2\mathbf{x}_2 + \cdots + a_k\mathbf{x}_k\}.$$

For example, the span of $\{\mathbf{e}_1, \mathbf{e}_2\}$ consists of all the points in the x-y plane. The span of the spherical basis vectors $\{\hat{\mathbf{r}}, \hat{\boldsymbol{\theta}}, \hat{\boldsymbol{\phi}}\}$ is $\mathbb{R}^3$.

Definition 1.6: Subspace.

A vector space S that is a subset of V is a subspace of V.

> **Definition 1.7: Basis and dimension**
>
> Let $S = \{\mathbf{x}_1, \mathbf{x}_2, \cdots, \mathbf{x}_n\}$ be a linearly independent subset of V. This set is a basis if $\text{span}(S) = V$. The number of vectors n in this set is called the dimension of the space.

1.1.4 Standard basis, length and normality.

> **Definition 1.8: Dot product.**
>
> Let $\mathbf{x}, \mathbf{y}$ be vectors from a vector space V. A dot product $\mathbf{x} \cdot \mathbf{y}$ is a mapping $V \times V \to \mathbb{R}$ with the following properties.
>
> | | |
> |---|---|
> | Symmetric | $\mathbf{x} \cdot \mathbf{y} = \mathbf{y} \cdot \mathbf{x}$ |
> | Bilinear | $(a\mathbf{x} + b\mathbf{y}) \cdot \mathbf{z} = a\mathbf{x} \cdot \mathbf{z} + b\mathbf{y} \cdot \mathbf{z}$, $\mathbf{x} \cdot (a\mathbf{y} + b\mathbf{z}) = a\mathbf{x} \cdot \mathbf{y} + b\mathbf{x} \cdot \mathbf{z}$ |
> | Positive length | $\mathbf{x} \cdot \mathbf{x} > 0, \mathbf{x} \neq 0$ |

> **Definition 1.9: Length**
>
> The length of a vector $\mathbf{x} \in V$ is defined as
>
> $$\|\mathbf{x}\| = \sqrt{\mathbf{x} \cdot \mathbf{x}}.$$

For example, $\mathbf{x} = \mathbf{e}_1 + \mathbf{e}_2$ has length $\|\mathbf{x}\| = \sqrt{2}$, and $\mathbf{x} = x\mathbf{e}_1 + y\mathbf{e}_2 + z\mathbf{e}_3$ has length $\|\mathbf{x}\| = \sqrt{(x^2 + y^2 + z^2)}$.

> **Definition 1.10: Unit vector**
>
> A vector $\mathbf{x}$ is called a unit vector if the dot product with itself is unity ($\mathbf{x} \cdot \mathbf{x} = 1$).

Examples of unit vectors include $\mathbf{e}_1, (\mathbf{e}_1 + \mathbf{e}_3)/\sqrt{3}, (2\mathbf{e}_1 - \mathbf{e}_2 - \mathbf{e}_3)/\sqrt{6}$, and any vector $\mathbf{x} = \alpha\mathbf{e}_1 + \beta\mathbf{e}_2 + \gamma\mathbf{e}_3$, where α, β, γ are direction cosines satisfying $\alpha^2 + \beta^2 + \gamma^2 = 1$.

> **Definition 1.11: Orthogonal**
>
> Two vectors $\mathbf{x}, \mathbf{y} \in V$ are orthogonal if their dot product is zero, $\mathbf{x} \cdot \mathbf{y} = 0$.

Examples of orthogonal vectors include $\mathbf{x}, \mathbf{y}$ where

$$\begin{aligned} \mathbf{x} &= \mathbf{e}_1 + \mathbf{e}_2 \\ \mathbf{y} &= \mathbf{e}_1 - \mathbf{e}_2, \end{aligned} \quad (1.5)$$

and $\mathbf{x}, \mathbf{y}, \mathbf{z}$ where

$$\begin{aligned} \mathbf{x} &= \mathbf{e}_1 + \mathbf{e}_2 + \mathbf{e}_3 \\ \mathbf{y} &= 2\mathbf{e}_1 - \mathbf{e}_2 - \mathbf{e}_3 \\ \mathbf{z} &= \mathbf{e}_3 - \mathbf{e}_2. \end{aligned} \quad (1.6)$$

> **Definition 1.12: Orthonormal**
>
> Two vectors $\mathbf{x}, \mathbf{y} \in V$ are orthonormal if they are both unit vectors and orthogonal to each other. A set of vectors $\{\mathbf{x}, \mathbf{y}, \cdots, \mathbf{z}\}$ is an orthonormal set if all pairs of vectors in that set are orthonormal.

Examples of orthonormal vectors include $\mathbf{x}, \mathbf{y}$ where

$$\begin{aligned} \mathbf{x} &= \frac{1}{\sqrt{2}} (\mathbf{e}_1 + \mathbf{e}_2) \\ \mathbf{y} &= \frac{1}{\sqrt{2}} (\mathbf{e}_1 - \mathbf{e}_2), \end{aligned} \quad (1.7)$$

and $\mathbf{x}, \mathbf{y}, \mathbf{z}$ where

$$\begin{aligned} \mathbf{x} &= \frac{1}{\sqrt{3}} (\mathbf{e}_1 + \mathbf{e}_2 + \mathbf{e}_3) \\ \mathbf{y} &= \frac{1}{\sqrt{6}} (2\mathbf{e}_1 - \mathbf{e}_2 - \mathbf{e}_3) \\ \mathbf{z} &= \frac{1}{\sqrt{2}} (\mathbf{e}_3 - \mathbf{e}_2). \end{aligned} \quad (1.8)$$

> **Definition 1.13: Standard basis.**
>
> A basis $\{\mathbf{e}_1, \mathbf{e}_2, \cdots, \mathbf{e}_N\}$ is called a standard basis if that set is orthonormal.

Any number of possible standard bases are possible, each differing by combinations of rotations and reflections. For example, given a standard basis $\{\mathbf{e}_1, \mathbf{e}_2, \mathbf{e}_3\}$, the set $\{\mathbf{x}, \mathbf{y}, \mathbf{z}\}$ from eq. (1.8) is also a standard basis.

> **Definition 1.14: Metric.**
>
> Given a basis $B = \{\mathbf{x}_1, \mathbf{x}_2, \cdots \mathbf{x}_N\}$, the metric of the space with respect to B is the (symmetric) matrix G with elements $g_{ij} = \mathbf{x}_i \cdot \mathbf{x}_j$.

For example, with a basis $B = \{\mathbf{x}_1, \mathbf{x}_2\}$ where $\mathbf{x}_1 = \mathbf{e}_1 + \mathbf{e}_2, \mathbf{x}_2 = 2\mathbf{e}_1 - \mathbf{e}_2$, the metric is

$$G = \begin{bmatrix} 2 & 1 \\ 1 & 5 \end{bmatrix}. \tag{1.9}$$

The metric with respect to a standard basis is just the identity matrix.

In relativisitic geometric algebra, the positive definite property of definition 1.8 is considered optional. In this case, the definition of length must be modified, and one would say the length of a vector $\mathbf{x}$ is $\sqrt{|\mathbf{x} \cdot \mathbf{x}|}$, and that $\mathbf{x}$ is a unit vector if $\mathbf{x} \cdot \mathbf{x} = \pm 1$. Such relativisitic dot products will not be used in this book, but they are ubiquitous in the geometric algebra literature, so it is worth knowing that the geometric algebra literature may use a weaker defition of dot product than typical. The metric for a relativistic vector space is not a positive definite matrix. In particular, the metric with respect to a relativistic standard basis is zero off diagonal, and has diagonals valued $(1, -1, -1, -1)$ or $(-1, 1, 1, 1)$. A space is called Euclidean, when the metric with respect to a standard basis is the identity matrix, that is $\mathbf{e}_i \cdot \mathbf{e}_j = \delta_{ij}$ for all standard basis elements $\mathbf{e}_i, \mathbf{e}_j$, and called non-Euclidean if $\mathbf{e}_i \cdot \mathbf{e}_i = -1$ for at least one standard basis vector $\mathbf{e}_i$.

1.2 MULTIVECTORS.

Geometric algebra builds upon a vector space by adding two additional operations, a vector multiplication operation, and a generalized addition operation that extends vector addition to include addition of scalars and products of vectors. Multiplication of vectors is indicated by juxtaposition, for example, if $\mathbf{x}, \mathbf{y}, \mathbf{e}_1, \mathbf{e}_2, \mathbf{e}_3, \cdots$ are vectors, then some vector products are

$$\mathbf{xy}, \mathbf{xyx}, \mathbf{xyxy},$$
$$\mathbf{e}_1\mathbf{e}_2, \mathbf{e}_2\mathbf{e}_1, \mathbf{e}_2\mathbf{e}_3, \mathbf{e}_3\mathbf{e}_2, \mathbf{e}_3\mathbf{e}_1, \mathbf{e}_1\mathbf{e}_3,$$
$$\mathbf{e}_1\mathbf{e}_2\mathbf{e}_3, \mathbf{e}_3\mathbf{e}_1\mathbf{e}_2, \mathbf{e}_2\mathbf{e}_3\mathbf{e}_1, \mathbf{e}_3\mathbf{e}_2\mathbf{e}_1, \mathbf{e}_2\mathbf{e}_1\mathbf{e}_3, \mathbf{e}_1\mathbf{e}_3\mathbf{e}_2, \quad (1.10)$$
$$\mathbf{e}_1\mathbf{e}_2\mathbf{e}_3\mathbf{e}_1, \mathbf{e}_1\mathbf{e}_2\mathbf{e}_3\mathbf{e}_1\mathbf{e}_3\mathbf{e}_2, \cdots$$

Products of vectors may be scalars, vectors, or other entities that represent higher dimensional oriented objects such as planes and volumes. Vector multiplication is constrained by a rule, called the contraction axiom, that gives a meaning to the square of a vector (a scalar equal to the squared length of the vector), and indirectly imposes an anti-commutative relationship between orthogonal vector products. The product of two vectors is not a vector, and may include a scalar component as well as an irreducible product of two orthogonal vectors called a bivector. With vectors and their products living in different spaces, geometric algebra allows scalars, vectors, or any products of vectors to be added, forming a larger closed space of more general objects. Such a sum is called a multivector, an example of which is

$$1 + 2\mathbf{e}_1 + 3\mathbf{e}_1\mathbf{e}_2 + 4\mathbf{e}_1\mathbf{e}_2\mathbf{e}_3. \quad (1.11)$$

In this example, we have added a

- scalar (or 0-vector) 1, to a
- vector (or 1-vector) $2\mathbf{e}_1$, to a
- bivector (or 2-vector) $3\mathbf{e}_1\mathbf{e}_2$, to a
- trivector (or 3-vector) $4\mathbf{e}_1\mathbf{e}_2\mathbf{e}_3$.

Geometric algebra uses vector multiplication to build up a hierarchy of geometrical objects, representing points, lines, planes, volumes and hypervolumes (in higher dimensional spaces.) Those objects are enumerated below to give an idea where we are headed before stating the formal definition of a multivector space.

Scalar. A scalar, also called a 0-vector, is a zero-dimensional object with sign, and a magnitude.

1.2 MULTIVECTORS.

Vector. A vector, also called a 1-vector, is a one-dimensional object with a sign, a magnitude, and a rotational attitude within the space it is embedded.

Bivector. A bivector, also called a 2-vector, is a 2 dimensional object representing a signed plane segment with magnitude and orientation. Assuming a vector product (with properties to be specified shortly), a bivector has the following algebraic description.

Definition 1.15: Bivector.

A bivector, or 2-vector, is a sum of products of pairs of orthogonal vectors. Given an N dimensional vector space V with an orthogonal basis $\{\mathbf{x}_1, \mathbf{x}_2, \cdots, \mathbf{x}_N\}$, a general bivector can be expressed as

$$\sum_{i \neq j} B_{ij} \mathbf{x}_i \mathbf{x}_j,$$

where B_{ij} is a scalar.

Given orthogonal vectors $\mathbf{x}, \mathbf{y}, \mathbf{z}$ and standard basis elements $\mathbf{e}_1, \mathbf{e}_2, \cdots$, examples of bivectors are $\mathbf{xy}, \mathbf{yz}, 3\mathbf{xy} - \mathbf{yz}, \mathbf{e}_1\mathbf{e}_2$, and $\mathbf{e}_1\mathbf{e}_2 + \mathbf{e}_2\mathbf{e}_3 + \mathbf{e}_3\mathbf{e}_1$.

The reader can check that bivectors specified by definition 1.15 form a vector space according to definition 1.1.

If a bivector is formed from the product of just two orthogonal vectors[1], that bivector is said to represent the plane containing those two vectors. Bivectors that represent the same plane can be summed by simply adding the respective (signed) areas, as illustrated in fig. 1.4. Note that the shape of a bivector's area is not significant, only the magnitude of the area and the sign of the bivector, which is represented as an oriented arc in the plane.

Addition of arbitrarily oriented bivectors in $\mathbb{R}^3$ or other higher dimensional spaces, requires decomposition of the bivector into a set of orthogonal planes, an operation best performed algebraically. The sum of a set of bivectors may not represent the same plane as any of the summands, as is crudely illustrated in fig. 1.5, where red + blue = green, where all bivectors have a different rotational attitude in space.

---

[1] Bivectors generated from $\mathbb{R}^2$, and $\mathbb{R}^3$ vectors can always be factored into a single product of orthogonal vectors, and therefore represent a plane. Such a factorization may not be possible in higher dimensional spaces.

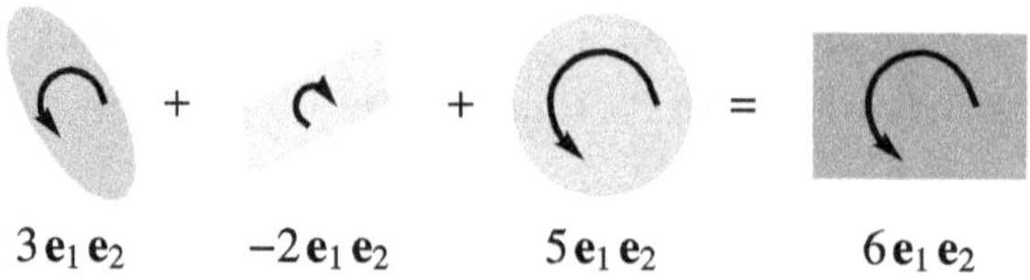

12 GEOMETRIC ALGEBRA.

$3\mathbf{e}_1\mathbf{e}_2$ $-2\mathbf{e}_1\mathbf{e}_2$ $5\mathbf{e}_1\mathbf{e}_2$ $6\mathbf{e}_1\mathbf{e}_2$

Figure 1.4: Graphical representation of bivector addition in the plane.

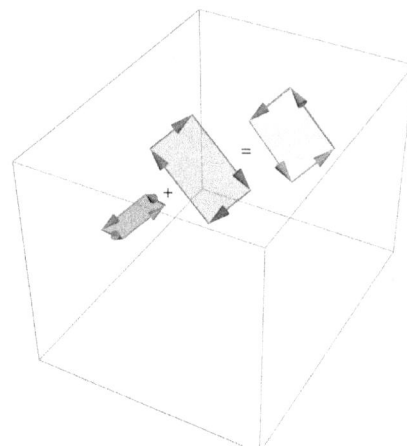

Figure 1.5: Bivector addition.

The bivector provides a structure that can encode plane oriented quantities such as torque, angular momentum, or a general plane of rotation. A quantity like angular momentum can be represented as a magnitude times a quantity that represents the orientation of the plane of rotation. In conventional vector algebra we use the normal of the plane to describe this orientation, but that is problematic in higher dimensional spaces where there is no unique normal. Use of the normal to represent a plane is unsatisfactory in two dimensional spaces, which have to be extended to three dimensions to use normal centric constructs like the cross product. A bivector representation of a plane can eliminate the requirement to utilize a third (normal) dimension, which may not be relevant in the problem, and can allow some concepts (like the cross product) to be generalized to dimensions other than three when desirable.

Later we will see that permutations of the orders of orthogonal vector products are not independent. In particular given a pair of orthogonal vectors $\mathbf{x}, \mathbf{y}$, that dependence is $\mathbf{xy} + \mathbf{yx} = 0$, or $\mathbf{yx} = -\mathbf{xy}$. This means that $\{\mathbf{e}_1\mathbf{e}_2, \mathbf{e}_2\mathbf{e}_1\}$ is not a basis for the $\mathbb{R}^2$ bivector space (those bivectors are not linearly independent), but that either $\{\mathbf{e}_1\mathbf{e}_2\}$ or $\{\mathbf{e}_2\mathbf{e}_1\}$ is an $\mathbb{R}^2$ bivector basis. Similarly, for $\mathbb{R}^3$, we may pick a set such as $R = \{\mathbf{e}_1\mathbf{e}_2, \mathbf{e}_2\mathbf{e}_3, \mathbf{e}_3\mathbf{e}_1\}$ for the bivector basis[2]. If $\mathbf{x}, \mathbf{y}$ are orthonormal vectors, the bivector product $\mathbf{xy}$ or $\mathbf{yx}$ will be called a unit bivector. The basis R is illustrated in fig. 1.6 with two different shaped representations of the "unit" bivector elements of this basis. In both cases, the sign of the bivector is represented graphically with an oriented arc.

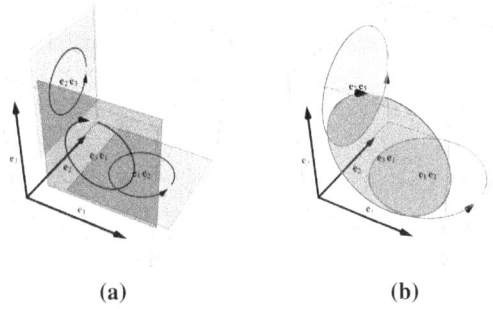

(a) (b)

Figure 1.6: Unit bivectors for $\mathbb{R}^3$

Trivector. A trivector, also called a 3-vector, is a 3 dimensional object representing a signed volume segment with magnitude and orientation. Assuming a vector product (with properties to be specified shortly), a trivector has the following algebraic description.

Definition 1.16: Trivector.

A trivector, or 3-vector, is a sum of products of triplets of mutually orthogonal vectors. Given an N dimensional vector space V with an orthogonal basis $\{\mathbf{x}_1, \mathbf{x}_2, \cdots, \mathbf{x}_N\}$, a trivector is any value

$$\sum_{i \neq j \neq k} T_{ijk} \mathbf{x}_i \mathbf{x}_j \mathbf{x}_k,$$

where T_{ijk} is a scalar.

---

2 R is a "right handed" choice of basis, as it is related to the right handed vector basis $\{\mathbf{e}_1, \mathbf{e}_2, \mathbf{e}_3\}$ in a fundamental way. Observe that the indexes i, j of each bivector $\mathbf{e}_i\mathbf{e}_j$ in R are cyclical permutations of $i, j = 1, 2$. Examples of other bivector basis choices include $\{\mathbf{e}_1\mathbf{e}_2, \mathbf{e}_2\mathbf{e}_3, \mathbf{e}_1\mathbf{e}_3\}$, the set of all pairs of bivectors $\mathbf{e}_i\mathbf{e}_j$ where $i < j$, or a "left handed" bivector basis $\{\mathbf{e}_2\mathbf{e}_1, \mathbf{e}_3\mathbf{e}_2, \mathbf{e}_1\mathbf{e}_3\}$.

For example $\mathbf{e}_1\mathbf{e}_2\mathbf{e}_3, 3\mathbf{e}_2\mathbf{e}_1\mathbf{e}_3, 5\mathbf{e}_1\mathbf{e}_2\mathbf{e}_4, -2\mathbf{e}_1\mathbf{e}_5\mathbf{e}_3$ are all trivectors. However, in $\mathbb{R}^3$, it turns out that all trivectors are scalar multiples of $\mathbf{e}_1\mathbf{e}_2\mathbf{e}_3$. Like scalars, there is no direction to such a quantity, and like scalars trivectors may be signed. The magnitude of a trivector may be interpreted as a volume. A geometric interpretation of the sign of a trivector will be deferred until integration theory is tackled.

K-vector. Scalars, vectors, bivectors, trivectors, or any higher dimensional analogues are all examples of a single algebraic structure composed of zero, one, two, three, or "k" products of orthogonal vectors. These are generally called k-vectors and defined as follows.

Definition 1.17: K-vector and grade.

A k-vector is a sum of products of k mutually orthogonal vectors. Given an N dimensional vector space with an orthonormal basis

$$\{\mathbf{x}_1, \mathbf{x}_2, \cdots, \mathbf{x}_N\},$$

a general k-vector can be expressed as

$$\sum_{r \neq s \neq \cdots \neq t} K_{rs\cdots t} \mathbf{x}_r \mathbf{x}_s \cdots \mathbf{x}_t,$$

where $K_{rs\cdots t}$ is a scalar, indexed by k indexes $r, s, \cdots, t$.

The number k of orthogonal vector factors of a k-vector is called the grade.

A 0-vector is a scalar.

Illustrating with some examples

- 1 is a 0-vector with grade 0
- $\mathbf{e}_1$ is a 1-vector with grade 1
- $\mathbf{e}_1\mathbf{e}_2, \mathbf{e}_2\mathbf{e}_3$, and $\mathbf{e}_3\mathbf{e}_1$ are 2-vectors with grade 2, and
- $\mathbf{e}_1\mathbf{e}_2\mathbf{e}_3$ is a 3-vector with grade 3.

The highest grade for a k-vector in an N dimensional vector space is N.

Multivector.

1.2 MULTIVECTORS.

> **Definition 1.18: Multivector.**
>
> Given an N dimensional (generating) vector space V and a vector multiplication operation represented by juxtaposition, a multivector is a sum of scalars, vectors, or products of vectors.

Any k-vector or sum of k-vectors is also a multivector. Examples:

- $\mathbf{e}_1\mathbf{e}_4, \mathbf{e}_1\mathbf{e}_2 + \mathbf{e}_2\mathbf{e}_3$. These are bivectors, and also multivectors with only grade 2 components.

- $\mathbf{e}_1\mathbf{e}_2\mathbf{e}_3, \mathbf{e}_2\mathbf{e}_3\mathbf{e}_4$. These are trivectors, and also multivectors with only grade 3 components.

- $1 + \mathbf{e}_1\mathbf{e}_2$ This is not a k-vector as there is no single grade, but is a multivector. In this case, it is a sum of a scalar (0-vector) and a bivector (2-vector).

- $0, 7, -3$. These are scalars (0-vectors), and also multivectors.

A k-vector was a sum of orthogonal products, but a multivector may also include arbitrary sums of any vector products, not all of which have to be orthogonal. Examples include

- $\mathbf{e}_1\mathbf{e}_1, \mathbf{e}_1\mathbf{e}_2\mathbf{e}_1\mathbf{e}_2$,

- $\mathbf{e}_1\mathbf{e}_2\mathbf{e}_1, \mathbf{e}_1\mathbf{e}_2\mathbf{e}_3\mathbf{e}_1\mathbf{e}_2$,

- $\mathbf{e}_1\mathbf{e}_2\mathbf{e}_1\mathbf{e}_3, \mathbf{e}_1\mathbf{e}_2\mathbf{e}_1\mathbf{e}_3\mathbf{e}_1\mathbf{e}_2$,

- $\mathbf{e}_1\mathbf{e}_2\mathbf{e}_1\mathbf{e}_3\mathbf{e}_1, \mathbf{e}_2\mathbf{e}_1\mathbf{e}_2\mathbf{e}_1\mathbf{e}_3\mathbf{e}_1\mathbf{e}_2$.

Once the definition of vector multiplication has been made more precise, we will be able to see that these multivectors are scalars, vectors, bivectors, and trivectors respectively.

Multivector space. Bivectors, trivectors, k-vectors, and multivectors all assumed that suitable multiplication and addition operations for vectors and vector products had been defined. The definition of a multivector space makes this more precise.

Definition 1.19: Multivector space.

Given an N dimensional (generating) vector space V, a multivector space generated by V is a set $M = \{x, y, z, \cdots\}$ of multivectors (sums of scalars, vectors, or products of vectors), where the following axioms are satisfied

Contraction	$\mathbf{x}^2 = \mathbf{x} \cdot \mathbf{x}, \forall \mathbf{x} \in V$
M is closed under addition	$x + y \in M$
M is closed under multiplication	$xy \in M$
Addition is associative	$(x + y) + z = x + (y + z)$
Addition is commutative	$y + x = x + y$
There exists a zero element $0 \in M$	$x + 0 = x$
For all $x \in M$ there exists a negative additive inverse $-x \in M$	$x + (-x) = 0$
Multiplication is distributive	$x(y + z) = xy + xz,$ $(x + y)z = xz + yz$
Multiplication is associative	$(xy)z = x(yz)$
There exists a multiplicative identity $1 \in M$	$1x = x$

The contraction axiom is arguably the most important of the multivector space axioms, as it allows for multiplicative closure. Another implication of the contraction axiom is that vector multiplication is not generally commutative (order matters). The multiplicative closure property and the commutative and non-commutative conditions for vector multiplication will be examined next.

Observe that the axioms of a multivector space are almost that of a field (i.e. real numbers, complex numbers, ...). However, a field also requires a multiplicative inverse element for all elements of the space. Such a multiplicative inverse exists for some multivector subspaces, but not in general.

The reader should compare definition 1.19 with definition 1.1 the specification of a vector space, and observe the similarities and differences.

1.3 COLINEAR VECTORS.

It was pointed out that the vector multiplication operation was not assumed to be commutative (order matters). The only condition for which the product of two vectors is order independent, is when those vectors are colinear.

Theorem 1.1: Vector commutation.

Given two vectors $\mathbf{x}, \mathbf{y}$, if $\mathbf{y} = \alpha \mathbf{x}$ for some scalar α, then $\mathbf{x}$ and $\mathbf{y}$ commute

$$\mathbf{xy} = \mathbf{yx}.$$

Proof.

$$\mathbf{yx} = \alpha \mathbf{xx}$$
$$\mathbf{xy} = \mathbf{x}\alpha\mathbf{x} = \alpha\mathbf{xx}. \qquad \square$$

The contraction axiom ensures that the product of two colinear vectors is a scalar. In particular, the square of a unit vector, say $\mathbf{u}$ is unity. This should be highlighted explicitly, because this property will be used again and again

$$\boxed{\mathbf{u}^2 = 1.} \qquad (1.12)$$

For example, the squares of any orthonormal basis vectors are unity $(\mathbf{e}_1)^2 = (\mathbf{e}_2)^2 = (\mathbf{e}_3)^2 = 1$.

A corollary of eq. (1.12) is

$$\boxed{1 = \mathbf{uu},} \qquad (1.13)$$

for any unit vector $\mathbf{u}$. Such a factorization trick will be used repeatedly in this book.

1.4 ORTHOGONAL VECTORS.

Theorem 1.2: Anticommutation of orthogonal vectors

Let **u**, and **v** be two orthogonal vectors, the product of which **uv** is a bivector. Changing the order of these products toggles the sign of the bivector.

uv = −vu.

This sign change on interchange is called anticommutation.

Proof. Let **u**, **v** be a pair of orthogonal vectors, such as those of fig. 1.7. The squared length of the sum **u** + **v** can be expressed in using the contraction axiom, or by explicit expansion (taking care to maintain the order of products)

$$(\mathbf{u}+\mathbf{v})^2 = (\mathbf{u}+\mathbf{v})(\mathbf{u}+\mathbf{v}) = \mathbf{u}^2 + \mathbf{uv} + \mathbf{vu} + \mathbf{v}^2$$
$$(\mathbf{u}+\mathbf{v})^2 = \|\mathbf{u}+\mathbf{v}\|^2 = \mathbf{u}^2 + \mathbf{v}^2.$$

Comparing the two expansions and rearranging completes the proof[3]. □

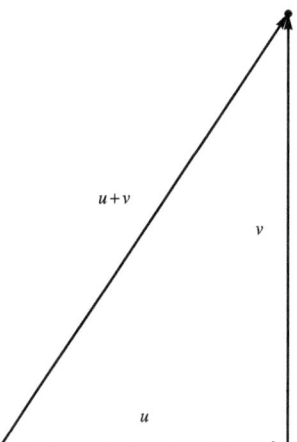

Figure 1.7: Sum of orthogonal vectors.

---

[3] We will see later (theorem 1.3) that the converse of this theorem is also true: If the product of two vectors is a bivector, those vectors are orthogonal.

Some examples of anticommuting pairs include, $\mathbf{e}_2\mathbf{e}_1 = -\mathbf{e}_1\mathbf{e}_2$, $\mathbf{e}_3\mathbf{e}_2 = -\mathbf{e}_2\mathbf{e}_3$, and $\mathbf{e}_1\mathbf{e}_3 = -\mathbf{e}_3\mathbf{e}_1$. This theorem can also be applied to any pairs of orthogonal vectors in a arbitrary k-vector, for example

$$\begin{aligned}
\mathbf{e}_3\mathbf{e}_2\mathbf{e}_1 &= (\mathbf{e}_3\mathbf{e}_2)\mathbf{e}_1 \\
&= -(\mathbf{e}_2\mathbf{e}_3)\mathbf{e}_1 \\
&= -\mathbf{e}_2(\mathbf{e}_3\mathbf{e}_1) \\
&= +\mathbf{e}_2(\mathbf{e}_1\mathbf{e}_3) \\
&= +(\mathbf{e}_2\mathbf{e}_1)\mathbf{e}_3 \\
&= -\mathbf{e}_1\mathbf{e}_2\mathbf{e}_3,
\end{aligned} \quad (1.14)$$

showing that reversal of all the factors in a trivector such as $\mathbf{e}_1\mathbf{e}_2\mathbf{e}_3$ toggles the sign.

1.5 SOME NOMENCLATURE.

The workhorse operator of geometric algebra is called grade selection, defined as

---
Definition 1.20: Grade selection operator

Given a set of k-vectors $M_k, k \in [0, N]$, and any multivector of their sum

$$M = \sum_{i=0}^{N} M_i,$$

the grade selection operator is defined as

$$\langle M \rangle_k = M_k.$$

Due to its importance, selection of the (scalar) zero grade is given the shorthand

$$\langle M \rangle = \langle M \rangle_0 = M_0.$$

The grade selection operator will be used to define a generalized dot product between multivectors, and the wedge product, which generalizes the cross product (and is related to the cross product in $\mathbb{R}^3$).

To illustrate grade selection by example, given a multivector $M = 3 - \mathbf{e}_3 + 2\mathbf{e}_1\mathbf{e}_2 + 7\mathbf{e}_1\mathbf{e}_2\mathbf{e}_4$, then

$$\begin{aligned} \langle M \rangle_0 &= \langle M \rangle = 3 \\ \langle M \rangle_1 &= -\mathbf{e}_3 \\ \langle M \rangle_2 &= 2\mathbf{e}_1\mathbf{e}_2 \\ \langle M \rangle_3 &= 7\mathbf{e}_1\mathbf{e}_2\mathbf{e}_4. \end{aligned} \quad (1.15)$$

Definition 1.21: Orthonormal product shorthand.

Given an orthonormal basis $\{\mathbf{e}_1, \mathbf{e}_2, \cdots\}$, a multiple indexed quantity $\mathbf{e}_{ij\cdots k}$ should be interpreted as the product (in the same order) of the basis elements with those indexes

$$\mathbf{e}_{ij\cdots k} = \mathbf{e}_i \mathbf{e}_j \cdots \mathbf{e}_k.$$

For example,

$$\begin{aligned} \mathbf{e}_{12} &= \mathbf{e}_1\mathbf{e}_2 \\ \mathbf{e}_{123} &= \mathbf{e}_1\mathbf{e}_2\mathbf{e}_3 \\ \mathbf{e}_{23121} &= \mathbf{e}_2\mathbf{e}_3\mathbf{e}_1\mathbf{e}_2\mathbf{e}_1. \end{aligned} \quad (1.16)$$

Definition 1.22: Pseudoscalar.

If $\{\mathbf{x}_1, \mathbf{x}_2, \cdots, \mathbf{x}_k\}$ is an orthogonal basis for a k-dimensional (sub)space, then the product $\mathbf{x}_1 \mathbf{x}_2 \cdots \mathbf{x}_k$ is called a pseudoscalar for that (sub)space. A pseudoscalar that squares to ± 1 is called a unit pseudoscalar.

A pseudoscalar is the highest grade k-vector in the algebra, so in $\mathbb{R}^2$ any bivector is a pseudoscalar, and in $\mathbb{R}^3$ any trivector is a pseudoscalar. In $\mathbb{R}^2$, $\mathbf{e}_1\mathbf{e}_2$ is a pseudoscalar, as is $3\mathbf{e}_2\mathbf{e}_1$, both of which are related by a constant factor. In $\mathbb{R}^3$ the trivector $\mathbf{e}_3\mathbf{e}_1\mathbf{e}_2$ is a pseudoscalar, as is $-7\mathbf{e}_3\mathbf{e}_1\mathbf{e}_2$, and both of these can also be related by a constant factor. For the subspace span $\mathbf{e}_1, \mathbf{e}_2 + \mathbf{e}_3$, one pseudoscalar is $\mathbf{e}_1(\mathbf{e}_2 + \mathbf{e}_3)$.

If all the vector factors of a pseudoscalar are not just orthogonal but orthonormal, then it is a unit pseudoscalar. It is conventional to refer to

$$\boxed{\mathbf{e}_{12} = \mathbf{e}_1\mathbf{e}_2,} \quad (1.17)$$

as "the pseudoscalar" for $\mathbb{R}^2$, and to

$$\boxed{\mathbf{e}_{123} = \mathbf{e}_1\mathbf{e}_2\mathbf{e}_3,} \tag{1.18}$$

as "the pseudoscalar" for a three dimensional space.

We will see that geometric algebra allows for many quantities that have a complex imaginary nature, and that the pseudoscalars of eq. (1.17) and eq. (1.18) both square to -1.

For this reason, it is often convenient to use an imaginary notation for the $\mathbb{R}^2$ and $\mathbb{R}^3$ pseudoscalars

$$\begin{aligned} i &= \mathbf{e}_{12} \\ I &= \mathbf{e}_{123}. \end{aligned} \tag{1.19}$$

For three dimensional problems in this book, i will often be used as the unit pseudoscalar for whatever planar subspace is relevant to the problem, which may not be the x-y plane. The meaning of i in any such cases will always be defined explicitly.

Exercise 1.4   Permutations of the $\mathbb{R}^3$ pseudoscalar

Show that all the cyclic permutations of the $\mathbb{R}^3$ pseudoscalar are equal

$$I = \mathbf{e}_2\mathbf{e}_3\mathbf{e}_1 = \mathbf{e}_3\mathbf{e}_1\mathbf{e}_2 = \mathbf{e}_1\mathbf{e}_2\mathbf{e}_3.$$

1.6 TWO DIMENSIONS.

The multiplication table for the $\mathbb{R}^2$ geometric algebra can be computed with relative ease. Many of the interesting products involve $i = \mathbf{e}_1\mathbf{e}_2$, the unit pseudoscalar. The imaginary nature of the pseudoscalar can be demonstrated using theorem 1.2

$$\begin{aligned} (\mathbf{e}_1\mathbf{e}_2)^2 &= (\mathbf{e}_1\mathbf{e}_2)(\mathbf{e}_1\mathbf{e}_2) \\ &= -(\mathbf{e}_1\mathbf{e}_2)(\mathbf{e}_2\mathbf{e}_1) \\ &= -\mathbf{e}_1(\mathbf{e}_2^2)\mathbf{e}_1 \\ &= -\mathbf{e}_1^2 \\ &= -1. \end{aligned} \tag{1.22}$$

Like the (scalar) complex imaginary, the bivector $\mathbf{e}_1\mathbf{e}_2$ also squares to -1. The only non-trivial products left to fill in the $\mathbb{R}^2$ multiplication table are those of the unit vectors with i, products that are order dependent

$$\begin{aligned}
\mathbf{e}_1 i &= \mathbf{e}_1 \left(\mathbf{e}_1 \mathbf{e}_2\right) \\
&= \left(\mathbf{e}_1 \mathbf{e}_1\right) \mathbf{e}_2 \\
&= \mathbf{e}_2 \\
i\mathbf{e}_1 &= \left(\mathbf{e}_1 \mathbf{e}_2\right) \mathbf{e}_1 \\
&= \left(-\mathbf{e}_2 \mathbf{e}_1\right) \mathbf{e}_1 \\
&= -\mathbf{e}_2 \left(\mathbf{e}_1 \mathbf{e}_1\right) \\
&= -\mathbf{e}_2 \\
\mathbf{e}_2 i &= \mathbf{e}_2 \left(\mathbf{e}_1 \mathbf{e}_2\right) \\
&= \mathbf{e}_2 \left(-\mathbf{e}_2 \mathbf{e}_1\right) \\
&= -\left(\mathbf{e}_2 \mathbf{e}_2\right) \mathbf{e}_1 \\
&= -\mathbf{e}_1 \\
i\mathbf{e}_2 &= \left(\mathbf{e}_1 \mathbf{e}_2\right) \mathbf{e}_2 \\
&= \mathbf{e}_1 \left(\mathbf{e}_2 \mathbf{e}_2\right) \\
&= \mathbf{e}_1.
\end{aligned} \quad (1.23)$$

The multiplication table for the $\mathbb{R}^2$ multivector basis can now be tabulated

Table 1.2: 2D Multiplication table.

	1	$\mathbf{e}_1$	$\mathbf{e}_2$	$\mathbf{e}_1\mathbf{e}_2$
1	1	$\mathbf{e}_1$	$\mathbf{e}_2$	$\mathbf{e}_1\mathbf{e}_2$
$\mathbf{e}_1$	$\mathbf{e}_1$	1	$\mathbf{e}_1\mathbf{e}_2$	$\mathbf{e}_2$
$\mathbf{e}_2$	$\mathbf{e}_2$	$-\mathbf{e}_1\mathbf{e}_2$	1	$-\mathbf{e}_1$
$\mathbf{e}_1\mathbf{e}_2$	$\mathbf{e}_1\mathbf{e}_2$	$-\mathbf{e}_2$	$\mathbf{e}_1$	-1

It is important to point out that the pseudoscalar i does not commute with either basis vector, but anticommutes with both, since $i\mathbf{e}_1 = -\mathbf{e}_1 i$, and $i\mathbf{e}_2 = -\mathbf{e}_2 i$. By superposition i anticommutes with any vector in the x-y plane.

More generally, if $\mathbf{u}$ and $\mathbf{v}$ are orthonormal, and $\mathbf{x} \in \text{span}\{\mathbf{u}, \mathbf{v}\}$ then the bivector $\mathbf{uv}$ anticommutes with $\mathbf{x}$, or any other vector in this plane.

1.7 PLANE ROTATIONS.

Plotting eq. (1.23), as in fig. 1.8, shows that multiplication by i rotates the $\mathbb{R}^2$ basis vectors by $\pm\pi/2$ radians, with the rotation direction dependent on the order of multiplication.

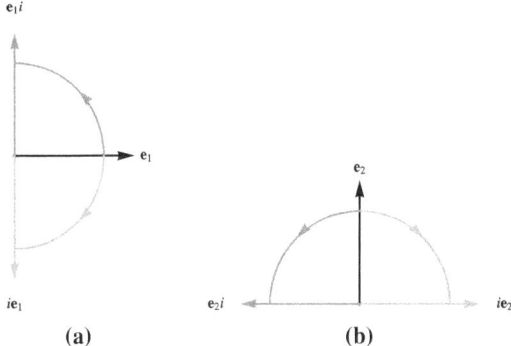

Figure 1.8: Multiplication by $e_1 e_2$.

Multiplying a polar vector representation

$$\mathbf{x} = \rho \left(\mathbf{e}_1 \cos \theta + \mathbf{e}_2 \sin \theta \right), \tag{1.24}$$

by i shows that a $\pi/2$ rotation is induced.

Multiplying the vector from the right by i gives

$$\begin{aligned}
\mathbf{x}i &= \mathbf{x}\mathbf{e}_1\mathbf{e}_2 \\
&= \rho \left(\mathbf{e}_1 \cos \theta + \mathbf{e}_2 \sin \theta \right) \mathbf{e}_1 \mathbf{e}_2 \\
&= \rho \left(\mathbf{e}_2 \cos \theta - \mathbf{e}_1 \sin \theta \right),
\end{aligned} \tag{1.25}$$

a counterclockwise rotation of $\pi/2$ radians, and multiplying the vector by i from the left gives

$$\begin{aligned}
i\mathbf{x} &= \mathbf{e}_1\mathbf{e}_2\mathbf{x} \\
&= \rho \mathbf{e}_1 \mathbf{e}_2 \left(\mathbf{e}_1 \cos \theta + \mathbf{e}_2 \sin \theta \right) \mathbf{e}_1 \mathbf{e}_2 \\
&= \rho \left(-\mathbf{e}_2 \cos \theta + \mathbf{e}_1 \sin \theta \right),
\end{aligned} \tag{1.26}$$

a clockwise rotation by $\pi/2$ radians (exercise 1.5).

The transformed vector $\mathbf{x}' = \mathbf{x}\mathbf{e}_1\mathbf{e}_2 = -\mathbf{e}_1\mathbf{e}_2\mathbf{x} (= \mathbf{x}i = -i\mathbf{x})$ has been rotated in the direction that takes $\mathbf{e}_1$ to $\mathbf{e}_2$, as illustrated in fig. 1.9.

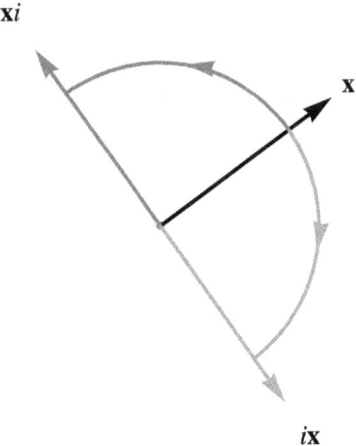

Figure 1.9: $\pi/2$ rotation in the plane using pseudoscalar multiplication.

In complex number theory the complex exponential $e^{i\theta}$ can be used as a rotation operator. Geometric algebra puts this rotation operator into the vector algebra toolbox, by utilizing Euler's formula

$$e^{i\theta} = \cos\theta + i\sin\theta, \qquad (1.27)$$

valid for this pseudoscalar imaginary representation too (exercise 1.6). By writing $\mathbf{e}_2 = \mathbf{e}_1\mathbf{e}_1\mathbf{e}_2$, a complex exponential can be factored directly out of the polar vector representation eq. (1.24)

$$\begin{aligned}
\mathbf{x} &= \rho\left(\mathbf{e}_1\cos\theta + \mathbf{e}_2\sin\theta\right) \\
&= \rho\left(\mathbf{e}_1\cos\theta + (\mathbf{e}_1\mathbf{e}_1)\mathbf{e}_2\sin\theta\right) \\
&= \rho\mathbf{e}_1\left(\cos\theta + \mathbf{e}_1\mathbf{e}_2\sin\theta\right) \\
&= \rho\mathbf{e}_1\left(\cos\theta + i\sin\theta\right) \\
&= \rho\mathbf{e}_1 e^{i\theta}.
\end{aligned} \qquad (1.28)$$

We end up with a complex exponential multivector factor on the right. Alternatively, since $\mathbf{e}_2 = \mathbf{e}_2\mathbf{e}_1\mathbf{e}_1$, a complex exponential can be factored out on the left

$$\begin{aligned}
\mathbf{x} &= \rho\left(\mathbf{e}_1\cos\theta + \mathbf{e}_2\sin\theta\right) \\
&= \rho\left(\mathbf{e}_1\cos\theta + \mathbf{e}_2(\mathbf{e}_1\mathbf{e}_1)\sin\theta\right) \\
&= \rho\left(\cos\theta - \mathbf{e}_1\mathbf{e}_2\sin\theta\right)\mathbf{e}_1 \\
&= \rho\left(\cos\theta - i\sin\theta\right)\mathbf{e}_1 \\
&= \rho e^{-i\theta}\mathbf{e}_1.
\end{aligned} \qquad (1.29)$$

Left and right exponential expressions have now been found for the polar representation

$$\rho \left(\mathbf{e}_1 \cos \theta + \mathbf{e}_2 \sin \theta \right) = \rho e^{-i\theta} \mathbf{e}_1 = \rho \mathbf{e}_1 e^{i\theta}. \tag{1.30}$$

This is essentially a recipe for rotation of a vector in the x-y plane. Such rotations are illustrated in fig. 1.10.

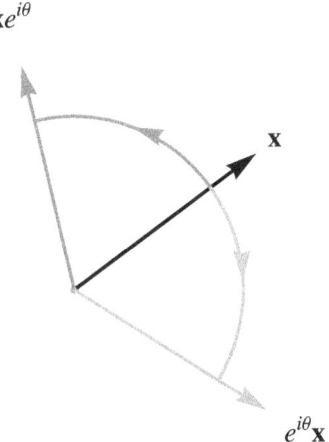

Figure 1.10: Rotation in a plane.

This generalizes to rotations of $\mathbb{R}^N$ vectors constrained to a plane. Given orthonormal vectors $\mathbf{u}, \mathbf{v}$ and any vector in the plane of these two vectors ($\mathbf{x} \in \text{span} \{\mathbf{u}, \mathbf{v}\}$), this vector is rotated θ radians in the direction of rotation that takes $\mathbf{u}$ to $\mathbf{v}$ by

$$\mathbf{x}' = \mathbf{x} e^{\mathbf{uv}\theta} = e^{-\mathbf{uv}\theta} \mathbf{x}. \tag{1.31}$$

The sense of rotation for the rotation $e^{\mathbf{uv}\theta}$ is opposite that of $e^{\mathbf{vu}\theta}$, which provides a first hint that bivectors can be characterized as having an orientation, somewhat akin to thinking of a vector as having a head and a tail.

Example 1.1: Velocity and acceleration in polar coordinates.

Complex exponential representations of rotations work very nicely for describing vectors in polar coordinates. A radial vector can be written as

$$\mathbf{r} = r\hat{\mathbf{r}}, \tag{1.32}$$

as illustrated in fig. 1.11. The polar representation of the radial and azimuthal unit vector are simply

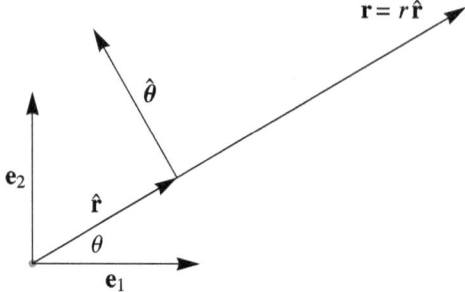

Figure 1.11: Radial vector in polar coordinates.

$$\hat{\mathbf{r}} = \mathbf{e}_1 e^{i\theta} = \mathbf{e}_1 \left(\cos\theta + \mathbf{e}_1\mathbf{e}_2 \sin\theta \right) = \mathbf{e}_1 \cos\theta + \mathbf{e}_2 \sin\theta$$
$$\hat{\theta} = \mathbf{e}_2 e^{i\theta} = \mathbf{e}_2 \left(\cos\theta + \mathbf{e}_1\mathbf{e}_2 \sin\theta \right) = \mathbf{e}_2 \cos\theta - \mathbf{e}_1 \sin\theta, \quad (1.33)$$

where $i = \mathbf{e}_{12}$ is the unit bivector for the x-y plane. We can easily show that these unit vectors are orthogonal

$$\begin{aligned}\hat{\mathbf{r}}\hat{\theta} &= \left(\mathbf{e}_1 e^{i\theta} \right)\left(e^{-i\theta} \mathbf{e}_2 \right) \\ &= \mathbf{e}_1 e^{i\theta} e^{-i\theta} \mathbf{e}_2 \\ &= \mathbf{e}_1 \mathbf{e}_2. \end{aligned} \quad (1.34)$$

By theorem 1.2, since the product of $\hat{\mathbf{r}}\hat{\theta}$ is a bivector, $\hat{\mathbf{r}}$ is orthogonal to $\hat{\theta}$.

We can find the velocity and acceleration by taking time derivatives

$$\begin{aligned}\mathbf{v} &= r'\hat{\mathbf{r}} + r\hat{\mathbf{r}}' \\ \mathbf{a} &= r''\hat{\mathbf{r}} + 2r'\hat{\mathbf{r}}' + r\hat{\mathbf{r}}'', \end{aligned} \quad (1.35)$$

but to make these more meaningful want to evaluate the $\hat{\mathbf{r}}, \hat{\boldsymbol{\theta}}$ derivatives explicitly. Those are

$$\mathbf{e}_1(\mathbf{e}_1\mathbf{e}_2) = (\mathbf{e}_1\mathbf{e}_1)\mathbf{e}_2$$

$$\begin{aligned}\hat{\mathbf{r}}' &= \left(\mathbf{e}_1 e^{i\theta}\right)' = \boxed{\mathbf{e}_1 i} e^{i\theta}\theta' = \mathbf{e}_2 e^{i\theta}\theta' = \hat{\boldsymbol{\theta}}\omega \\ \hat{\boldsymbol{\theta}}' &= \left(\mathbf{e}_2 e^{i\theta}\right)' = \boxed{\mathbf{e}_2 i} e^{i\theta}\theta' = -\mathbf{e}_1 e^{i\theta}\theta' = -\hat{\mathbf{r}}\omega,\end{aligned} \quad (1.36)$$

$$\mathbf{e}_2\mathbf{e}_1\mathbf{e}_2 = (-\mathbf{e}_1\mathbf{e}_2)\mathbf{e}_2$$

where $\omega = d\theta/dt$, and primes denote time derivatives. The velocity and acceleration vectors can now be written explicitly in terms of radial and azimuthal components. The velocity is

$$\mathbf{v} = r'\hat{\mathbf{r}} + r\omega\hat{\boldsymbol{\theta}}, \quad (1.37)$$

and the acceleration is

$$\begin{aligned}\mathbf{a} &= r''\hat{\mathbf{r}} + 2r'\omega\hat{\boldsymbol{\theta}} + r(\omega\hat{\boldsymbol{\theta}})' \\ &= r''\hat{\mathbf{r}} + 2r'\omega\hat{\boldsymbol{\theta}} + r\omega'\hat{\boldsymbol{\theta}} - r\omega^2\hat{\mathbf{r}},\end{aligned} \quad (1.38)$$

or

$$\mathbf{a} = \hat{\mathbf{r}}\left(r'' - r\omega^2\right) + \frac{1}{r}\hat{\boldsymbol{\theta}}\left(r^2\omega\right)'. \quad (1.39)$$

Using eq. (1.33), we also have the option of factoring out the rotation operation from the position vector or any of its derivatives

$$\begin{aligned}\mathbf{r} &= (r\mathbf{e}_1)\,e^{i\theta} \\ \mathbf{v} &= (r'\mathbf{e}_1 + r\omega\mathbf{e}_2)\,e^{i\theta} \\ \mathbf{a} &= \left(\left(r'' - r\omega^2\right)\mathbf{e}_1 + \frac{1}{r}\left(r^2\omega\right)'\mathbf{e}_2\right)e^{i\theta}.\end{aligned} \quad (1.40)$$

In particular, for uniform circular motion, each of the position, velocity and acceleration vectors can be represented by a vector that is fixed in space, subsequently rotated by an angle θ.

Exercise 1.5 $\mathbb{R}^2$ rotations.

Using familiar methods, such as rotation matrices, show that the counter-clockwise and clockwise rotations of eq. (1.24) are given by eq. (1.25) and eq. (1.26) respectively.

Exercise 1.6 Multivector Euler's formula and trig relations.

For a multivector x assume an infinite series representation of the exponential, sine and cosine functions and their hyperbolic analogues

$$e^x = \sum_{k=0}^{\infty} \frac{x^k}{k!}$$

$$\cos x = \sum_{k=0}^{\infty} (-1)^k \frac{x^{2k}}{(2k)!} \qquad \sin x = \sum_{k=0}^{\infty} (-1)^k \frac{x^{2k+1}}{(2k+1)!}$$

$$\cosh x = \sum_{k=0}^{\infty} \frac{x^{2k}}{(2k)!} \qquad \sinh x = \sum_{k=0}^{\infty} \frac{x^{2k+1}}{(2k+1)!}$$

a. Show that for scalar θ, and any multivectors J that satisfies $J^2 = -1$, and $K^2 = 1$, then hold for multivectors J, K satisfying $J^2 = -1$ and $K^2 = 1$ respectively.

$$\cosh(J\theta) = \cos\theta, \quad \cosh(K\theta) = \cosh\theta$$
$$\sinh(J\theta) = J\sin\theta, \quad \sinh(K\theta) = K\sinh\theta.$$

b. Show that the trigonometric and hyperbolic Euler formulas

$$e^{J\theta} = \cos\theta + J\sin\theta$$
$$e^{K\theta} = \cosh\theta + K\sinh\theta,$$

hold for multivectors J, K satisfying $J^2 = -1$ and $K^2 = 1$ respectively.

c. Given multivectors X, Y, show that $e^{X+Y} = e^X e^Y$ if X, Y commute. That is $XY = YX$.

Exercise 1.7 Exponential derivatives.

a. For real or complex x, we know that

$$(e^x)' = x'e^x,$$

but this is not generally true for multivectors x (or square matrices for that matter.) Expand e^x in Taylor series and take derivatives, and show that this identity requires that that x' commutes with x.

b. Given any j^2 = constant, and scalar constant θ, show that

$$\left(e^{j\theta}\right)' = j' j^{-1} \sinh(j\theta),$$

where derivatives are with respect to j.

We will see an example of such an exponential later when we construct an exponential representation of spherical polar vectors, setting $i = \mathbf{e}_{12}$, $j = \mathbf{e}_{31} e^{i\phi}$, $\mathbf{x} = r \mathbf{e}_3 e^{j\theta}$. Such a vector representation hides all the ϕ dependence away in the bivector j, and computation of $\partial \mathbf{x}/\partial \phi$ requires that we know how to correctly compute these sorts of exponential derivatives.

Hint: show that $jj' = -j'j$, and consider the effect of this on the $(j^k)'$ term in the Taylor series.

c. Let when $j^2 = -1$, show that

$$\left(e^{j\theta}\right)' = j' \sin(\theta).$$

Observe can also be found (much more easily) by first expanding the exponential as

$$e^{j\theta} = \cos\theta + j \sin\theta,$$

and then taking derivatives. Contrast this to an application of $(e^x)' = x' e^x$, which would lead us to believe that $(e^{j\theta})' = \theta j' e^{j\theta}$, which is incorrect.

1.8 DUALITY.

Pseudoscalar multiplication maps a subspace to its orthogonal complement, called the dual.

Definition 1.23: Dual

Given a multivector M and a unit pseudoscalar I for the space, the dual is designated M^*, and has the value $M^* = MI$.

For example, in $\mathbb{R}^2$ with $i = \mathbf{e}_{12}$, the dual of a vector $\mathbf{x} = x\mathbf{e}_1 + y\mathbf{e}_2$ is

$$\begin{aligned} \mathbf{x}i &= (x\mathbf{e}_1 + y\mathbf{e}_2)i \\ &= x\mathbf{e}_2 - y\mathbf{e}_1, \end{aligned} \quad (1.63)$$

which is perpendicular to **x**. This was also observed in eq. (1.25) and eq. (1.26) which showed that multiplying a vector in a plane by a unit pseudoscalar for that plane, maps a vector (say **x**) to a vector **x**i that is orthogonal to **x**. The direction that **x**i points depends on the orientation of the chosen pseudoscalar.

In three dimensions, a bivector can be factored into two orthogonal vector factors, say $B = \mathbf{ab}$, and pseudoscalar multiplication of $BI = \mathbf{c}$ produces a vector **c** that is orthogonal to the factors **a, b**. For example, the unit vectors and bivectors are related in the following fashion

$$\begin{aligned} \mathbf{e}_2\mathbf{e}_3 &= \mathbf{e}_1 I & \mathbf{e}_2\mathbf{e}_3 I &= -\mathbf{e}_1 \\ \mathbf{e}_3\mathbf{e}_1 &= \mathbf{e}_2 I & \mathbf{e}_3\mathbf{e}_1 I &= -\mathbf{e}_2 \\ \mathbf{e}_1\mathbf{e}_2 &= \mathbf{e}_3 I & \mathbf{e}_1\mathbf{e}_2 I &= -\mathbf{e}_3. \end{aligned} \quad (1.64)$$

For example, with $\mathbf{r} = a\mathbf{e}_1 + b\mathbf{e}_2$, the dual is

$$\begin{aligned} \mathbf{r}I &= (a\mathbf{e}_1 + b\mathbf{e}_2)\,\mathbf{e}_{123} \\ &= a\mathbf{e}_{23} + b\mathbf{e}_{31} \\ &= \mathbf{e}_3\left(-a\mathbf{e}_2 + b\mathbf{e}_1\right). \end{aligned} \quad (1.65)$$

Here $\mathbf{e}_3$ was factored out of the resulting bivector, leaving two factors both perpendicular to the original vector. Every vector that lies in the span of the plane represented by this bivector is perpendicular to the original vector. This is illustrated in fig. 1.12.

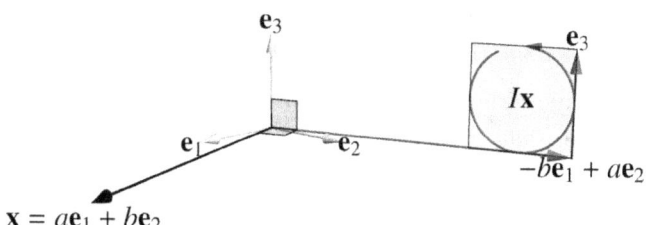

Figure 1.12: $\mathbb{R}^3$ duality illustration.

Some notes on duality

- The dual of any scalar is a pseudoscalar, whereas the dual of a pseudoscalar is a scalar.

1.9 VECTOR PRODUCT, DOT PRODUCT AND WEDGE PRODUCT.

- The dual of a k-vector in a N-dimensional space is an $(N-k)$-vector. For example, eq. (1.64) showed that the dual of a 1-vector in $\mathbb{R}^3$ was a $(3-1)$-vector, and the dual of a 2-vector is a $(3-2)$-vector. In $\mathbb{R}^7$, say, the dual of a 2-vector is a 5-vector, the dual of a 3-vector is a 4-vector, and so forth.

- All factors of the dual $(N-k)$-vector are orthogonal to all the factors of the k-vector. Looking to eq. (1.64) for examples, we see that the dual of the bivector $\mathbf{e}_2\mathbf{e}_3$ is $\mathbf{e}_1$, and both factors of the bivector $\mathbf{e}_2, \mathbf{e}_3$ are orthogonal to the dual of that bivector $\mathbf{e}_1$.

- Some authors use different sign conventions for duality, in particular, designating the dual as MI^{-1}, which can have a different sign. As one may choose pseudoscalars that differ in sign anyways, the duality convention doesn't matter too much, provided one is consistent.

1.9 VECTOR PRODUCT, DOT PRODUCT AND WEDGE PRODUCT.

The product of two colinear vectors is a scalar, and the product of two orthogonal vectors is a bivector. The product of two general vectors is a multivector with structure to be determined. In the process of exploring this structures we will prove the following theorems.

> **Theorem 1.3: Dot product as a scalar selection.**
>
> The dot product of two vectors $\mathbf{a}, \mathbf{b}$ can be computed by scalar grade selection
>
> $$\mathbf{a} \cdot \mathbf{b} = \langle \mathbf{ab} \rangle.$$

Proving theorem 1.3 will be deferred slightly. Computation of dot products using scalar grade selection will be used extensively in this book, as scalar grade selection of vector products will often be the easiest way to compute a dot product.

> **Theorem 1.4: Grades of a vector product.**

> The product of two vectors is a multivector that has only grades 0 and 2. That is
>
> $$\mathbf{ab} = \langle \mathbf{ab} \rangle + \langle \mathbf{ab} \rangle_2.$$

We've seen special cases of both theorem 1.3 and theorem 1.4 considering colinear and orthogonal vectors. The more general cases will be proven in two ways, first using a polar representation of two vectors in a plane, and then using a coordinate expansion of the vectors. This will also provide some insight about the bivector component of the product of two vectors.

Proof. Let $\{\mathbf{u}, \mathbf{v}\}$ be an orthonormal basis for the plane containing two vectors $\mathbf{a}$ and $\mathbf{b}$, where the rotational sense of $\mathbf{u} \to \mathbf{v}$ is in the same direction as the shortest rotation that takes $\mathbf{a}/\|\mathbf{a}\| \to \mathbf{b}/\|\mathbf{b}\|$, as illustrated in fig. 1.13.

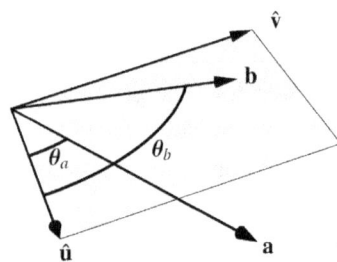

Figure 1.13: Two vectors in a plane.

Let $i_{\mathbf{uv}} = \mathbf{uv}$ designate the unit pseudoscalar for the plane, so that a polar representation of $\mathbf{a}, \mathbf{b}$ is

$$\begin{aligned} \mathbf{a} &= \|\mathbf{a}\| \, \mathbf{u} e^{i_{\mathbf{uv}} \theta_a} = \|\mathbf{a}\| \, e^{-i_{\mathbf{uv}} \theta_a} \mathbf{u} \\ \mathbf{b} &= \|\mathbf{b}\| \, \mathbf{u} e^{i_{\mathbf{uv}} \theta_b} = \|\mathbf{b}\| \, e^{-i_{\mathbf{uv}} \theta_b} \mathbf{u}, \end{aligned} \qquad (1.66)$$

1.9 VECTOR PRODUCT, DOT PRODUCT AND WEDGE PRODUCT.

The vector product of these two vectors is

$$\begin{aligned} \mathbf{ab} &= \left(\|\mathbf{a}\| \, e^{-i_{\mathbf{uv}}\theta_a} \mathbf{u} \right) \left(\|\mathbf{b}\| \, \mathbf{u} e^{i_{\mathbf{uv}}\theta_b} \right) \\ &= \|\mathbf{a}\| \, \|\mathbf{b}\| \, e^{-i_{\mathbf{uv}}\theta_a} (\mathbf{uu}) e^{i_{\mathbf{uv}}\theta_b} \\ &= \|\mathbf{a}\| \, \|\mathbf{b}\| \, e^{i_{\mathbf{uv}}(\theta_b - \theta_a)}. \\ &= \|\mathbf{a}\| \, \|\mathbf{b}\| \, \left(\cos(\theta_b - \theta_a) + i_{\mathbf{uv}} \sin(\theta_b - \theta_a) \right). \end{aligned} \qquad (1.67)$$

This completes the proof of theorem 1.4, as we see that the product of two vectors is a multivector with only grades 0 and 2. It is also clear that the scalar grade of the end result of eq. (1.67) is the $\mathbb{R}^N$ dot product, completing the proof of theorem 1.3. $\square$

The grade 2 component of the vector product is something new that requires a name, which we call the wedge product.

Definition 1.24: Wedge product of two vectors.

Given two vectors $\mathbf{a}, \mathbf{b}$, the wedge product of the vectors is defined as a grade-2 selection operation of their vector product and written

$$\mathbf{a} \wedge \mathbf{b} \equiv \langle \mathbf{ab} \rangle_2.$$

Given this notation, the product of two vectors can be written

$$\mathbf{ab} = \mathbf{a} \cdot \mathbf{b} + \mathbf{a} \wedge \mathbf{b}.$$

The split of a vector product into dot and wedge product components is also important. However, to utilize it, the properties of the wedge product have to be determined.

Summarizing eq. (1.67) with our new operators, where $i_{\mathbf{uv}} = \mathbf{uv}$, and $\mathbf{u}, \mathbf{v}$ are orthonormal vectors in the plane of $\mathbf{a}, \mathbf{b}$ with the same sense of the smallest rotation that takes $\mathbf{a}$ to $\mathbf{b}$, the vector, dot and wedge products are

$$\begin{aligned} \mathbf{ab} &= \|\mathbf{a}\| \, \|\mathbf{b}\| \exp\left(i_{\mathbf{uv}}(\theta_b - \theta_a) \right) \\ \mathbf{a} \cdot \mathbf{b} &= \|\mathbf{a}\| \, \|\mathbf{b}\| \cos(\theta_b - \theta_a) \\ \mathbf{a} \wedge \mathbf{b} &= i_{\mathbf{uv}} \|\mathbf{a}\| \, \|\mathbf{b}\| \sin(\theta_b - \theta_a). \end{aligned} \qquad (1.68)$$

Example 1.2: Products of two unit vectors.

To develop some intuition about the vector product, let's consider product of two unit vectors **a**, **b** in the equilateral triangle of fig. 1.14, where

$$\mathbf{a} = \frac{1}{\sqrt{2}} (\mathbf{e}_3 + \mathbf{e}_1) = \mathbf{e}_3 \exp(\mathbf{e}_{31}\pi/4)$$
$$\mathbf{b} = \frac{1}{\sqrt{2}} (\mathbf{e}_3 + \mathbf{e}_2) = \mathbf{e}_3 \exp(\mathbf{e}_{32}\pi/4).$$
(1.69)

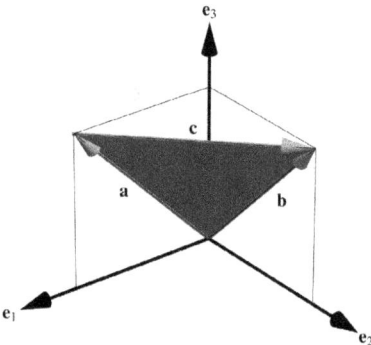

Figure 1.14: Equilateral triangle in $\mathbb{R}^3$.

The product of these vectors is

$$\begin{aligned}\mathbf{ab} &= \frac{1}{2} (\mathbf{e}_3 + \mathbf{e}_1)(\mathbf{e}_3 + \mathbf{e}_2) \\ &= \frac{1}{2} (1 + \mathbf{e}_{32} + \mathbf{e}_{13} + \mathbf{e}_{12}) \\ &= \frac{1}{2} + \frac{\sqrt{3}}{2} \frac{\mathbf{e}_{32} + \mathbf{e}_{13} + \mathbf{e}_{12}}{\sqrt{3}}.\end{aligned}$$
(1.70)

Let the bivector factor be designated

$$j = \frac{\mathbf{e}_{32} + \mathbf{e}_{13} + \mathbf{e}_{12}}{\sqrt{3}}.$$
(1.71)

The reader can check (exercise 1.11) that j is a unit bivector (i.e. it squares to -1), allowing us to write

$$\begin{aligned}\mathbf{ab} &= \frac{1}{2} + \frac{\sqrt{3}}{2}j \\ &= \cos(\pi/3) + j\sin(\pi/3) \\ &= \exp(j\pi/3).\end{aligned} \quad (1.72)$$

Since both vector factors were unit length, this "complex" exponential has no leading scalar factor contribution from $\|\mathbf{a}\| \|\mathbf{b}\|$.

Now, let's calculate the vector product using the polar form, which gives

$$\begin{aligned}\mathbf{ab} &= (\exp(-\mathbf{e}_{31}\pi/4)\mathbf{e}_3)(\mathbf{e}_3 \exp(\mathbf{e}_{32}\pi/4)) \\ &= \exp(-\mathbf{e}_{31}\pi/4)\exp(\mathbf{e}_{32}\pi/4).\end{aligned} \quad (1.73)$$

The product of two unit vectors, each with a component in the z-axis direction, results in a product of complex exponential rotation operators, each a grade $(0, 2)$-multivectors. The product of these complex exponentials is another grade $(0, 2)$-multivector. This is a specific example of the product of two rotation operators producing a rotation operator for the composition of rotations, as follows

$$\exp(\mathbf{e}_{13}\pi/4)\exp(\mathbf{e}_{32}\pi/4) = \exp(j\pi/3). \quad (1.74)$$

The rotation operator that describes the composition of rotations has a different rotational plane, and rotates through a different rotation angle.

We are left with a geometric interpretation for the vector product. The product of two vectors can be interpreted as a rotation and scaling operator. The product of two unit length vectors can be interpreted as a pure rotation operator.

Two wedge product properties can be immediately deduced from the polar representation of eq. (1.68)

1. $\mathbf{b} \wedge \mathbf{a} = -\mathbf{a} \wedge \mathbf{b}$.

2. $\mathbf{a} \wedge (\alpha \mathbf{a}) = 0, \quad \forall \alpha \in \mathbb{R}$.

We have now had a few hints that the wedge product might be related to the cross product. Given two vectors $\mathbf{a}, \mathbf{b}$ both the wedge and the cross product contain a $\|\mathbf{a}\| \|\mathbf{b}\| \sin \Delta\theta$ factor, and both the wedge and cross product are antisymmetric operators. The cross product is a bilinear operator $(\mathbf{a}+\mathbf{b}) \times (\mathbf{c}+\mathbf{d}) = \mathbf{a} \times \mathbf{c} + \mathbf{a} \times \mathbf{d} + \mathbf{b} \times \mathbf{c} + \mathbf{b} \times \mathbf{d}$. To see whether this is the case for the wedge product, let's examine the coordinate expansion of the wedge product. Let

$$\begin{aligned} \mathbf{a} &= \sum_i a_i \mathbf{e}_i \\ \mathbf{b} &= \sum_i b_i \mathbf{e}_i. \end{aligned} \tag{1.75}$$

The product of these vectors is

$$\begin{aligned} \mathbf{ab} &= \left(\sum_i a_i \mathbf{e}_i\right)\left(\sum_j b_j \mathbf{e}_j\right) \\ &= \sum_{ij} a_i b_j \mathbf{e}_i \mathbf{e}_j \\ &= \sum_{i=j} a_i b_j \mathbf{e}_i \mathbf{e}_j + \sum_{i \neq j} a_i b_j \mathbf{e}_i \mathbf{e}_j. \end{aligned} \tag{1.76}$$

Since $\mathbf{e}_i \mathbf{e}_i = 1$, we see again that the scalar component of the product is the dot product $\sum_i a_i b_i$. The remaining grade 2 components are the wedge product, for which the coordinate expansion can be simplified further

$$\begin{aligned} \mathbf{a} \wedge \mathbf{b} &= \sum_{i \neq j} a_i b_j \mathbf{e}_i \mathbf{e}_j \\ &= \sum_{i<j} a_i b_j \mathbf{e}_i \mathbf{e}_j + \sum_{j<i} a_i b_j \mathbf{e}_i \mathbf{e}_j \\ &= \sum_{i<j} a_i b_j \mathbf{e}_i \mathbf{e}_j + \sum_{i<j} a_j b_i \mathbf{e}_j \mathbf{e}_i \\ &= \sum_{i<j} (a_i b_j - a_j b_i) \mathbf{e}_i \mathbf{e}_j. \end{aligned} \tag{1.77}$$

The scalar factors can be written as 2x2 determinants

$$\boxed{\mathbf{a} \wedge \mathbf{b} = \sum_{i<j} \begin{vmatrix} a_i & a_j \\ b_i & b_j \end{vmatrix} \mathbf{e}_i \mathbf{e}_j.} \tag{1.78}$$

It is now straightforward to show that the wedge product is distributive and bilinear (exercise 1.10). It is also simple to use eq. (1.78) to show that $\mathbf{b} \wedge \mathbf{a} = -\mathbf{a} \wedge \mathbf{b}$ and $\mathbf{a} \wedge \mathbf{a} = 0$.

1.9 VECTOR PRODUCT, DOT PRODUCT AND WEDGE PRODUCT.

For $\mathbb{R}^2$ there is only one term in eq. (1.78)

$$\mathbf{a} \wedge \mathbf{b} = \begin{vmatrix} a_1 & a_2 \\ b_1 & b_2 \end{vmatrix} \mathbf{e}_1 \mathbf{e}_2. \tag{1.79}$$

We are used to writing the cross product as a 3×3 determinant, which can also be done with the coordinate expansion of the $\mathbb{R}^3$ wedge product

$$\mathbf{a} \wedge \mathbf{b} = \sum_{ij \in \{12,13,23\}} \begin{vmatrix} a_i & a_j \\ b_i & b_j \end{vmatrix} \mathbf{e}_i \mathbf{e}_j = \begin{vmatrix} \mathbf{e}_2 \mathbf{e}_3 & \mathbf{e}_3 \mathbf{e}_1 & \mathbf{e}_1 \mathbf{e}_2 \\ a_1 & a_2 & a_3 \\ b_1 & b_2 & b_3 \end{vmatrix}. \tag{1.80}$$

Let's summarize the wedge product properties and relations we have found so far, comparing the $\mathbb{R}^3$ wedge product to the cross product

Table 1.3: Cross product and $\mathbb{R}^3$ wedge product comparison.

Property	Cross product	Wedge product
Same vectors	$\mathbf{a} \times \mathbf{a} = 0$	$\mathbf{a} \wedge \mathbf{a} = 0$
Antisymmetry	$\mathbf{b} \times \mathbf{a} = -\mathbf{a} \times \mathbf{b}$	$\mathbf{b} \wedge \mathbf{a} = -\mathbf{a} \wedge \mathbf{b}$
Linear	$\mathbf{a} \times (\alpha \mathbf{b}) = \alpha(\mathbf{a} \times \mathbf{b})$	$\mathbf{a} \wedge (\alpha \mathbf{b}) = \alpha(\mathbf{a} \wedge \mathbf{b})$
Distributive	$\mathbf{a} \times (\mathbf{b} + \mathbf{c}) = \mathbf{a} \times \mathbf{b} + \mathbf{a} \times \mathbf{c}$	$\mathbf{a} \wedge (\mathbf{b} + \mathbf{c}) = \mathbf{a} \wedge \mathbf{b} + \mathbf{a} \wedge \mathbf{c}$
Determinant expansion	$\mathbf{a} \times \mathbf{b} = \begin{vmatrix} \mathbf{e}_1 & \mathbf{e}_2 & \mathbf{e}_3 \\ a_1 & a_2 & a_3 \\ b_1 & b_2 & b_3 \end{vmatrix}$	$\mathbf{a} \wedge \mathbf{b} = \begin{vmatrix} \mathbf{e}_2 \mathbf{e}_3 & \mathbf{e}_3 \mathbf{e}_1 & \mathbf{e}_1 \mathbf{e}_2 \\ a_1 & a_2 & a_3 \\ b_1 & b_2 & b_3 \end{vmatrix}$
Coordinate expansion	$\mathbf{a} \times \mathbf{b} = \sum_{i<j} \begin{vmatrix} a_i & a_j \\ b_i & b_j \end{vmatrix} \mathbf{e}_i \times \mathbf{e}_j$	$\mathbf{a} \wedge \mathbf{b} = \sum_{i<j} \begin{vmatrix} a_i & a_j \\ b_i & b_j \end{vmatrix} \mathbf{e}_i \mathbf{e}_j$
Polar form	$\hat{\mathbf{n}} \|\mathbf{a}\| \|\mathbf{b}\| \sin(\theta_b - \theta_a)$	$i \|\mathbf{a}\| \|\mathbf{b}\| \sin(\theta_b - \theta_a)$

All the wedge properties except the determinant expansion above are valid in any dimension. Comparing eq. (1.80) to the determinant representation of the cross product, and referring to eq. (1.64), shows that the $\mathbb{R}^3$

wedge product is related to the cross product by a duality transformation $i = I\hat{n}$, or

$$\mathbf{a} \wedge \mathbf{b} = I(\mathbf{a} \times \mathbf{b}). \tag{1.81}$$

The direction of the cross product $\mathbf{a} \times \mathbf{b}$ is orthogonal to the plane represented by the bivector $\mathbf{a} \wedge \mathbf{b}$. The magnitude of both (up to a sign) is the area of the parallelogram spanned by the two vectors.

Example 1.3: Wedge and cross product relationship.

To take some of the abstraction from eq. (1.81) let's consider a specific example. Let

$$\begin{aligned} \mathbf{a} &= \mathbf{e}_1 + 2\mathbf{e}_2 + 3\mathbf{e}_3 \\ \mathbf{b} &= 4\mathbf{e}_1 + 5\mathbf{e}_2 + 6\mathbf{e}_3. \end{aligned} \tag{1.82}$$

The reader should check that the cross product of these two vectors is

$$\mathbf{a} \times \mathbf{b} = -3\mathbf{e}_1 + 6\mathbf{e}_2 - 3\mathbf{e}_3. \tag{1.83}$$

By direct computation, we find that the wedge and the cross products are related by a $\mathbb{R}^3$ pseudoscalar factor

$$\begin{aligned} \mathbf{a} \wedge \mathbf{b} &= (\mathbf{e}_1 + 2\mathbf{e}_2 + 3\mathbf{e}_3) \wedge (4\mathbf{e}_1 + 5\mathbf{e}_2 + 6\mathbf{e}_3) \\ &= \cancel{\mathbf{e}_1 \wedge (4\mathbf{e}_1)} + \cancel{(2\mathbf{e}_2) \wedge (5\mathbf{e}_2)} + \cancel{(3\mathbf{e}_3) \wedge (6\mathbf{e}_3)} \\ &\quad + 5\mathbf{e}_{12} + 6\mathbf{e}_{13} + 8\mathbf{e}_{21} + 12\mathbf{e}_{23} + 12\mathbf{e}_{31} + 15\mathbf{e}_{32} \\ &= (5-8)\mathbf{e}_{12} + (6-12)\mathbf{e}_{13} + (12-15)\mathbf{e}_{23} \\ &= -3\mathbf{e}_{12} - 6\mathbf{e}_{13} - 3\mathbf{e}_{23} \\ &= \mathbf{e}_{123}(-3\mathbf{e}_3) + \mathbf{e}_{132}(-6\mathbf{e}_2) + \mathbf{e}_{231}(-3\mathbf{e}_1) \\ &= \mathbf{e}_{123}(-3\mathbf{e}_3 + 6\mathbf{e}_2 - 3\mathbf{e}_1) \\ &= I(\mathbf{a} \times \mathbf{b}). \end{aligned} \tag{1.84}$$

The relationship between the wedge and cross products allows us to express the $\mathbb{R}^3$ vector product as a multivector combination of the dot and cross products

$$\mathbf{ab} = \mathbf{a} \cdot \mathbf{b} + I(\mathbf{a} \times \mathbf{b}). \tag{1.85}$$

1.9 VECTOR PRODUCT, DOT PRODUCT AND WEDGE PRODUCT.

This is a very important relationship.

In particular, for electromagnetism, eq. (1.85) can be used with $\mathbf{a} = \nabla$ to combine pairs of Maxwell's equations to form pairs of multivector gradient equations, which can be merged further. The resulting multivector equation will be called Maxwell equation (singular), and will be the starting point of all our electromagnetic analysis.

We are used to expressing the dot and cross product components of eq. (1.85) separately, for example, as

$$\begin{aligned} \mathbf{a} \cdot \mathbf{b} &= \|\mathbf{a}\| \|\mathbf{b}\| \cos(\Delta\theta) \\ \mathbf{a} \times \mathbf{b} &= \hat{\mathbf{n}} \|\mathbf{a}\| \|\mathbf{b}\| \sin(\Delta\theta), \end{aligned} \quad (1.86)$$

Introducing a unit bivector i_{ab} normal to the unit normal $\hat{\mathbf{n}}$ (i.e. $i_{ab}\hat{\mathbf{n}} = \mathbf{e}_{123}$), we can assemble eq. (1.86) into a $\cos + i \sin$ form using eq. (1.85)

$$\begin{aligned} \mathbf{ab} &= \|\mathbf{a}\| \|\mathbf{b}\| \left(\cos(\Delta\theta) + I\hat{\mathbf{n}} \sin(\Delta\theta) \right) \\ &= \|\mathbf{a}\| \|\mathbf{b}\| \exp(i_{ab}\Delta\theta). \end{aligned} \quad (1.87)$$

Exercise 1.8 Wedge product of colinear vectors.

Given $\mathbf{b} = \alpha\mathbf{a}$, use eq. (1.78) to show that the wedge product of any pair of colinear vectors is zero.

Exercise 1.9 Wedge product antisymmetry.

Prove that the wedge product is antisymmetric using using eq. (1.78).

Exercise 1.10 Wedge product distributivity and bilinearity.

For vectors $\mathbf{a}, \mathbf{b}, \mathbf{c}$ and $\mathbf{d}$, and scalars α, β use eq. (1.78) to show that

a. the wedge product is distributive

$$(\mathbf{a} + \mathbf{b}) \wedge (\mathbf{c} + \mathbf{d}) = \mathbf{a} \wedge \mathbf{c} + \mathbf{a} \wedge \mathbf{d} + \mathbf{b} \wedge \mathbf{c} + \mathbf{b} \wedge \mathbf{d},$$

b. and show that the wedge product is bilinear

$$(\alpha\mathbf{a}) \wedge (\beta\mathbf{b}) = (\alpha\beta)(\mathbf{a} \wedge \mathbf{b}).$$

Note that these imply the wedge product also has the cross product filtering property $\mathbf{a} \wedge (\mathbf{b} + \alpha\mathbf{a}) = \mathbf{a} \wedge \mathbf{b}$.

Exercise 1.11 Unit bivector.

Verify by explicit multiplication that the bivector of eq. (1.71) squares to -1.

1.10 REVERSE.

> **Definition 1.25: Reverse**
>
> Let A be a multivector with j multivector factors, $A = B_1 B_2 \cdots B_j$, not necessarily orthogonal. The reverse $A^\dagger$, or reversion, of this multivector A is
>
> $$A^\dagger = B_j^\dagger B_{j-1}^\dagger \cdots B_1^\dagger.$$
>
> Scalars and vectors are their own reverse, and the reverse of a sum of multivectors is the sum of the reversions of its summands.

Examples:

$$\begin{aligned}(1 + 2\mathbf{e}_{12} + 3\mathbf{e}_{321})^\dagger &= 1 + 2\mathbf{e}_{21} + 3\mathbf{e}_{123} \\ ((1+\mathbf{e}_1)(\mathbf{e}_{23} - \mathbf{e}_{12}))^\dagger &= (\mathbf{e}_{32} + \mathbf{e}_{12})(1 + \mathbf{e}_1).\end{aligned} \quad (1.97)$$

A useful relation for k-vectors that are composed entirely of products of orthogonal vectors exists. We call such k-vectors blades

> **Definition 1.26: Blade.**
>
> A product of k orthogonal vectors is called a k-blade, or a blade of grade k. A grade zero blade is a scalar.
>
> The notation $F \in \bigwedge^k$ is used in the literature to indicate that F is a blade of grade k.

Any k-blade is also a k-vector, but not all k-vectors are k-blades. For example in $\mathbb{R}^4$ the bivector $\mathbf{e}_{12} + \mathbf{e}_{34}$ is not a 2-blade, since it cannot be factored into orthogonal products, whereas any $\mathbb{R}^3$ bivector, such as $\mathbf{e}_{12} + \mathbf{e}_{23} + \mathbf{e}_{31}$ is a blade (exercise 1.19). This will be relevant when formulating rotations since bivectors that are blades can be used to simply describe rotations or Lorentz boosts [4] whereas it is not easily possible to compute an exponential of a non-blade bivector argument.

---

[4] A rotation in spacetime.

> **Theorem 1.5: Reverse of k-blade.**
>
> The reverse of a k-blade $A_k = \mathbf{a}_1 \mathbf{a}_2 \cdots \mathbf{a}_k$ is given by
>
> $$A_k^\dagger = (-1)^{k(k-1)/2} A_k.$$

Proof. We prove by successive interchange of factors.

$$\begin{aligned}
A_k^\dagger &= \mathbf{a}_k \mathbf{a}_{k-1} \cdots \mathbf{a}_1 \\
&= (-1)^{k-1} \mathbf{a}_1 \mathbf{a}_k \mathbf{a}_{k-1} \cdots \mathbf{a}_2 \\
&= (-1)^{k-1}(-1)^{k-2} \mathbf{a}_1 \mathbf{a}_2 \mathbf{a}_k \mathbf{a}_{k-1} \cdots \mathbf{a}_3 \\
&\vdots \\
&= (-1)^{k-1}(-1)^{k-2} \cdots (-1)^1 \mathbf{a}_1 \mathbf{a}_2 \cdots \mathbf{a}_k \\
&= (-1)^{k(k-1)/2} \mathbf{a}_1 \mathbf{a}_2 \cdots \mathbf{a}_k \\
&= (-1)^{k(k-1)/2} A_k.
\end{aligned}$$
$\square$

A special, but important case, is the reverse of the $\mathbb{R}^3$ pseudoscalar, which is negated by reversion

$$\boxed{I^\dagger = -I.} \tag{1.98}$$

1.11 COMPLEX REPRESENTATIONS.

We've seen that bivectors like $\mathbf{e}_{12}$ square to minus one. Geometric algebra has infinitely many such imaginary numbers, which can be utilized to introduce problem specific "complex planes" as desired. In three dimensional and higher spaces, imaginary representations (such as the $\mathbb{R}^3$ pseudoscalar) with grades higher than two are also possible.

Using the reversion relationship of eq. (1.98), we can see that the I behaves as an imaginary

$$\begin{aligned}
I^2 &= I(-I^\dagger) \\
&= -(\mathbf{e}_1 \mathbf{e}_2 \mathbf{e}_3)(\mathbf{e}_3 \mathbf{e}_2 \mathbf{e}_1) \quad = -\mathbf{e}_1 \mathbf{e}_2 \mathbf{e}_2 \mathbf{e}_1 \\
&= -\mathbf{e}_1 \mathbf{e}_1 \\
&= -1.
\end{aligned} \tag{1.99}$$

Given many possible imaginary representations, complex and complex-like numbers can be represented in GA for any k-vector i that satisfies $i^2 = -1$ since the multivector

$$z = x + iy, \qquad (1.100)$$

will have all the required properties of a complex number.

For example, in Euclidean spaces we could use either of

$$\begin{aligned} i &= \frac{\mathbf{u} \wedge \mathbf{v}}{\sqrt{-(\mathbf{u} \wedge \mathbf{v})^2}} \\ I &= \frac{\mathbf{u} \wedge \mathbf{v} \wedge \mathbf{w}}{\sqrt{-(\mathbf{u} \wedge \mathbf{v} \wedge \mathbf{w})^2}}, \end{aligned} \qquad (1.101)$$

provided $\mathbf{u}, \mathbf{v}, \mathbf{w}$ are linearly independent vectors. Given a set of orthonormal vectors $\mathbf{u}, \mathbf{v}, \mathbf{w}$, then

$$\begin{aligned} i &= \mathbf{uv} \\ I &= \mathbf{uvw}, \end{aligned} \qquad (1.102)$$

are also suitable as imaginaries. Note that in eq. (1.102), the bivector i differs from the unit $\mathbb{R}^2$ pseudoscalar only by a sign ($i = \pm \mathbf{e}_{12}$), and the trivector I, also differs from the $\mathbb{R}^3$ unit pseudoscalar only by a sign ($I = \pm \mathbf{e}_{123}$).

Other complex number like representations are also possible with GA. Quaternions, which are often used in computer graphics to represent rotations, are the set $q \in \{a + x\mathbf{i} + y\mathbf{j} + z\mathbf{k} \mid a, x, y, z \in \mathbb{R}\}$ where

$$\begin{aligned} \mathbf{i}^2 &= \mathbf{j}^2 = \mathbf{k}^2 = -1 \\ \mathbf{ij} &= \mathbf{k} = -\mathbf{ji} \\ \mathbf{jk} &= \mathbf{i} = -\mathbf{kj} \\ \mathbf{ki} &= \mathbf{j} = -\mathbf{ik}. \end{aligned} \qquad (1.103)$$

Like complex numbers, quaternions can be represented in GA as grade $(0, 2)$-multivectors, but require three imaginaries instead of one.

Exercise 1.12 Quaternions.

Show that the relations eq. (1.103) are satisfied by the unit bivectors $\mathbf{i} = \mathbf{e}_{32}, \mathbf{j} = \mathbf{e}_{13}, \mathbf{k} = \mathbf{e}_{21}$, demonstrating that quaternions, like complex numbers, may be represented as multivector subspaces.

1.12 MULTIVECTOR DOT PRODUCT.

In general the product of two k-vectors is a multivector, with a selection of different grades. For example, the product of two bivectors may have grades 0, 2, or 4

$$\mathbf{e}_{12}\left(\mathbf{e}_{21} + \mathbf{e}_{23} + \mathbf{e}_{34}\right) = 1 + \mathbf{e}_{13} + \mathbf{e}_{1234}. \tag{1.107}$$

Similarly, the product of a vector and bivector generally has grades 1 and 3

$$\mathbf{e}_1\left(\mathbf{e}_{12} + \mathbf{e}_{23}\right) = \mathbf{e}_2 + \mathbf{e}_{123}. \tag{1.108}$$

The dot product was identified with scalar grade selection, which picks out the lowest grade of their product. This motivates the definition of a general multivector dot product

> **Definition 1.27: Multivector dot product**
>
> The dot (or inner) product of two multivectors $A = \sum_{i=0}^{N} \langle A \rangle_i$, $B = \sum_{i=0}^{N} \langle B \rangle_i$ is defined as
>
> $$A \cdot B \equiv \sum_{i,j=0}^{N} \langle A_i B_j \rangle_{|i-j|}.$$

If A, B are k-vectors with equal grade, then the dot product is just the scalar selection of their product

$$A \cdot B = \langle AB \rangle, \tag{1.109}$$

and if A, B are a k-vectors with grades $r \neq s$ respectively, then their dot product is a single grade selection

$$A \cdot B = \langle AB \rangle_{|r-s|}. \tag{1.110}$$

> **Example 1.4: Multivector dot products.**

The most common and useful multivector dot products are for pairs of multivectors that are each entirely a specific grade, such as a vector-bivector dot product

$$(\mathbf{e}_1 + 2\mathbf{e}_2) \cdot (\mathbf{e}_{12} + \mathbf{e}_{23}) = \langle (\mathbf{e}_1 + 2\mathbf{e}_2)(\mathbf{e}_{12} + \mathbf{e}_{23}) \rangle_1$$
$$= \mathbf{e}_2 - 2\mathbf{e}_1 + 2\mathbf{e}_3, \quad (1.111)$$

or a vector-trivector dot product

$$(\mathbf{e}_1 + 2\mathbf{e}_2) \cdot \mathbf{e}_{123} = \langle (\mathbf{e}_1 + 2\mathbf{e}_2) \mathbf{e}_{123} \rangle_2$$
$$= \mathbf{e}_{23} + 2\mathbf{e}_{31}. \quad (1.112)$$

Should the products be of mixed grade, then we sum all the individual dot products

$$(1 + \mathbf{e}_1 + 2\mathbf{e}_{23}) \cdot (\mathbf{e}_2 - \mathbf{e}_{31})$$
$$= \langle 1\mathbf{e}_2 \rangle_1 + \langle \mathbf{e}_1\mathbf{e}_2 \rangle + 2\langle \mathbf{e}_{23}\mathbf{e}_2 \rangle_1$$
$$- \langle 1\mathbf{e}_{31} \rangle_2 - \langle \mathbf{e}_1\mathbf{e}_{31} \rangle_1 - 2\langle \mathbf{e}_{23}\mathbf{e}_{31} \rangle \quad (1.113)$$
$$= \mathbf{e}_2 - 2\mathbf{e}_3 + \mathbf{e}_{13} + \mathbf{e}_3.$$

Unfortunately, the notation for the multivector dot product is not standardized. In particular, some authors [7] prefer left and right contraction operations that omit the absolute value in the grade selections. A dot product like operator for scalar selection is also common.

Definition 1.28: Alternate dot products.

The left and right contraction operations are respectively defined as

$$A \rfloor B = \sum_{i,j=0}^{N} \langle A_i B_j \rangle_{j-i}$$

$$A \lfloor B = \sum_{i,j=0}^{N} \langle A_i B_j \rangle_{i-j},$$

where any selection of a negative grade is taken to be zero. The scalar product is defined as

$$A * B = \sum_{i,j=0}^{N} \langle A_i B_j \rangle$$

In an attempt to avoid inundating the reader with too many new operators, this book will stick to the dot, wedge and grade selection operators. However, these alternates are common enough that they deserve mentioning.

1.12.1 *Dot product of a vector and bivector*

An important example of the generalized dot product is the dot product of a vector and bivector. Unlike the dot product of two vectors, a vector-bivector dot product is order dependent.

The vector dot product is zero when the two vectors are orthogonal. This is also true if the vector and bivector are orthogonal, that is, having no common factor, as in

$$\mathbf{e}_1 \cdot \mathbf{e}_{23} = \langle \mathbf{e}_{123} \rangle_1 = 0. \tag{1.114}$$

On the other hand, a non-zero vector-bivector dot product requires the vector to have some overlap with the bivector. A bivector formed from the product of two orthogonal vectors $B = \mathbf{ab}$, where $\mathbf{a} \cdot \mathbf{b} = 0$, will have a non-zero dot product with any vector that lies in span $\{\mathbf{a}, \mathbf{b}\}$

$$(\alpha \mathbf{a} + \beta \mathbf{b}) \cdot (\mathbf{ab}) = \alpha \|\mathbf{a}\|^2 \mathbf{b} - \beta \|\mathbf{b}\|^2 \mathbf{a}. \tag{1.115}$$

It is often useful to be able to expand a vector-bivector dot product. A useful identity for such an expansion is

---

Theorem 1.6: Dot product of vector and wedge product.

The dot product of a vector $\mathbf{a}$ with the wedge product of two vectors $\mathbf{b}, \mathbf{c}$ distributes as

$$\mathbf{a} \cdot (\mathbf{b} \wedge \mathbf{c}) = (\mathbf{c} \wedge \mathbf{b}) \cdot \mathbf{a} = (\mathbf{a} \cdot \mathbf{b})\mathbf{c} - (\mathbf{a} \cdot \mathbf{c})\mathbf{b}.$$

Before proving this theorem, let's take a look at what it implies. This shows that only vectors with some component in the span of the plane represented by the bivector will result in a non-zero vector-bivector dot product. We also know that when a vector that lies entirely in the span of a bivector is multiplied by that bivector, the result is rotated by $\pm\pi/2$. This means that a vector-bivector dot product is orthogonal to the vector that is dotted with the bivector. This can also be seen algebraically since

$$\begin{aligned}\mathbf{a} \cdot (\mathbf{a} \cdot (\mathbf{b} \wedge \mathbf{c})) &= \mathbf{a} \cdot ((\mathbf{a} \cdot \mathbf{b})\mathbf{c} - (\mathbf{a} \cdot \mathbf{c})\mathbf{b}) \\ &= (\mathbf{a} \cdot \mathbf{c})(\mathbf{a} \cdot \mathbf{b}) - (\mathbf{a} \cdot \mathbf{c})(\mathbf{a} \cdot \mathbf{b}) \\ &= 0.\end{aligned} \quad (1.116)$$

A vector-bivector dot product selects only the component of the vector that lies in the plane of the bivector, and rotates that component by $\pm\pi/2$ in that plane.

Proof. There are (somewhat tricky) coordinate free ways to prove theorem 1.6, but a straightforward expansion in coordinates also does the job.

$$\begin{aligned}\mathbf{a} \cdot (\mathbf{b} \wedge \mathbf{c}) &= \sum_{i,j,k} a_i b_j c_k \mathbf{e}_i \cdot (\mathbf{e}_j \wedge \mathbf{e}_k) = \sum_{i,j\neq k} a_i b_j c_k \langle \mathbf{e}_i \mathbf{e}_j \mathbf{e}_k \rangle_1, \\ (\mathbf{c} \wedge \mathbf{b}) \cdot \mathbf{a} &= \sum_{i,j,k} a_i b_j c_k (\mathbf{e}_k \wedge \mathbf{e}_j) \cdot \mathbf{e}_i = \sum_{i,j\neq k} a_i b_j c_k \langle \mathbf{e}_k \mathbf{e}_j \mathbf{e}_i \rangle_1.\end{aligned} \quad (1.117)$$

If all of i, j, k are unique then $\langle \mathbf{e}_i \mathbf{e}_j \mathbf{e}_k \rangle_1 = 0$, so the vector selection is non-zero only when i equals one of j, k. For example

$$\begin{aligned}\langle \mathbf{e}_1 \mathbf{e}_1 \mathbf{e}_2 \rangle_1 &= \mathbf{e}_2 \\ \langle \mathbf{e}_1 \mathbf{e}_2 \mathbf{e}_1 \rangle_1 &= -\mathbf{e}_2.\end{aligned} \quad (1.118)$$

Given $j \neq k$, and $i = j$ or $i = k$, then it is simple to show (exercise 1.13) that

$$\langle \mathbf{e}_i \mathbf{e}_j \mathbf{e}_k \rangle_1 = \langle \mathbf{e}_k \mathbf{e}_j \mathbf{e}_i \rangle_1, \quad (1.119)$$

so $\mathbf{a} \cdot (\mathbf{b} \wedge \mathbf{c}) = (\mathbf{c} \wedge \mathbf{b}) \cdot \mathbf{a}$. Additionally, again if $j \neq k$ (exercise 1.14)

$$\langle \mathbf{e}_i \mathbf{e}_j \mathbf{e}_k \rangle_1 = \mathbf{e}_k (\mathbf{e}_j \cdot \mathbf{e}_i) - \mathbf{e}_j (\mathbf{e}_k \cdot \mathbf{e}_i). \quad (1.120)$$

Plugging eq. (1.120) back into eq. (1.117) proves the theorem

$$\begin{aligned}\mathbf{a} \cdot (\mathbf{b} \wedge \mathbf{c}) &= \sum_{i,j,k} a_i b_j c_k (\mathbf{e}_k (\mathbf{e}_j \cdot \mathbf{e}_i) - \mathbf{e}_j (\mathbf{e}_k \cdot \mathbf{e}_i)) \\ &= (\mathbf{a} \cdot \mathbf{b}) \mathbf{c} - (\mathbf{a} \cdot \mathbf{c}) \mathbf{b}.\end{aligned} \quad (1.121)$$

The RHS of eq. (1.121) shows that the vector-bivector dot product has the following relation to the $\mathbb{R}^3$ vector triple product

> **Theorem 1.7: Triple cross product.**
>
> For vectors in $\mathbb{R}^3$, the dot product of a vector and vector wedge product can be expressed as a vector triple product
>
> $$\mathbf{a} \cdot (\mathbf{b} \wedge \mathbf{c}) = (\mathbf{b} \times \mathbf{c}) \times \mathbf{a}.$$

Proof. The lazy proof, not related to geometric algebra at all, would be to invoke the well known distribution identity for the vector triple product ([15])

$$\mathbf{a} \times (\mathbf{b} \times \mathbf{c}) = (\mathbf{a} \cdot \mathbf{c})\mathbf{b} - (\mathbf{a} \cdot \mathbf{b})\mathbf{c}. \tag{1.122}$$

We can prove this result directly by applying the identity $\mathbf{a} \wedge \mathbf{b} = I(\mathbf{a} \times \mathbf{b})$ to the vector-bivector product, and then selecting the vector grade

$$\begin{aligned}
\mathbf{a}(\mathbf{b} \wedge \mathbf{c}) &= \mathbf{a}I(\mathbf{b} \times \mathbf{c}) \\
&= I(\mathbf{a} \cdot (\mathbf{b} \times \mathbf{c})) + I(\mathbf{a} \wedge (\mathbf{b} \times \mathbf{c})) \\
&= I(\mathbf{a} \cdot (\mathbf{b} \times \mathbf{c})) + I^2 \mathbf{a} \times (\mathbf{b} \times \mathbf{c}) \\
&= I(\mathbf{a} \cdot (\mathbf{b} \times \mathbf{c})) + (\mathbf{b} \times \mathbf{c}) \times \mathbf{a}.
\end{aligned} \tag{1.123}$$

This multivector has a pseudoscalar (grade 3) component, and a vector component. Selecting the grade one component, and invoking definition 1.27 to express this grade selection as a dot product, completes the proof

$$\langle \mathbf{a}(\mathbf{b} \wedge \mathbf{c}) \rangle_1 = \mathbf{a} \cdot (\mathbf{b} \wedge \mathbf{c}) = (\mathbf{b} \times \mathbf{c}) \times \mathbf{a}. \tag{1.124}$$

□

1.12.2 Bivector dot product.

Being able to compute the generalized dot product of two bivectors will also have a number of applications. When those bivectors are wedge products, there is a useful distribution identity for this dot product.

> **Theorem 1.8: Dot product distribution over wedge products.**
>
> Given two sets of wedge products $\mathbf{a} \wedge \mathbf{b}$, and $\mathbf{c} \wedge \mathbf{d}$, their dot product is
>
> $$(\mathbf{a} \wedge \mathbf{b}) \cdot (\mathbf{c} \wedge \mathbf{d}) = ((\mathbf{a} \wedge \mathbf{b}) \cdot \mathbf{c}) \cdot \mathbf{d} = (\mathbf{b} \cdot \mathbf{c})(\mathbf{a} \cdot \mathbf{d}) - (\mathbf{a} \cdot \mathbf{c})(\mathbf{b} \cdot \mathbf{d}).$$

Proof. To prove this, select the scalar grade of the product $(\mathbf{a} \wedge \mathbf{b})(\mathbf{c} \wedge \mathbf{d})$

$$\begin{aligned}(\mathbf{a} \wedge \mathbf{b})(\mathbf{c} \wedge \mathbf{d}) &= (\mathbf{a} \wedge \mathbf{b})(\mathbf{c}\mathbf{d} - \mathbf{c} \cdot \mathbf{d}) \\ &= (\mathbf{a} \wedge \mathbf{b})\mathbf{c}\mathbf{d} - (\mathbf{a} \wedge \mathbf{b})(\mathbf{c} \cdot \mathbf{d}).\end{aligned} \quad (1.125)$$

The second term, a bivector, is not of interest since it will be killed by the scalar selection operation. The remainder can be expanded in grades, first making use of the fact that a bivector-vector product has only grade 1 and 3 components

$$(\mathbf{a} \wedge \mathbf{b})\mathbf{c} = (\mathbf{a} \wedge \mathbf{b}) \cdot \mathbf{c} + \langle (\mathbf{a} \wedge \mathbf{b})\mathbf{c} \rangle_3. \quad (1.126)$$

Multiplication of the trivector term by $\mathbf{d}$ produces only grades $2, 4$, which will be discarded when we apply a scalar grade selection, so we ignore those. The product of $(\mathbf{a} \wedge \mathbf{b}) \cdot \mathbf{c}$, a vector, with $\mathbf{d}$ is a grade $(0, 2)$-multivector, of which only the scalar grade is of interest. That is

$$\begin{aligned}(\mathbf{a} \wedge \mathbf{b}) \cdot (\mathbf{c} \wedge \mathbf{d}) &= \langle (\mathbf{a} \wedge \mathbf{b})(\mathbf{c} \wedge \mathbf{d}) \rangle \\ &= ((\mathbf{a} \wedge \mathbf{b}) \cdot \mathbf{c}) \cdot \mathbf{d}.\end{aligned} \quad (1.127)$$

To complete the proof, we apply theorem 1.6

$$\begin{aligned}((\mathbf{a} \wedge \mathbf{b}) \cdot \mathbf{c}) \cdot \mathbf{d} &= \big(\mathbf{a}(\mathbf{b} \cdot \mathbf{c}) - \mathbf{b}(\mathbf{a} \cdot \mathbf{c})\big) \cdot \mathbf{d} \\ &= (\mathbf{a} \cdot \mathbf{d})(\mathbf{b} \cdot \mathbf{c}) - (\mathbf{b} \cdot \mathbf{d})(\mathbf{a} \cdot \mathbf{c}).\end{aligned} \quad (1.128)$$

□

Identity eq. (1.128) has the following $\mathbb{R}^3$ cross product equivalent.

> **Theorem 1.9: Dot products of wedges as cross products.**
>
> The dot product of two $\mathbb{R}^3$ wedge products can be expressed as cross products
>
> $$(\mathbf{a} \wedge \mathbf{b}) \cdot (\mathbf{c} \wedge \mathbf{d}) = -(\mathbf{a} \times \mathbf{b}) \cdot (\mathbf{c} \times \mathbf{d}).$$

Proof. This follows by scalar grade selection

$$\begin{aligned}(\mathbf{a} \wedge \mathbf{b}) \cdot (\mathbf{c} \wedge \mathbf{d}) &= \langle (\mathbf{a} \wedge \mathbf{b})(\mathbf{c} \wedge \mathbf{d}) \rangle \\ &= \langle I(\mathbf{a} \times \mathbf{b})I(\mathbf{c} \times \mathbf{d}) \rangle \\ &= -(\mathbf{a} \times \mathbf{b}) \cdot (\mathbf{c} \times \mathbf{d}).\end{aligned} \quad (1.129)$$

□

Table 1.4: Comparison of distribution identities.

	Geometric algebra	Vector algebra
vector-bivector dot product (scalar triple cross product)	$\mathbf{a} \cdot (\mathbf{b} \wedge \mathbf{c})$	$(\mathbf{b} \times \mathbf{c}) \times \mathbf{a}$
bivector dot product (dot product of cross products)	$(\mathbf{a} \wedge \mathbf{b}) \cdot (\mathbf{c} \wedge \mathbf{d})$	$-(\mathbf{a} \times \mathbf{b}) \cdot (\mathbf{c} \times \mathbf{d})$

1.12.3 Problems.

Exercise 1.13   Index permutation in vector selection.

Prove eq. (1.119). That is, given $j \neq k$, and $i = j$ or $i = k$, show that

$$\langle \mathbf{e}_i \mathbf{e}_j \mathbf{e}_k \rangle_1 = \langle \mathbf{e}_k \mathbf{e}_j \mathbf{e}_i \rangle_1, \quad (1.130)$$

Exercise 1.14   Dot product of unit vector with unit bivector.

Prove eq. (1.120). That is, given $j \neq k$, show that

$$\langle \mathbf{e}_i \mathbf{e}_j \mathbf{e}_k \rangle_1 = \mathbf{e}_k (\mathbf{e}_j \cdot \mathbf{e}_i) - \mathbf{e}_j (\mathbf{e}_k \cdot \mathbf{e}_i). \quad (1.133)$$

1.13 PERMUTATION WITHIN SCALAR SELECTION.

As scalar selection is at the heart of the generalized dot product, it is worth knowing some of the ways that such a selection operation can be manipulated.

> **Theorem 1.10: Permutation within scalar selection.**
>
> The factors within a scalar grade selection of a pair of multivector products may be permuted or may be cyclically permuted
>
> $$\langle AB \rangle = \langle BA \rangle$$
> $$\langle AB \cdots YZ \rangle = \langle ZAB \cdots Y \rangle.$$

Proof. It is sufficient to prove just the two multivector permutation case. One simple, but inelegant method, is to first expand the pair of multivectors in coordinates. Let

$$A = a_0 + \sum_i a_i \mathbf{e}_i + \sum_{i<j} a_{ij} \mathbf{e}_{ij} + \cdots$$
$$B = b_0 + \sum_i b_i \mathbf{e}_i + \sum_{i<j} b_{ij} \mathbf{e}_{ij} + \cdots \quad (1.138)$$

Only the products of equal unit k-vectors $\mathbf{e}_{ij}, \mathbf{e}_{ijk}, \cdots$ can contribute scalar components to the sum, so the scalar selection of the products must have the form

$$\langle AB \rangle = a_0 b_0 + \sum_i a_i b_i \mathbf{e}_i^2 + \sum_{i<j} a_{ij} b_{ij} \mathbf{e}_{ij}^2 + \cdots \quad (1.139)$$

This sum is also clearly equal to $\langle BA \rangle$. □

1.14 MULTIVECTOR WEDGE PRODUCT.

We've identified the vector wedge product of two vectors with the selection of the highest grade of their product. Looking back to the multivector products of eq. (1.107), and eq. (1.108) as motivation, a generalized wedge product can be defined that selects the highest grade terms of a given multivector product

> **Definition 1.29: Multivector wedge product.**

1.14 MULTIVECTOR WEDGE PRODUCT.

For the multivectors A, B defined in definition 1.27, the wedge (or outer) product is defined as

$$A \wedge B \equiv \sum_{i,j=0}^{N} \langle A_i B_j \rangle_{i+j}.$$

If A, B are a k-vectors with grades r, s respectively, then their wedge product is a single grade selection

$$A \wedge B = \langle AB \rangle_{r+s}. \tag{1.140}$$

The most important example of the generalized wedge is the wedge product of a vector with wedge of two vectors

Theorem 1.11: Wedge of three vectors.

The wedge product of three vectors is associative

$$(\mathbf{a} \wedge \mathbf{b}) \wedge \mathbf{c} = \mathbf{a} \wedge (\mathbf{b} \wedge \mathbf{c}),$$

so can be written simply as $\mathbf{a} \wedge \mathbf{b} \wedge \mathbf{c}$.

Proof. The proof follows directly from the definition

$$\begin{aligned}
(\mathbf{a} \wedge \mathbf{b}) \wedge \mathbf{c} &= \langle (\mathbf{a} \wedge \mathbf{b})\mathbf{c} \rangle_3 \\
&= \langle (\mathbf{ab} - \mathbf{a} \cdot \mathbf{b})\mathbf{c} \rangle_3 \\
&= \langle \mathbf{abc} \rangle_3 - (\mathbf{a} \cdot \mathbf{b})\langle \mathbf{c} \rangle_3 \\
&= \langle \mathbf{abc} \rangle_3,
\end{aligned} \tag{1.141}$$

where the grade-3 selection of a vector is zero by definition. Similarly

$$\begin{aligned}
\mathbf{a} \wedge (\mathbf{b} \wedge \mathbf{c}) &= \langle \mathbf{a}(\mathbf{b} \wedge \mathbf{c}) \rangle_3 \\
&= \langle \mathbf{a}(\mathbf{bc} - \mathbf{b} \cdot \mathbf{c}) \rangle_3 \\
&= \langle \mathbf{abc} \rangle_3 - (\mathbf{b} \cdot \mathbf{c})\langle \mathbf{a} \rangle_3 \\
&= \langle \mathbf{abc} \rangle_3.
\end{aligned} \tag{1.142}$$

□

It is simple to show that the wedge of three vectors is completely anti-symmetric (any interchange of vectors changes the sign), and that cyclic permutation $\mathbf{a} \to \mathbf{b} \to \mathbf{c} \to \mathbf{a}$ of the vectors leaves it unchanged (exercise 1.15). These properties are also common to the scalar triple product of $\mathbb{R}^3$ vector algebra, because both the scalar triple product and the wedge of three vectors has a determinant structure, which can be shown by direct expansion in coordinates

$$\begin{aligned} \mathbf{a} \wedge \mathbf{b} \wedge \mathbf{c} &= \langle a_i b_j c_k \mathbf{e}_i \mathbf{e}_j \mathbf{e}_k \rangle_3 \\ &= \sum_{i \neq j \neq k} a_i b_j c_k \mathbf{e}_i \mathbf{e}_j \mathbf{e}_k \\ &= \sum_{i<j<k} \begin{vmatrix} a_i & a_j & a_k \\ b_i & b_j & b_k \\ c_i & c_j & c_k \end{vmatrix} \mathbf{e}_{ijk}. \end{aligned} \tag{1.143}$$

In $\mathbb{R}^3$ this sum has only one term

$$\mathbf{a} \wedge \mathbf{b} \wedge \mathbf{c} = \begin{vmatrix} a_1 & a_2 & a_3 \\ b_1 & b_2 & b_3 \\ c_1 & c_2 & c_3 \end{vmatrix} I, \tag{1.144}$$

in which the determinant is recognizable as the scalar triple product. This shows that the $\mathbb{R}^3$ wedge of three vectors is the scalar triple product times the pseudoscalar

$$\boxed{\mathbf{a} \wedge \mathbf{b} \wedge \mathbf{c} = (\mathbf{a} \cdot (\mathbf{b} \times \mathbf{c})) I.} \tag{1.145}$$

Note that the wedge of n vectors is also associative. A full proof is possible by induction, which won't be done here. Instead, as a hint of how to proceed if desired, consider the coordinate expansion of a trivector wedged with a vector

$$\begin{aligned} (\mathbf{a} \wedge \mathbf{b} \wedge \mathbf{c}) \wedge \mathbf{d} &= \sum_{i \neq j \neq k, l} \langle a_i b_j c_k \mathbf{e}_i \mathbf{e}_j \mathbf{e}_k d_l \mathbf{e}_l \rangle_4 \\ &= \sum_{i \neq j \neq k \neq l} a_i b_j c_k d_l \mathbf{e}_i \mathbf{e}_j \mathbf{e}_k \mathbf{e}_l. \end{aligned} \tag{1.146}$$

This can be rewritten with any desired grouping $((\mathbf{a} \wedge \mathbf{b}) \wedge \mathbf{c}) \wedge \mathbf{d} = (\mathbf{a} \wedge \mathbf{b}) \wedge (\mathbf{c} \wedge \mathbf{d}) = \mathbf{a} \wedge (\mathbf{b} \wedge \mathbf{c} \wedge \mathbf{d}) = \cdots$. Observe that this can also be put into a determinant form like that of eq. (1.143). Whenever the number of vectors matches the dimension of the underlying vector space, this will be a single determinant of all the coordinates of the vectors multiplied by the unit pseudoscalar for the vector space.

1.14.1 Problems.

Exercise 1.15   Properties of the wedge of three vectors.

Show that the wedge product of three vectors is completely antisymmetric, and show that the wedge product of three vectors is invariant with respect to cyclic permutation.

Exercise 1.16 $\mathbb{R}^4$ wedge of a non-blade with itself.

While the wedge product of a blade with itself is always zero, this is not generally true of the wedge products of arbitrary k-vectors in higher dimensional spaces. To demonstrate this, show that the wedge of the bivector $B = \mathbf{e}_1\mathbf{e}_2 + \mathbf{e}_3\mathbf{e}_4$ with itself is non-zero. Why is this bivector not a blade?

1.15 PROJECTION AND REJECTION.

Let's now look at how the dot plus wedge product decomposition of the vector product can be applied to compute vector projection and rejection, which are defined as

---

Definition 1.30: Vector projection and rejection.

Given a vector $\mathbf{x}$ and vector $\mathbf{u}$ the projection of $\mathbf{x}$ onto the direction of $\mathbf{u}$ is defined as

$$\text{Proj}_\mathbf{u}(\mathbf{x}) = (\mathbf{x} \cdot \hat{\mathbf{u}})\hat{\mathbf{u}},$$

where $\hat{\mathbf{u}} = \mathbf{u}/\|\mathbf{u}\|$. The rejection of $\mathbf{x}$ by $\mathbf{u}$ is defined as the component of $\mathbf{x}$ that is orthogonal to $\mathbf{u}$

$$\text{Rej}_\mathbf{u}(\mathbf{x}) = \mathbf{x} - \text{Proj}_\mathbf{u}(\mathbf{x}).$$

An example of projection and rejection with respect to a direction vector $\mathbf{u}$ is illustrated in fig. 1.15.

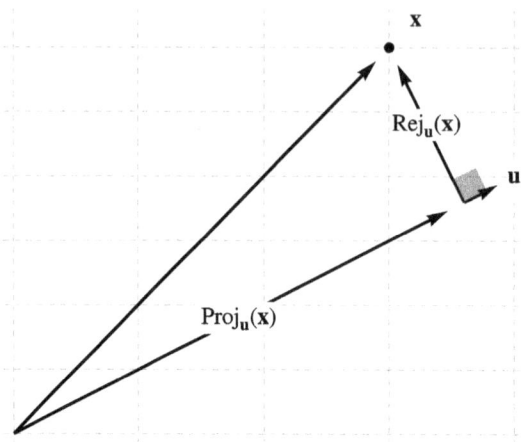

Figure 1.15: Projection and rejection illustrated.

Computation of the projective and rejective components of a vector **x** relative to a direction $\hat{\mathbf{u}}$ requires little more than a multiplication by $1 = \hat{\mathbf{u}}\hat{\mathbf{u}}$, and some rearrangement

$$\begin{aligned}
\mathbf{x} &= \mathbf{x}\hat{\mathbf{u}}\hat{\mathbf{u}} \\
&= (\mathbf{x}\hat{\mathbf{u}})\hat{\mathbf{u}} \\
&= \left(\mathbf{x}\cdot\hat{\mathbf{u}} + \mathbf{x}\wedge\hat{\mathbf{u}}\right)\hat{\mathbf{u}} \\
&= (\mathbf{x}\cdot\hat{\mathbf{u}})\hat{\mathbf{u}} + (\mathbf{x}\wedge\hat{\mathbf{u}})\hat{\mathbf{u}}.
\end{aligned} \quad (1.152)$$

The vector **x** is split nicely into its projection and rejective components in a complementary fashion

$$\text{Proj}_{\mathbf{u}}(\mathbf{x}) = (\mathbf{x}\cdot\hat{\mathbf{u}})\hat{\mathbf{u}} \quad (1.153\text{a})$$

$$\text{Rej}_{\mathbf{u}}(\mathbf{x}) = (\mathbf{x}\wedge\hat{\mathbf{u}})\hat{\mathbf{u}}. \quad (1.153\text{b})$$

By construction, $(\mathbf{x}\wedge\hat{\mathbf{u}})\hat{\mathbf{u}}$ must be a vector, despite any appearance of a multivector nature.

The utility of this multivector rejection formula is not for hand or computer algebra calculations, where it will generally be faster and simpler to compute $\mathbf{x} - (\mathbf{x}\cdot\hat{\mathbf{u}})\hat{\mathbf{u}}$, than to use eq. (1.153b). Instead this will come in handy as a new abstract algebraic tool.

1.15 PROJECTION AND REJECTION.

When it is desirable to perform this calculation explicitly, it can be done more efficiently using a no-op grade selection operation. In particular, a vector can be written as its own grade-1 selection

$$\mathbf{x} = \langle \mathbf{x} \rangle_1, \tag{1.154}$$

so the rejection can be re-expressed using definition 1.27 as a generalized bivector-vector dot product

$$\text{Rej}_{\hat{\mathbf{u}}}(\mathbf{x}) = \langle (\mathbf{x} \wedge \hat{\mathbf{u}}) \hat{\mathbf{u}} \rangle_1 = (\mathbf{x} \wedge \hat{\mathbf{u}}) \cdot \hat{\mathbf{u}}. \tag{1.155}$$

In $\mathbb{R}^3$, using theorem 1.7, the rejection operation can also be expressed as a vector triple product

$$\text{Rej}_{\hat{\mathbf{u}}}(\mathbf{x}) = \hat{\mathbf{u}} \times (\mathbf{x} \times \hat{\mathbf{u}}). \tag{1.156}$$

To help establish some confidence with these new additions to our toolbox, here are a pair of illustrative examples using eq. (1.153b), and eq. (1.155) respectively.

Example 1.5: An $\mathbb{R}^2$ rejection.

Let $\mathbf{x} = a\mathbf{e}_1 + b\mathbf{e}_2$ and $\mathbf{u} = \mathbf{e}_1$ for which the wedge is $\mathbf{x} \wedge \hat{\mathbf{u}} = b\mathbf{e}_2\mathbf{e}_1$. Using eq. (1.153b) the rejection of $\mathbf{x}$ by $\mathbf{u}$ is

$$\begin{aligned}
\text{Rej}_{\hat{\mathbf{u}}}(\mathbf{x}) &= (\mathbf{x} \wedge \hat{\mathbf{u}}) \hat{\mathbf{u}} \\
&= (b\mathbf{e}_2\mathbf{e}_1)\mathbf{e}_1 \\
&= b\mathbf{e}_2(\mathbf{e}_1\mathbf{e}_1) \\
&= b\mathbf{e}_2,
\end{aligned} \tag{1.157}$$

as expected.

This example provides some guidance about what is happening geometrically in eq. (1.153b). The wedge operation produces a pseudoscalar for the plane spanned by $\{\mathbf{x}, \mathbf{u}\}$ that is scaled as $\sin\theta$ where θ is the angle between $\mathbf{x}$ and $\mathbf{u}$. When that pseudoscalar is multiplied by $\hat{\mathbf{u}}$, $\hat{\mathbf{u}}$ is rotated in the plane by $\pi/2$ radians towards $\mathbf{x}$, yielding the normal component of the vector $\mathbf{x}$.

Here's a slightly less trivial $\mathbb{R}^3$ example

Example 1.6: An $\mathbb{R}^3$ rejection.

Let $\mathbf{x} = a\mathbf{e}_2 + b\mathbf{e}_3$ and $\hat{\mathbf{u}} = (\mathbf{e}_1 + \mathbf{e}_2)/\sqrt{2}$ for which the wedge product is

$$\begin{aligned}
\mathbf{x} \wedge \hat{\mathbf{u}} &= \frac{1}{\sqrt{2}} \begin{vmatrix} \mathbf{e}_{23} & \mathbf{e}_{31} & \mathbf{e}_{12} \\ 0 & a & b \\ 1 & 1 & 0 \end{vmatrix} \\
&= \frac{1}{\sqrt{2}} \left(\mathbf{e}_{23}(-b) - \mathbf{e}_{31}(-b) + \mathbf{e}_{12}(-a) \right) \\
&= \frac{1}{\sqrt{2}} \left(b(\mathbf{e}_{32} + \mathbf{e}_{31}) + a\mathbf{e}_{21} \right).
\end{aligned} \tag{1.158}$$

Using eq. (1.155) the rejection of $\mathbf{x}$ by $\mathbf{u}$ is

$$(\mathbf{x} \wedge \hat{\mathbf{u}}) \cdot \hat{\mathbf{u}} = \frac{1}{2} \left(b(\mathbf{e}_{32} + \mathbf{e}_{31}) + a\mathbf{e}_{21} \right) \cdot (\mathbf{e}_1 + \mathbf{e}_2). \tag{1.159}$$

Each of these bivector-vector dot products has the form $\mathbf{e}_{rs} \cdot \mathbf{e}_t = \langle \mathbf{e}_{rst} \rangle_1$ which is zero whenever the indexes r, s, t are unique, and is a vector whenever one of indexes are repeated ($r = t$, or $s = t$). This leaves

$$\begin{aligned}
(\mathbf{x} \wedge \hat{\mathbf{u}}) \cdot \hat{\mathbf{u}} &= \frac{1}{2} \left(b\mathbf{e}_3 + a\mathbf{e}_2 + b\mathbf{e}_3 - a\mathbf{e}_1 \right) \\
&= b\mathbf{e}_3 + \frac{a}{2}(\mathbf{e}_2 - \mathbf{e}_1).
\end{aligned} \tag{1.160}$$

Example 1.7: Velocity and acceleration in polar coordinates.

In eq. (1.37), and eq. (1.38) we found the polar representation of the velocity and acceleration vectors associated with the radial parameterization $\mathbf{r}(r, \theta) = r\hat{\mathbf{r}}(\theta)$.

We can alternatively compute the radial and azimuthal components of these vectors in terms of their projective and rejective components

$$\begin{aligned}
\mathbf{v} &= \mathbf{v}\hat{\mathbf{r}}\hat{\mathbf{r}} = (\mathbf{v} \cdot \hat{\mathbf{r}} + \mathbf{v} \wedge \hat{\mathbf{r}})\hat{\mathbf{r}} \\
\mathbf{a} &= \mathbf{a}\hat{\mathbf{r}}\hat{\mathbf{r}} = (\mathbf{a} \cdot \hat{\mathbf{r}} + \mathbf{a} \wedge \hat{\mathbf{r}})\hat{\mathbf{r}},
\end{aligned} \tag{1.161}$$

so

$$\mathbf{v} \cdot \hat{\mathbf{r}} = r'$$
$$\mathbf{v} \wedge \hat{\mathbf{r}} = r\omega\hat{\boldsymbol{\theta}} \wedge \hat{\mathbf{r}} = \omega\hat{\boldsymbol{\theta}} \wedge \mathbf{r}$$
$$\mathbf{a} \cdot \hat{\mathbf{r}} = r'' - r\omega^2 \qquad (1.162)$$
$$\mathbf{a} \wedge \hat{\mathbf{r}} = \frac{1}{r}\left(r^2\omega\right)' \hat{\boldsymbol{\theta}} \wedge \hat{\mathbf{r}}.$$

We see that it is natural to introduce angular velocity and acceleration bivectors. These both lie in the $\hat{\boldsymbol{\theta}} \wedge \hat{\mathbf{r}}$ plane. Of course, it is also possible to substitute the cross product for the wedge product, but doing so requires the introduction of a normal direction that may not intrinsically be part of the problem (i.e. two dimensional problems).

In the GA literature the projection and rejection operations are usually written using the vector inverse.

Definition 1.31: Vector inverse.

Define the inverse of a vector $\mathbf{x}$, when it exists, as

$$\mathbf{x}^{-1} = \frac{\mathbf{x}}{\|\mathbf{x}\|^2}.$$

This inverse satisfies $\mathbf{x}^{-1}\mathbf{x} = \mathbf{x}\mathbf{x}^{-1} = 1$.

The vector inverse may not exist in a non-Euclidean vector space where $\mathbf{x}^2$ can be zero for non-zero vectors $\mathbf{x}$.

In terms of the vector inverse, the projection and rejection operations with respect to $\mathbf{u}$ can be written without any reference to the unit vector $\hat{\mathbf{u}} = \mathbf{u}/\|\mathbf{u}\|$ that lies along that vector

$$\begin{aligned} \text{Proj}_{\mathbf{u}}(\mathbf{x}) &= (\mathbf{x} \cdot \mathbf{u})\frac{1}{\mathbf{u}} \\ \text{Rej}_{\mathbf{u}}(\mathbf{x}) &= (\mathbf{x} \wedge \mathbf{u})\frac{1}{\mathbf{u}} = (\mathbf{x} \wedge \mathbf{u}) \cdot \frac{1}{\mathbf{u}}. \end{aligned} \qquad (1.163)$$

It was claimed in the definition of rejection that the rejection is orthogonal to the projection. This can be shown trivially without any use of GA (exercise 1.17). This also follows naturally using the grade selection operator representation of the dot product

$$\begin{align}
\text{Rej}_{\mathbf{u}}(\mathbf{x}) \cdot \text{Proj}_{\mathbf{u}}(\mathbf{x}) &= \langle \text{Rej}_{\mathbf{u}}(\mathbf{x}) \text{Proj}_{\mathbf{u}}(\mathbf{x}) \rangle \\
&= \langle (\mathbf{x} \wedge \hat{\mathbf{u}}) \hat{\mathbf{u}} (\mathbf{x} \cdot \hat{\mathbf{u}}) \hat{\mathbf{u}} \rangle \\
&= (\mathbf{x} \cdot \hat{\mathbf{u}}) \langle (\mathbf{x} \wedge \hat{\mathbf{u}}) \hat{\mathbf{u}}^2 \rangle \tag{1.164} \\
&= (\mathbf{x} \cdot \hat{\mathbf{u}}) \langle \mathbf{x} \wedge \hat{\mathbf{u}} \rangle \\
&= 0.
\end{align}$$

This is zero because the scalar grade of a wedge product, a bivector, is zero by definition.

Exercise 1.17 Rejection orthogonality.

Prove, without any use of GA, that $\mathbf{x} - \text{Proj}_{\mathbf{u}}(\mathbf{x})$ is orthogonal to $\mathbf{u}$, as claimed in definition 1.30.

Exercise 1.18 Rejection example.

a. Repeat example 1.6 by calculating $(\mathbf{x} \wedge \hat{\mathbf{u}})\hat{\mathbf{u}}$ and show that all the grade three components of this multivector product vanish.
b. Compute $\mathbf{x} - (\mathbf{x} \cdot \hat{\mathbf{u}})\hat{\mathbf{u}}$ and show that this matches eq. (1.160).

1.16 NORMAL FACTORIZATION OF THE WEDGE PRODUCT.

A general bivector has the form

$$B = \sum_{i \neq j} a_{ij} \mathbf{e}_{ij}, \tag{1.170}$$

which is not necessarily a blade. On the other hand, a wedge product is always a blade [5]

Theorem 1.12: Wedge product normal factorization

The wedge product of any two non-colinear vectors $\mathbf{a}, \mathbf{b}$ always has a orthogonal (2-blade) factorization

$$\mathbf{a} \wedge \mathbf{b} = \mathbf{u}\mathbf{v}, \quad \mathbf{u} \cdot \mathbf{v} = 0.$$

---

[5] In $\mathbb{R}^3$ any bivector is also a blade [1]

This can be proven by construction. Pick $\mathbf{u} = \mathbf{a}$ and $\mathbf{v} = \text{Rej}_{\mathbf{a}}(\mathbf{b})$, then

$$\begin{aligned} \mathbf{a}\,\text{Rej}_{\mathbf{a}}(\mathbf{b}) &= \mathbf{a} \cdot \text{Rej}_{\mathbf{a}}(\mathbf{b}) + \mathbf{a} \wedge \text{Rej}_{\mathbf{a}}(\mathbf{b}) \\ &= \mathbf{a} \wedge \left(\mathbf{b} - \frac{\mathbf{b} \cdot \mathbf{a}}{\|\mathbf{a}\|^2}\mathbf{a}\right) \\ &= \mathbf{a} \wedge \mathbf{b}, \end{aligned} \quad (1.171)$$

since $\mathbf{a} \wedge (\alpha \mathbf{a}) = 0$ for any scalar α.

The significance of theorem 1.12 is that the square of any wedge product is negative

$$\begin{aligned} (\mathbf{uv})^2 &= (\mathbf{uv})(-\mathbf{vu}) \\ &= -\mathbf{u}(\mathbf{v}^2)\mathbf{u} \\ &= -|\mathbf{u}|^2|\mathbf{v}|^2, \end{aligned} \quad (1.172)$$

which in turn means that exponentials with wedge product arguments can be used as rotation operators.

Exercise 1.19 $\mathbb{R}^3$ bivector factorization.

Find some orthogonal factorizations for the $\mathbb{R}^3$ bivector $\mathbf{e}_{12} + \mathbf{e}_{23} + \mathbf{e}_{31}$.

1.17 THE WEDGE PRODUCT AS AN ORIENTED AREA.

The coordinate representation of the $\mathbb{R}^2$ wedge product (eq. (1.79)) had a single $\mathbf{e}_{12}$ bivector factor, whereas the expansion in coordinates for the general $\mathbb{R}^N$ wedge product was considerably messier (eq. (1.78)). This difference can be eliminated by judicious choice of basis.

A simpler coordinate representation for the $\mathbb{R}^N$ wedge product follows by choosing an orthonormal basis for the planar subspace spanned by the wedge vectors. Given vectors $\mathbf{a}, \mathbf{b}$, let $\{\hat{\mathbf{u}}, \hat{\mathbf{v}}\}$ be an orthonormal basis for the plane subspace $P = \text{span}\{\mathbf{a}, \mathbf{b}\}$. The coordinate representations of $\mathbf{a}, \mathbf{b}$ in this basis are

$$\begin{aligned} \mathbf{a} &= (\mathbf{a} \cdot \hat{\mathbf{u}})\hat{\mathbf{u}} + (\mathbf{a} \cdot \hat{\mathbf{v}})\hat{\mathbf{v}} \\ \mathbf{b} &= (\mathbf{b} \cdot \hat{\mathbf{u}})\hat{\mathbf{u}} + (\mathbf{b} \cdot \hat{\mathbf{v}})\hat{\mathbf{v}}. \end{aligned} \quad (1.179)$$

The wedge of these vectors is

$$\mathbf{a} \wedge \mathbf{b} = \Big((\mathbf{a}\cdot\hat{\mathbf{u}})\hat{\mathbf{u}} + (\mathbf{a}\cdot\hat{\mathbf{v}})\hat{\mathbf{v}}\Big) \wedge \Big((\mathbf{b}\cdot\hat{\mathbf{u}})\hat{\mathbf{u}} + (\mathbf{b}\cdot\hat{\mathbf{v}})\hat{\mathbf{v}}\Big)$$
$$= \Big((\mathbf{a}\cdot\hat{\mathbf{u}})(\mathbf{b}\cdot\hat{\mathbf{v}}) - (\mathbf{a}\cdot\hat{\mathbf{v}})(\mathbf{b}\cdot\hat{\mathbf{u}})\Big)\hat{\mathbf{u}}\hat{\mathbf{v}} \quad (1.180)$$
$$= \begin{vmatrix} \mathbf{a}\cdot\hat{\mathbf{u}} & \mathbf{a}\cdot\hat{\mathbf{v}} \\ \mathbf{b}\cdot\hat{\mathbf{u}} & \mathbf{b}\cdot\hat{\mathbf{v}} \end{vmatrix} \hat{\mathbf{u}}\hat{\mathbf{v}}.$$

We see that this basis allows for the most compact (single term) coordinate representation of the wedge product.

If a counterclockwise rotation by $\pi/2$ takes $\hat{\mathbf{u}}$ to $\hat{\mathbf{v}}$ the determinant will equal the area of the parallelogram spanned by $\mathbf{a}$ and $\mathbf{b}$. Let that area be designated

$$A = \begin{vmatrix} \mathbf{a}\cdot\hat{\mathbf{u}} & \mathbf{a}\cdot\hat{\mathbf{v}} \\ \mathbf{b}\cdot\hat{\mathbf{u}} & \mathbf{b}\cdot\hat{\mathbf{v}} \end{vmatrix}. \quad (1.181)$$

A given wedge product may have any number of other wedge or orthogonal product representations

$$\begin{aligned}
\mathbf{a} \wedge \mathbf{b} &= (\mathbf{a} + \beta\mathbf{b}) \wedge \mathbf{b} \\
&= \mathbf{a} \wedge (\mathbf{b} + \alpha\mathbf{a}) \\
&= (A\hat{\mathbf{u}}) \wedge \hat{\mathbf{v}} \\
&= \hat{\mathbf{u}} \wedge (A\hat{\mathbf{v}}) \\
&= (\alpha A\hat{\mathbf{u}}) \wedge \frac{\hat{\mathbf{v}}}{\alpha} \\
&= (\beta A\hat{\mathbf{u}}') \wedge \frac{\hat{\mathbf{v}}'}{\beta}
\end{aligned} \quad (1.182)$$

These equivalencies can be thought of as different geometrical representations of the same object. Since the spanned area and relative ordering of the wedged vectors remains constant. Some different parallelogram representations of a wedge products are illustrated in fig. 1.16.

As there are many possible orthogonal factorizations for a given wedge product, and also many possible wedge products that produce the same value bivector, we can say that a wedge product represents an area with a specific cyclic orientation, but any such area is a valid representation. This is illustrated in fig. 1.17.

1.17 THE WEDGE PRODUCT AS AN ORIENTED AREA. 61

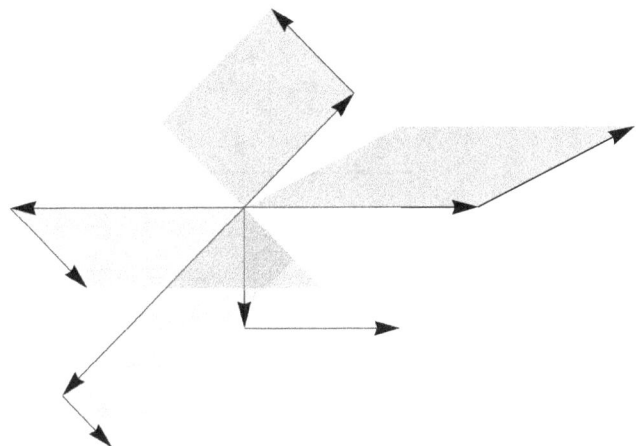

Figure 1.16: Parallelogram representations of wedge products.

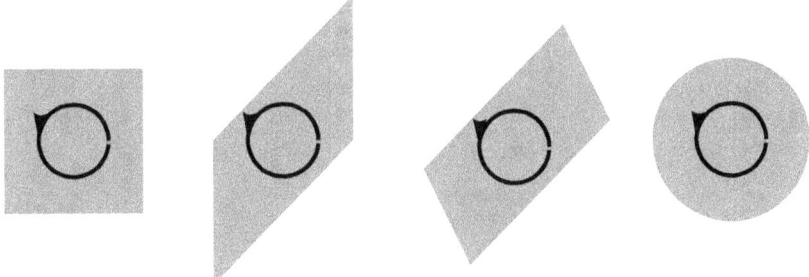

Figure 1.17: Different shape representations of a wedge product.

Exercise 1.20 Parallelogram area.

Show that the area A of the parallelogram spanned by vectors $\mathbf{a}, \mathbf{b}$ as illustrated in fig. 1.18,

$$\mathbf{a} = a_1\mathbf{e}_1 + a_2\mathbf{e}_2$$
$$\mathbf{b} = b_1\mathbf{e}_1 + b_2\mathbf{e}_2,$$

is

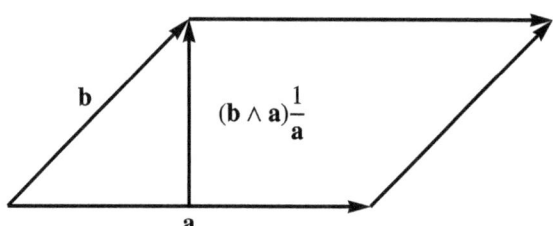

Figure 1.18: Parallelogram area.

$$A = \pm \begin{vmatrix} b_1 & b_2 \\ a_1 & a_2 \end{vmatrix},$$

where we adjust the sign to make the end result come out positive.

1.18 GENERAL ROTATION.

Equation (1.23) showed that the $\mathbb{R}^2$ pseudoscalar anticommutes with any vector $\mathbf{x} \in \mathbb{R}^2$,

$$\mathbf{x}i = -i\mathbf{x}, \tag{1.188}$$

and that the sign of the bivector exponential argument must be negated to maintain the value of the vector $\mathbf{x} \in \mathbb{R}^2$ on interchange

$$\mathbf{x}e^{i\theta} = e^{-i\theta}\mathbf{x}. \tag{1.189}$$

The higher dimensional generalization of these results are

Theorem 1.13: Bivector exponential properties.

Given two non-colinear vectors $\mathbf{a}, \mathbf{b}$, let the planar subspace formed by their span be designated $S = \text{span}\{\mathbf{a}, \mathbf{b}\}$.

(a) Any vector $\mathbf{p} \in S$ anticommutes with the wedge product $\mathbf{a} \wedge \mathbf{b}$

$$\mathbf{p}(\mathbf{a} \wedge \mathbf{b}) = -(\mathbf{a} \wedge \mathbf{b})\mathbf{p}.$$

(b) Any vector $\mathbf{n}$ orthogonal to this plane ($\mathbf{n} \cdot \mathbf{a} = \mathbf{n} \cdot \mathbf{b} = 0$) commutes with this wedge product

$$\mathbf{n}(\mathbf{a} \wedge \mathbf{b}) = (\mathbf{a} \wedge \mathbf{b})\mathbf{n}.$$

(c) Reversing the order of multiplication of a vector $\mathbf{p} \in S$ with an exponential $e^{\mathbf{a} \wedge \mathbf{b}}$, requires the sign of the exponential argument to be negated

$$\mathbf{p} e^{\mathbf{a} \wedge \mathbf{b}} = e^{-\mathbf{a} \wedge \mathbf{b}} \mathbf{p}.$$

This sign change on interchange will be called conjugation.

(d) Any orthogonal vectors $\mathbf{n}$ commute with a such a complex exponential

$$\mathbf{n} e^{\mathbf{a} \wedge \mathbf{b}} = e^{\mathbf{a} \wedge \mathbf{b}} \mathbf{n}.$$

Proof. The proof relies on the fact that a orthogonal factorization of the wedge product is possible. If $\mathbf{p}$ is one of those factors, then the other is uniquely determined by the multivector equation $\mathbf{a} \wedge \mathbf{b} = \mathbf{p}\mathbf{q}$, for which we must have $\mathbf{q} = \frac{1}{\mathbf{x}}(\mathbf{a} \wedge \mathbf{b}) \in S$ and $\mathbf{p} \cdot \mathbf{q} = 0$ [6]. Then

$$\begin{aligned}\mathbf{p}(\mathbf{a} \wedge \mathbf{b}) &= \mathbf{p}(\mathbf{p}\mathbf{q}) \\ &= \mathbf{p}(-\mathbf{q}\mathbf{p}) \\ &= -(\mathbf{p}\mathbf{q})\mathbf{p} \\ &= -(\mathbf{a} \wedge \mathbf{b})\mathbf{p}.\end{aligned} \quad (1.190)$$

---

[6] The identities required to show that $\mathbf{q}$ above has no trivector grades, and to evaluate it explicitly in terms of $\mathbf{a}, \mathbf{b}, \mathbf{x}$, will be derived later.

Any orthogonal vectors **n** must also be perpendicular to the factors **p, q**, with $\mathbf{n} \cdot \mathbf{p} = \mathbf{n} \cdot \mathbf{q} = 0$, so

$$\begin{aligned}\mathbf{n}(\mathbf{a} \wedge \mathbf{b}) &= \mathbf{n}(\mathbf{p}\mathbf{q}) \\ &= (-\mathbf{p}\mathbf{n})\mathbf{q} \\ &= -\mathbf{p}(-\mathbf{q}\mathbf{n}) \\ &= (\mathbf{p}\mathbf{q})\mathbf{n} \\ &= (\mathbf{a} \wedge \mathbf{b})\mathbf{n}.\end{aligned} \quad (1.191)$$

For the complex exponentials, introduce a unit pseudoscalar for the plane $i = \hat{\mathbf{p}}\hat{\mathbf{q}}$ satisfying $i^2 = -1$ and a scalar rotation angle $\theta = (\mathbf{a} \wedge \mathbf{b})/i$, then for vectors $\mathbf{p} \in S$

$$\begin{aligned}\mathbf{p}e^{\mathbf{a} \wedge \mathbf{b}} &= \mathbf{p}e^{i\theta} \\ &= \mathbf{p}\left(\cos\theta + i\sin\theta\right) \\ &= \left(\cos\theta - i\sin\theta\right)\mathbf{p} \\ &= e^{-i\theta}\mathbf{p} \\ &= e^{-\mathbf{a} \wedge \mathbf{b}}\mathbf{p},\end{aligned} \quad (1.192)$$

and for vectors **n** orthogonal to S

$$\begin{aligned}\mathbf{n}e^{\mathbf{a} \wedge \mathbf{b}} &= \mathbf{n}e^{i\theta} \\ &= \mathbf{n}\left(\cos\theta + i\sin\theta\right) \\ &= \left(\cos\theta + i\sin\theta\right)\mathbf{n} \\ &= e^{i\theta}\mathbf{n} \\ &= e^{\mathbf{a} \wedge \mathbf{b}}\mathbf{n}.\end{aligned} \quad (1.193)$$

□

The point of this somewhat abstract seeming theorem is to prepare for the statement of a general $\mathbb{R}^N$ rotation, which is

---

Definition 1.32: General rotation

Let $B = \{\hat{\mathbf{p}}, \hat{\mathbf{q}}\}$ be an orthonormal basis for a planar subspace with unit pseudoscalar $i = \hat{\mathbf{p}}\hat{\mathbf{q}}$ where $i^2 = -1$. The rotation of a vector **x** through an angle θ with respect to this plane is

$$R_\theta(\mathbf{x}) = e^{-i\theta/2}\mathbf{x}e^{i\theta/2}.$$

Here the rotation sense is that of the $\pi/2$ rotation from $\hat{\mathbf{p}}$ to $\hat{\mathbf{q}}$ in the subspace $S = \text{span } B$.

This statement did not make any mention of an orthogonal direction. Such an orthogonal direction is not unique for dimensions higher than 3, nor defined for two dimensions. Instead the rotational sense is defined by the ordering of the factors in the bivector i.

To check that this operation has the desired semantics, let $\mathbf{x} = \mathbf{x}_\| + \mathbf{x}_\perp$, where $\mathbf{x}_\| \in S$ and $\mathbf{x}_\perp \cdot \mathbf{p} = 0 \; \forall \mathbf{p} \in S$. Then

$$\begin{aligned} R_\theta(\mathbf{x}) &= e^{-i\theta/2} \mathbf{x} e^{i\theta/2} \\ &= e^{-i\theta/2} \left(\mathbf{x}_\| + \mathbf{x}_\perp \right) e^{i\theta/2} \\ &= \mathbf{x}_\| e^{i\theta} + \mathbf{x}_\perp e^{-i\theta/2} e^{i\theta/2} \\ &= \mathbf{x}_\| e^{i\theta} + \mathbf{x}_\perp. \end{aligned} \quad (1.194)$$

As desired, this rotation operation rotates components of the vector that lies in the planar subspace S by θ, while leaving the components of the vector orthogonal to the plane unchanged, as illustrated in fig. 1.19. This is what we can call rotation around a normal in $\mathbb{R}^3$.

1.19 SYMMETRIC AND ANTISYMMETRIC VECTOR SUMS.

Theorem 1.14: Symmetric and antisymmetric vector products.

1. The dot product of vectors $\mathbf{x}, \mathbf{y}$ can be written as

$$\mathbf{x} \cdot \mathbf{y} = \frac{1}{2} \left(\mathbf{x}\mathbf{y} + \mathbf{y}\mathbf{x} \right).$$

This sum, including all permutations of the products of $\mathbf{x}$ and $\mathbf{y}$ is called a completely symmetric sum. A useful variation of this relationship is

$$\mathbf{y}\mathbf{x} = 2(\mathbf{x} \cdot \mathbf{y}) - \mathbf{x}\mathbf{y}.$$

2. The wedge product of vectors $\mathbf{x}, \mathbf{y}$ can be written as

$$\mathbf{x} \wedge \mathbf{y} = \frac{1}{2} \left(\mathbf{x}\mathbf{y} - \mathbf{y}\mathbf{x} \right).$$

66 GEOMETRIC ALGEBRA.

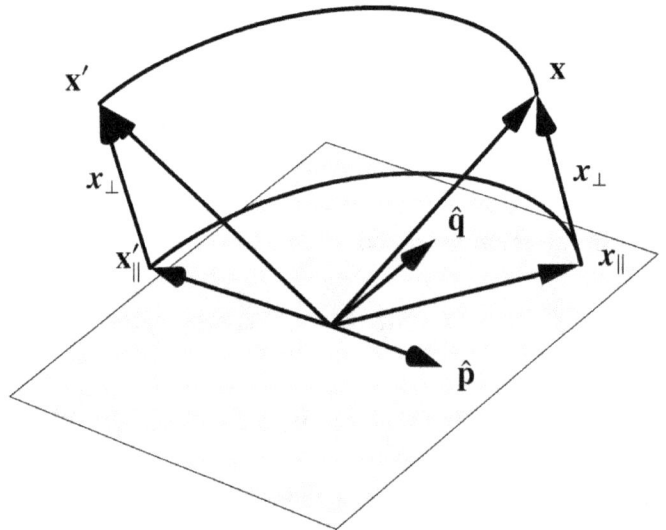

Figure 1.19: Rotation with respect to the plane of a pseudoscalar.

> This sum, including all permutations of the products **x** and **y**, with a sign change for any interchange, is called a completely antisymmetric sum.

Proof. These identities highlight the symmetric and antisymmetric nature of the respective dot and wedge products in a coordinate free form, and will be useful in the manipulation of various identities. The proof follows by direct computation after first noting that the respect vector products are

$$\mathbf{xy} = \mathbf{x} \cdot \mathbf{y} + \mathbf{x} \wedge \mathbf{y} \tag{1.195a}$$

$$\begin{aligned} \mathbf{yx} &= \mathbf{y} \cdot \mathbf{x} + \mathbf{y} \wedge \mathbf{x} \\ &= \mathbf{x} \cdot \mathbf{y} - \mathbf{x} \wedge \mathbf{y}. \end{aligned} \tag{1.195b}$$

In eq. (1.195b) the interchange utilized the respective symmetric and antisymmetric nature of the dot and wedge products.

Adding and subtracting eq. (1.195) proves the result. □

1.20 REFLECTION.

Geometrically the reflection of a vector **x** across a line directed along **u** is the difference of the projection and rejection

$$\begin{aligned} \mathbf{x}' &= (\mathbf{x} \cdot \mathbf{u}) \frac{1}{\mathbf{u}} - (\mathbf{x} \wedge \mathbf{u}) \frac{1}{\mathbf{u}} \\ &= (\mathbf{x} \cdot \mathbf{u} - \mathbf{x} \wedge \mathbf{u}) \frac{1}{\mathbf{u}}. \end{aligned} \tag{1.196}$$

Using the symmetric and antisymmetric sum representations of the dot and wedge products from theorem 1.14 the reflection can be expressed as vector products

$$\mathbf{x}' = \frac{1}{2} \left(\mathbf{xu} + \mathbf{ux} - \mathbf{xu} + \mathbf{ux} \right) \frac{1}{\mathbf{u}}, \tag{1.197}$$

yielding a remarkably simple form in terms of vector products

$$\boxed{\mathbf{x}' = \mathbf{ux} \frac{1}{\mathbf{u}}.} \tag{1.198}$$

As an illustration, here is a sample CliffordBasic reflection computation

```
In[1]:=   ClearAll[u, x, uu, invu, i, o, proj, rej, ux,
            uxu]
          u = 4 e[1] + 2 e[2];
          x = 3 e[1] + 3 e[2];
          uu = InnerProduct[u, u];
          invu = u / uu;
          i = InnerProduct[x, u];
          o = OuterProduct[x, u];
          proj = i invu // N // Simplify
          rej = GeometricProduct[o, invu] // N // Simplify
          ux = GeometricProduct[u, x]

Out[1]=   3.6 e[1] + 1.8 e[2]

Out[2]=   -0.6 e[1] + 1.2 e[2]

Out[3]=   18 + 6 e[1,2]

Out[4]=   4.2 e[1] + 0.6 e[2]
```

the results of which are plotted in fig. 1.20.

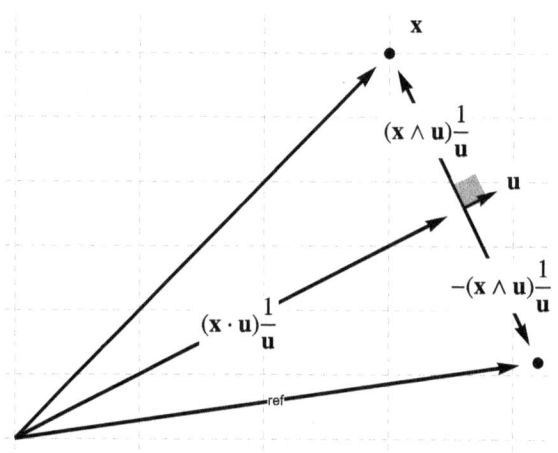

Figure 1.20: Reflection.

Table 1.5: Comparision of geometric identities.

	Geometric algebra	Traditional algebra
Projection	$(\mathbf{x} \cdot \hat{\mathbf{u}})\hat{\mathbf{u}}$	$(\mathbf{x} \cdot \hat{\mathbf{u}})\hat{\mathbf{u}}$
Rejection	$(\mathbf{x} \wedge \hat{\mathbf{u}})\hat{\mathbf{u}}$	$(\hat{\mathbf{u}} \times \mathbf{x}) \times \hat{\mathbf{u}}$
2D rotation	$\mathbf{x}e^{i\theta}, \quad i = \mathbf{e}_{12}$	$\begin{bmatrix} \cos\theta & \sin\theta \\ -\sin\theta & \cos\theta \end{bmatrix} \begin{bmatrix} x \\ y \end{bmatrix}$
3D rotation in the plane of $\hat{\mathbf{u}}, \hat{\mathbf{v}}$, where $\hat{\mathbf{u}} \cdot \hat{\mathbf{v}} = 0$	$e^{-\hat{\mathbf{u}}\hat{\mathbf{v}}\theta/2} \mathbf{x} e^{\hat{\mathbf{u}}\hat{\mathbf{v}}\theta/2}$	$(\mathbf{x} \cdot \hat{\mathbf{u}})(\hat{\mathbf{u}}\cos\theta + \hat{\mathbf{v}}\sin\theta) + (\mathbf{x} \cdot \hat{\mathbf{v}})(\hat{\mathbf{v}}\cos\theta - \hat{\mathbf{u}}\sin\theta) + (\hat{\mathbf{u}} \times \mathbf{x}) \times \hat{\mathbf{u}}$
Reflection	$\hat{\mathbf{u}}\mathbf{x}\hat{\mathbf{u}}$	$(\mathbf{x} \cdot \hat{\mathbf{u}})\hat{\mathbf{u}} + \hat{\mathbf{u}} \times (\hat{\mathbf{u}} \times \mathbf{x})$

1.21 LINEAR SYSTEMS.

Theorem 1.15: Best fit solution of linear system.

Given k linearly independent vectors $\mathbf{a}_1, \mathbf{a}_2, \cdots \mathbf{a}_k$, and the projection $\mathbf{b}_{\parallel}$ of a vector $\mathbf{b}$ onto the hypervolume spanned by $\{\mathbf{a}_1, \cdots, \mathbf{a}_k\}$

$$\mathbf{b}_{\parallel} = i^{-1}(i \cdot \mathbf{b}),$$

where $i = \mathbf{a}_1 \wedge \mathbf{a}_2 \wedge \cdots \wedge \mathbf{a}_k$, is a pseudoscalar for that hypervolume, then the system

$$\mathbf{a}_1 x_1 + \mathbf{a}_2 x_2 \cdots + \mathbf{a}_k x_k = \mathbf{b}_{\parallel},$$

is solved by

$$x_1 = i^{-1} \cdot (\mathbf{b} \wedge \mathbf{a}_2 \wedge \cdots \wedge \mathbf{a}_k)$$
$$x_2 = i^{-1} \cdot (\mathbf{a}_1 \wedge \mathbf{b} \wedge \cdots \wedge \mathbf{a}_k)$$
$$\vdots$$
$$x_n = i^{-1} \cdot (\mathbf{a}_1 \wedge \mathbf{a}_2 \wedge \cdots \wedge \mathbf{b}).$$

If $\mathbf{b} \in \text{span}\{\mathbf{a}_1, \cdots, \mathbf{a}_k\}$, so that $\mathbf{b}_{\parallel} = \mathbf{b}$, then the dot products between the k-blades above may be dropped.

This is equivalent to a Moore-Penrose or SVD pseudoinverse solution for the system

$$\begin{bmatrix} \mathbf{a}_1 & \cdots & \mathbf{a}_n \end{bmatrix} \begin{bmatrix} x_1 \\ \vdots \\ x_n \end{bmatrix} = \mathbf{b}. \quad (1.199)$$

Furthermore, also when the system is exact, if the dimension of the vectors $\mathbf{a}_i$ is k, then this solution is equivalent to Cramer's rule.

Rather than formally trying to prove this theorem, we can tackle it informally, starting with some examples.

1.21.0.1 Example: two variable system.

The simplest example is that of a two variable system

$$\mathbf{a}x + \mathbf{b}y = \mathbf{c}. \quad (1.200)$$

Let's proceed to solve this using the wedge product, assuming to start with that the system has an exact solution (i.e.: that $\mathbf{c}$ is a linear combination of $\mathbf{a}, \mathbf{b}$.)

To solve for x simply wedge with $\mathbf{b}$, and to solve for y wedge with $\mathbf{a}$

$$\begin{aligned} (\mathbf{a}x + \cancel{\mathbf{b}y}) \wedge \mathbf{b} &= \mathbf{c} \wedge \mathbf{b} \\ \mathbf{a} \wedge (\cancel{\mathbf{a}x} + \mathbf{b}y) &= \mathbf{a} \wedge \mathbf{c}, \end{aligned} \quad (1.201)$$

so, the solution, if it exists, is given by

$$\begin{aligned} x &= \frac{1}{\mathbf{a} \wedge \mathbf{b}} \mathbf{c} \wedge \mathbf{b} \\ y &= \frac{1}{\mathbf{a} \wedge \mathbf{b}} \mathbf{a} \wedge \mathbf{c}. \end{aligned} \quad (1.202)$$

1.21.0.2 Example: exact k variable system.

This idea generalizes trivially to higher order systems can be solved, simply requiring wedging more times to eliminate all terms other than the one of interest.

For example, if the k variable system

$$\mathbf{a}_1 x_1 + \mathbf{a}_2 x_2 \cdots + \mathbf{a}_k x_k = \mathbf{b}, \quad (1.203)$$

has a solution, we can solve for any of the x_i's by wedging repeatedly. For example, we can find x_1 by wedging with all $\mathbf{a}_2, \cdots \mathbf{a}_k$, to find

$$x_1 \left(\mathbf{a}_1 \wedge \mathbf{a}_2 \wedge \cdots \wedge \mathbf{a}_k \right) = \mathbf{b} \wedge \mathbf{a}_2 \wedge \cdots \wedge \mathbf{a}_k, \tag{1.204}$$

or

$$x_1 = \frac{1}{\mathbf{a}_1 \wedge \mathbf{a}_2 \wedge \cdots \wedge \mathbf{a}_k} \left(\mathbf{b} \wedge \mathbf{a}_2 \wedge \cdots \wedge \mathbf{a}_k \right) \tag{1.205}$$

If this system has no solution, then these k-vector ratios will not be scalars.

It's fairly easy to see that to solve for x_j, we start switch the numerator to the pseudoscalar i, with $\mathbf{b}$ taking the place of $\mathbf{a}_j$.

1.21.0.3 Example: $\mathbb{R}^3$ Cramer's rule.

If this sounds like Cramer's rule, that is because the two are equivalent when the dimension of the vector equals the number of variables in the linear system. For example, consider the solution for x_1 of eq. (1.203) for an $\mathbb{R}^3$ system, with $\mathbf{a}_1 = \mathbf{u}, \mathbf{a}_2 = \mathbf{v}, \mathbf{a}_3 = \mathbf{w}$

$$x_1 = \frac{\mathbf{b} \wedge \mathbf{v} \wedge \mathbf{w}}{\mathbf{u} \wedge \mathbf{v} \wedge \mathbf{w}} = \frac{\begin{vmatrix} b_1 & v_1 & w_1 \\ b_2 & v_2 & w_2 \\ b_3 & v_3 & w_3 \end{vmatrix} \mathbf{e}_1 \mathbf{e}_2 \mathbf{e}_3}{\begin{vmatrix} u_1 & v_1 & w_1 \\ u_2 & v_2 & w_2 \\ u_3 & v_3 & w_3 \end{vmatrix} \mathbf{e}_1 \mathbf{e}_2 \mathbf{e}_3}, \tag{1.206}$$

which is exactly the ratio of determinants found in the Cramer's rule solution of this problem. We get Cramer's rule for free due to the antisymmetric structure of the wedge product.

Cramer's rule doesn't apply to cases where the dimension of the space exceeds the number of variables, but a wedge product solution does not have that restriction.

1.21.0.4 Example: Some $\mathbb{R}^4$ vectors.

As an example, consider the two variable system eq. (1.200) for vectors in $\mathbb{R}^4$ as follows

$$\mathbf{a} = \begin{bmatrix} 1 \\ 1 \\ 0 \\ 0 \end{bmatrix}, \quad \mathbf{b} = \begin{bmatrix} 1 \\ 0 \\ 0 \\ 1 \end{bmatrix}, \quad \mathbf{c} = \begin{bmatrix} 1 \\ 2 \\ 0 \\ -1 \end{bmatrix}. \tag{1.207}$$

Here's a (Mathematica) computation of the wedge products for the solution [7]

```
In[5]:=  ClearAll[a, b, c, iab, aWedgeB, cWedgeB, aWedgeC,
          x, y]
         a = e[1] + e[2];
         b = e[1] + e[4];
         c = e[1] + 2 e[2] - e[4];

         aWedgeB = OuterProduct[a, b];
         cWedgeB = OuterProduct[c, b];
         aWedgeC = OuterProduct[a, c];

         (* 1/aWedgeB *)
         iab = aWedgeB / GeometricProduct[aWedgeB,
           aWedgeB];
         x = GeometricProduct[iab, cWedgeB];
         y = GeometricProduct[iab, aWedgeC];

         {{a ∧ b = , aWedgeB},{c ∧ b = , cWedgeB},
          {a ∧ c = , aWedgeC},{" = " x},{y = " y}

Out[5]=  a ∧ b =      -e[1,2] + e[1,4] + e[2,4]
         c ∧ b =      -2 e[1,2] + 2 e[1,4] + 2 e[2,4]
         a ∧ c =      e[1,2] - e[1,4] - e[2,4]
         x =    2
```

which shows that $2\mathbf{a} - \mathbf{b} = \mathbf{c}$.

1.21.0.5 Example: intersection of two lines.

As a concrete example, let's solve the intersection of two lines problem illustrated in fig. 1.21.

---
[7] Using the CliffordBasic.m geometric algebra module from [2].

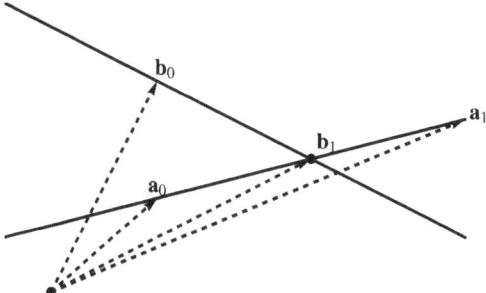

Figure 1.21: Intersection of two lines.

In parametric form, the lines in this problem are

$$\begin{aligned}\mathbf{r}_1(s) &= \mathbf{a}_0 + s(\mathbf{a}_1 - \mathbf{a}_0) \\ \mathbf{r}_2(t) &= \mathbf{b}_0 + t(\mathbf{b}_1 - \mathbf{b}_0),\end{aligned} \quad (1.208)$$

so the solution, if it exists, is found at the point satisfying the equality

$$\mathbf{a}_0 + s(\mathbf{a}_1 - \mathbf{a}_0) = \mathbf{b}_0 + t(\mathbf{b}_1 - \mathbf{b}_0). \quad (1.209)$$

With

$$\begin{aligned}\mathbf{u}_1 &= \mathbf{a}_1 - \mathbf{a}_0 \\ \mathbf{u}_2 &= \mathbf{b}_1 - \mathbf{b}_0 \\ \mathbf{d} &= \mathbf{a}_0 - \mathbf{b}_0,\end{aligned} \quad (1.210)$$

the desired equation to solve is

$$\mathbf{d} + s\mathbf{u}_1 = t\mathbf{u}_2. \quad (1.211)$$

As with any linear system, we can solve for s or t by wedging both sides with one of $\mathbf{u}_1$ or $\mathbf{u}_2$

$$\begin{aligned}\mathbf{d} \wedge \mathbf{u}_1 &= t\mathbf{u}_2 \wedge \mathbf{u}_1 \\ \mathbf{d} \wedge \mathbf{u}_2 + s\mathbf{u}_1 \wedge \mathbf{u}_2 &= 0.\end{aligned} \quad (1.212)$$

In $\mathbb{R}^2$ these equations have a solution if $\mathbf{u}_1 \wedge \mathbf{u}_2 \neq 0$, and in $\mathbb{R}^N$ these have solutions if the bivectors on each sides of the equations describe the same plane (i.e. the bivectors on each side of eq. (1.212) are related by a scalar factor). Put another way, these have solutions when s and t are scalars with the values

$$\begin{aligned}s &= \frac{\mathbf{u}_2 \wedge \mathbf{d}}{\mathbf{u}_1 \wedge \mathbf{u}_2} \\ t &= \frac{\mathbf{u}_1 \wedge \mathbf{d}}{\mathbf{u}_1 \wedge \mathbf{u}_2}.\end{aligned} \quad (1.213)$$

Exercise 1.21   Intersection of a line and plane.

Let a line be parameterized by

$$\mathbf{r}(a) = \mathbf{p} + \alpha \mathbf{a},$$

and a plane be parameterized by

$$\mathbf{r}(b, c) = \mathbf{q} + \beta \mathbf{b} + \gamma \mathbf{c}.$$

a. For the intersection of the two, state the vector equation to be solved, and its solution for a in terms of a ratio of wedge products.
b. State the conditions for which the solution exist in $\mathbb{R}^3$ and $\mathbb{R}^N$.
c. In terms of coordinates in $\mathbb{R}^3$ write out the ratio of wedge products as determinants and compare to the Cramer's rule solution.

Exercise 1.22   Find static load in two member configuration.

In introductory physics or mechanics classes, systems such as those illustrated in fig. 1.22, are common. All of these have a static load under gravity, and with supporting members (rigid beams or wire lines), which can be under compression, or tension, depending on the geometry.

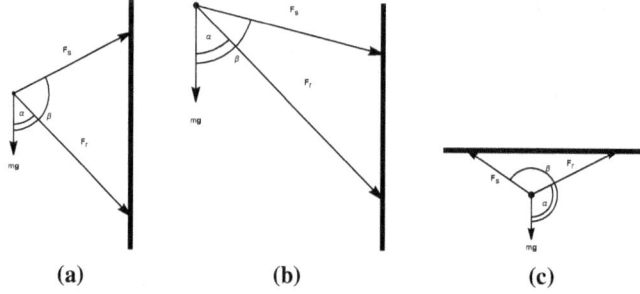

Figure 1.22: Static load with two members.

We seek the magnitudes of the forces in the two members, where the static configuration means that we have a force balance equation of the form

$$\mathbf{F}_s + \mathbf{F}_r + m\mathbf{g} = 0. \tag{1.222}$$

If each of these forces are expressed in complex exponential form

$$\mathbf{F}_r = f_r \mathbf{e}_1 e^{i\alpha}$$
$$\mathbf{F}_s = f_s \mathbf{e}_1 e^{i\beta} \tag{1.223}$$
$$\mathbf{g} = g\mathbf{e}_1,$$

then the system to be solved is

$$f_r \mathbf{e}_1 e^{i\alpha} + f_s \mathbf{e}_1 e^{i\beta} + mg\mathbf{e}_1 = 0. \tag{1.224}$$

By computing a ratio of wedge products, find the scalar magnitudes f_r, f_s for such a system.

1.21.0.6 *Example: Best fit solution for two variable system.*

Now, let's consider the case where the system cannot be solved exactly. It's sufficient to illustrate the ideas using just two variables.

Geometrically, the best we can do is to try to solve the related "least squares" problem

$$x\mathbf{a} + y\mathbf{b} = \mathbf{c}_{\|}, \tag{1.227}$$

where $\mathbf{c}_{\|}$ is the projection of $\mathbf{c}$ onto the plane spanned by $\mathbf{a}, \mathbf{b}$. Regardless of the value of $\mathbf{c}$, we can always find a solution to this problem. For example, solving for x, we have

$$\begin{aligned} x &= \frac{1}{\mathbf{a} \wedge \mathbf{b}} \mathbf{c}_{\|} \wedge \mathbf{b} \\ &= \frac{1}{\mathbf{a} \wedge \mathbf{b}} \cdot (\mathbf{c}_{\|} \wedge \mathbf{b}) \\ &= \frac{1}{\mathbf{a} \wedge \mathbf{b}} \cdot (\mathbf{c} \wedge \mathbf{b}) - \frac{1}{\mathbf{a} \wedge \mathbf{b}} \cdot (\mathbf{c}_{\perp} \wedge \mathbf{b}). \end{aligned} \tag{1.228}$$

The zero above follows because $\mathbf{c}_{\perp}$ is perpendicular to both $\mathbf{a}$ and $\mathbf{b}$ by construction. Geometrically, we are trying to dot two perpendicular bivectors, where $\mathbf{b}$ is a common factor of those two bivectors, as illustrated in fig. 1.23.

We see that the solution to this two variable linear system problem, is

$$x = \frac{1}{\mathbf{a} \wedge \mathbf{b}} \cdot (\mathbf{c} \wedge \mathbf{b}). \tag{1.229a}$$

$$y = \frac{1}{\mathbf{a} \wedge \mathbf{b}} \cdot (\mathbf{a} \wedge \mathbf{c}). \tag{1.229b}$$

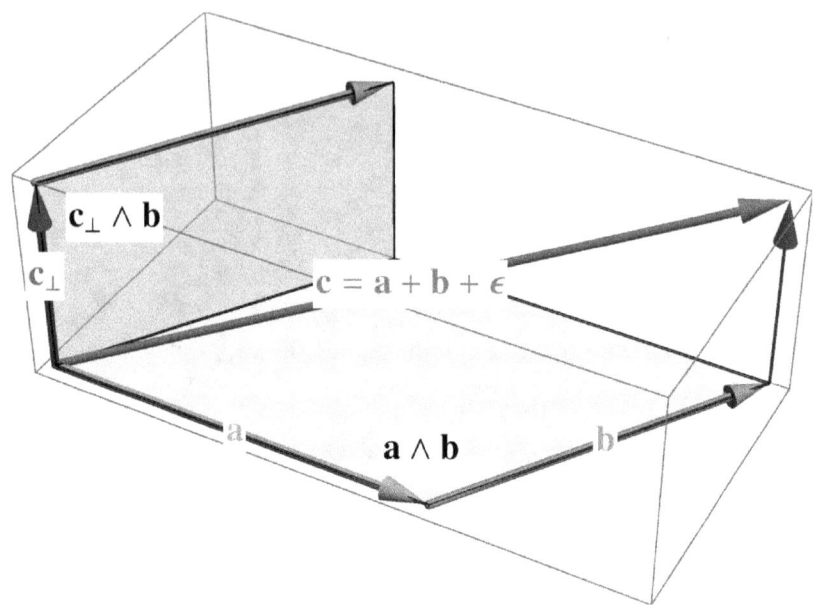

Figure 1.23: Perpendicular bivectors.

Exercise 1.23   Perpendicular blades.

Show algebraically, that the second term from eq. (1.228)

$$-\frac{1}{\mathbf{a} \wedge \mathbf{b}} \cdot (\mathbf{c}_\perp \wedge \mathbf{b}),$$

is zero.

Exercise 1.24   Two variable least squares problem.

We called the projection solution, a least-squares solution, without full justification. Justify this by finding the best fit solution to the two variable system

$$x\mathbf{a} + y\mathbf{b} = \mathbf{c},$$

by minimizing the squared error function

$$\epsilon = (\mathbf{c} - x\mathbf{a} - y\mathbf{b})^2. \qquad (1.232)$$

Show that the resulting solution is identical to eq. (1.229).

1.22 PROBLEM SOLUTIONS.

Answer for Exercise 1.3

The reader can check that with zero element $0 = \begin{bmatrix} 0 & 0 \\ 0 & 0 \end{bmatrix}$, and a scalar multiplicative identity 1, all the vector space properties are satisified.

For the coordinates observe that $\mathbf{x} = \begin{bmatrix} c & a - ib \\ a + ib & -c \end{bmatrix}$, and

$$\text{tr}(\sigma_1 \mathbf{x}) = \text{tr} \begin{bmatrix} a + ib & -c \\ c & a - ib \end{bmatrix} = 2a$$

$$\text{tr}(\sigma_2 \mathbf{x}) = \text{tr} \begin{bmatrix} -ia + b & ic \\ ic & ia + b \end{bmatrix} = 2b \qquad (1.4)$$

$$\text{tr}(\sigma_3 \mathbf{x}) = \text{tr} \begin{bmatrix} c & a - ib \\ a + ib & c \end{bmatrix} = 2c,$$

so $a = \text{tr}(\sigma_1 \mathbf{x})/2$, $b = \text{tr}(\sigma_2 \mathbf{x})/2$, $c = \text{tr}(\sigma_3 \mathbf{x})/2$.

Answer for Exercise 1.4

To verify, swap repeatedly, changing the sign with each swap. Any cyclic permutation requires exactly two swaps

$$\begin{aligned}
\mathbf{e}_2 \mathbf{e}_3 \mathbf{e}_1 &= \mathbf{e}_2 (\mathbf{e}_3 \mathbf{e}_1) \\
&= -\mathbf{e}_2 (\mathbf{e}_1 \mathbf{e}_3) \\
&= -(\mathbf{e}_2 \mathbf{e}_1) \mathbf{e}_3 \\
&= +\mathbf{e}_1 \mathbf{e}_2 \mathbf{e}_3,
\end{aligned} \qquad (1.20)$$

$$\begin{aligned}
\mathbf{e}_3 \mathbf{e}_1 \mathbf{e}_2 &= (\mathbf{e}_3 \mathbf{e}_1) \mathbf{e}_2 \\
&= -(\mathbf{e}_1 \mathbf{e}_3) \mathbf{e}_2 \\
&= -\mathbf{e}_1 (\mathbf{e}_3 \mathbf{e}_2) \\
&= +\mathbf{e}_1 \mathbf{e}_2 \mathbf{e}_3.
\end{aligned} \qquad (1.21)$$

Answer for Exercise 1.5

The 2D rotation matrix is

$$R_\theta = \begin{bmatrix} \cos\theta & -\sin\theta \\ \sin\theta & \cos\theta \end{bmatrix}, \tag{1.41}$$

so to rotate coordinates by $\pm\pi/2$, we multiply by

$$R_{\pm\pi/2} = \pm \begin{bmatrix} 0 & -1 \\ 1 & 0 \end{bmatrix}. \tag{1.42}$$

In particular

$$R_{\pm\pi/2} \begin{bmatrix} \rho\cos\theta \\ \rho\sin\theta \end{bmatrix} = \pm\pi/2 \begin{bmatrix} 0 & -1 \\ 1 & 0 \end{bmatrix} \begin{bmatrix} \rho\cos\theta \\ \rho\sin\theta \end{bmatrix} = \pm\rho \begin{bmatrix} -\sin\theta \\ \cos\theta \end{bmatrix}, \tag{1.43}$$

consistent with the results observed from left and right multiplication with the plane pseudoscalar $\mathbf{e}_1\mathbf{e}_2$.

Answer for Exercise 1.6

Solution Part a. Let χ be a multivector that squares to ± 1. Series expansion of $\cosh(\chi\theta)$, for scalar *theta* yields

$$\cosh(\chi\theta) = \sum_{k=0}^{\infty} \frac{(\chi\theta)^{2k}}{(2k)!} = \sum_{k=0}^{\infty} \frac{\chi^{2k}\theta^{2k}}{(2k)!}. \tag{1.44}$$

In particular, for $\chi = J, K$ respectively, we have

$$\cosh(J\theta) = \sum_{k=0}^{\infty} \frac{(-1)^k \theta^{2k}}{(2k)!} = \cos\theta$$

$$\cosh(K\theta) = \sum_{k=0}^{\infty} \frac{(+1)^k \theta^{2k}}{(2k)!} = \cosh\theta. \tag{1.45}$$

Similarly,

$$\sinh(\chi\theta) = \sum_{k=0}^{\infty} \frac{(\chi\theta)^{2k+1}}{(2k+1)!} = \chi \sum_{k=0}^{\infty} \frac{\chi^{2k}\theta^{2k+1}}{(2k+1)!}. \tag{1.46}$$

So, for $\chi = J, K$ respectively, we have

$$\sinh(J\theta) = J \sum_{k=0}^{\infty} \frac{(-1)^k \theta^{2k+1}}{(2k+1)!} = J\sin\theta$$

$$\sinh(K\theta) = K \sum_{k=0}^{\infty} \frac{(+1)^k \theta^{2k+1}}{(2k+1)!} = K\sinh\theta. \tag{1.47}$$

Solution Part b. Series expanding again, we may split the exponential into even and odd parts, for any multivector x

$$\begin{aligned} e^x &= \sum_{k=0}^{\infty} \frac{x^k}{k!} \\ &= \sum_{k=0}^{\infty} \frac{x^{2k}}{(2k)!} + \sum_{k=0}^{\infty} \frac{x^{2k+1}}{(2k+1)!} \\ &= \cosh(x) + \sinh(x). \end{aligned} \quad (1.48)$$

There is nothing in such a series expansion that cares about the type of x, only that we can take repeated powers. The remainder of the problem follows from our results above after substitution of $x = J\theta$ and $x = K\theta$ respectively.

Solution Part c. The exponential of a sum, such as $X + Y$, regardless of the types or characteristics of X and Y is

$$e^{X+Y} = \sum_{k=0}^{\infty} \frac{(X+Y)^k}{k!}. \quad (1.49)$$

Let's look at the powers of such a sum. For the square and cube we have

$$(X+Y)^2 = X^2 + XY + YX + Y^2, \quad (1.50)$$

$$(X+Y)^3 = X^3 + X^2Y + XYX + YX^2 + Y^2X + YXY + XY^2 + Y^3. \quad (1.51)$$

Observe that the conventional binomial series form for these powers is only possible if X and Y commute. If we have such commutation, then the exponential takes the form

$$\begin{aligned} e^{X+Y} &= \sum_{k=0}^{\infty} \sum_{j=0}^{k} \binom{k}{j} \frac{X^j Y^{k-j}}{k!} \\ &= \sum_{k=0}^{\infty} \sum_{j=0}^{k} \frac{X^j Y^{k-j}}{j!(k-j)!}. \end{aligned} \quad (1.52)$$

This is a sum over all points in a trianglular region of the first quadrant of indexes on the k, j axes. We can, however, sum over all the diagonals

$s = k - j$ = constant, and index our position on each of those diagonals by $u = j$, to find

$$e^{X+Y} = \sum_{s=0}^{\infty} \sum_{u=0}^{\infty} \frac{X^u Y^s}{u!\, s!}$$
$$= \sum_{u=0}^{\infty} \frac{X^u}{u!} \sum_{s=0}^{\infty} \frac{Y^s}{s!} \quad (1.53)$$
$$= e^X e^Y.$$

We see that commutation of variables is required for an exponential of a sum to equal the product of the exponentials. This is worth understanding since it shows us that we can factor exponentials of sums such as $Z = 1 + \mathbf{e}_1\mathbf{e}_2$, $Z = \mathbf{e}_1\mathbf{e}_2 + \mathbf{e}_3\mathbf{e}_4$, $Z = \mathbf{e}_1 + \mathbf{e}_1\mathbf{e}_2\mathbf{e}_3$, into the product of the exponentials of the summands of those multivectors, but cannot do so with multivectors like $Z = \mathbf{e}_1 + \mathbf{e}_2$, $Z = \mathbf{e}_1 + \mathbf{e}_1\mathbf{e}_2$, or $Z = \mathbf{e}_1\mathbf{e}_2 + \mathbf{e}_2\mathbf{e}_3$.

Answer for Exercise 1.7

Part a. From the power series representation of the exponential, we compute

$$\left(e^x\right)' = \sum_{k=1}^{\infty} \frac{(x^k)'}{k!} \quad (1.54)$$

If x is an arbitrary algebraic entity with unknown characteristics, we may only write

$$\left(x^k\right)' = x' x^{k-1} + x x' x^{k-2} + \cdots \quad (1.55)$$

It is only when we know a-priori that x and x' commute, can we reduce this in the usual combinatoric fashion, writing

$$\left(x^k\right)' = k x' x^{k-1}. \quad (1.56)$$

If (and only if) that is true, do we have

$$(e^x)' = \sum_{k=1}^{\infty} \frac{kx'x^{k-1}}{k!}$$

$$= x' \sum_{k=1}^{\infty} \frac{x^{k-1}}{(k-1)!} \qquad (1.57)$$

$$= x' \sum_{k=0}^{\infty} \frac{x^k}{k!}$$

$$= x'e^x.$$

Hiding in this identity is the assumption that x commutes with x', so it is not generally true for non-commutative objects with as multivectors, or square matrices.

Part b. If j^2 is a constant, then we must have

$$(j^2)' = jj' + j'j = 0, \qquad (1.58)$$

or

$$jj' = -j'j. \qquad (1.59)$$

For example, for the $j = \mathbf{e}_{31}e^{i\phi}$ in the spherical polar example that was mentioned, taking derivatives with respect to ϕ, we have $j'j = -jj' = \mathbf{e}_{12}$. We now set $x = j\theta$, and see that we have

$$\left((j\theta)^k\right)' = \left(j'j^{k-1} + jj'j^{k-2} + \cdots\right)\theta^k, \qquad (1.60)$$

but since j anticommutes with j' this is zero whenever k is even, and all but one term cancels out when k is odd. The exponential derivative is

$$\left(e^{j\theta}\right)' = \sum_{k=1, k\in\text{odd}}^{\infty} \frac{j'j^{k-1}\theta^k}{k!}$$

$$= j'j^{-1} \sum_{k=1, k\in\text{odd}}^{\infty} \frac{j^k\theta^k}{k!} \qquad (1.61)$$

$$= j'j^{-1} \sinh(j\theta).$$

Part c. When $j^2 = -1$, $\sinh(j\theta) = j\sin\theta$, so

$$\left(e^{j\theta}\right)' = j'j^{-1}j\sin\theta = j'\sin\theta. \tag{1.62}$$

Clearly, this is what we find if we first expand the exponential in its cis form.

Answer for Exercise 1.8

$$\mathbf{a} \wedge (\alpha\mathbf{a}) = \sum_{i<j} \begin{vmatrix} a_i & a_j \\ \alpha a_i & \alpha a_j \end{vmatrix} \mathbf{e}_i\mathbf{e}_j, \tag{1.88}$$

but

$$\begin{vmatrix} a_i & a_j \\ \alpha a_i & \alpha a_j \end{vmatrix} = \alpha \begin{vmatrix} a_i & a_j \\ a_i & a_j \end{vmatrix} = 0, \tag{1.89}$$

for all i, j, so $\mathbf{a} \wedge (\alpha\mathbf{a}) = 0$.

Answer for Exercise 1.9

Substitution gives

$$\mathbf{b} \wedge \mathbf{a} = \sum_{i<j} \begin{vmatrix} b_i & b_j \\ a_i & a_j \end{vmatrix} \mathbf{e}_i\mathbf{e}_j, \tag{1.90}$$

but

$$\begin{vmatrix} b_i & b_j \\ a_i & a_j \end{vmatrix} = -\begin{vmatrix} a_i & a_j \\ b_i & b_j \end{vmatrix}, \tag{1.91}$$

for all i, j, so $\mathbf{b} \wedge \mathbf{a} = -\mathbf{a} \wedge \mathbf{b}$.

Answer for Exercise 1.10

Solution Part a. Substitution gives

$$(\mathbf{a}+\mathbf{b}) \wedge (\mathbf{c}+\mathbf{d}) = \mathbf{a} \wedge \mathbf{b} = \sum_{i<j} \begin{vmatrix} (a_i+b_i) & (a_j+b_j) \\ (c_i+d_i) & (c_j+d_j) \end{vmatrix} \mathbf{e}_i\mathbf{e}_j, \tag{1.92}$$

1.22 PROBLEM SOLUTIONS. 83

but that determinant expands as

$$
\begin{aligned}
\begin{vmatrix} (a_i+b_i) & (a_j+b_j) \\ (c_i+d_i) & (c_j+d_j) \end{vmatrix}
&= (a_i+b_i)(c_j+d_j) - (a_j+b_j)(c_i+d_i) \\
&= a_ic_j - a_jc_i + b_ic_j - b_jc_i + a_id_j - a_jd_i + b_id_j - b_jd_i \\
&= \begin{vmatrix} a_i & a_j \\ c_i & c_j \end{vmatrix} + \begin{vmatrix} a_i & a_j \\ d_i & d_j \end{vmatrix} + \begin{vmatrix} b_i & b_j \\ c_i & c_j \end{vmatrix} + \begin{vmatrix} b_i & b_j \\ d_i & d_j \end{vmatrix}.
\end{aligned}
\qquad (1.93)
$$

Backsubstitution and comparison proves the result.

Solution Part b. Substitution gives

$$
(\alpha \mathbf{a}) \wedge (\beta \mathbf{b}) = \sum_{i<j} \begin{vmatrix} \alpha a_i & \alpha a_j \\ \beta b_i & \beta b_j \end{vmatrix} \mathbf{e}_i \mathbf{e}_j, \qquad (1.94)
$$

but

$$
\begin{vmatrix} \alpha a_i & \alpha a_j \\ \beta b_i & \beta b_j \end{vmatrix} = \alpha\beta \begin{vmatrix} a_i & a_j \\ b_i & b_j \end{vmatrix}, \qquad (1.95)
$$

proving the result.

Answer for Exercise 1.11

$$
\begin{aligned}
j^2 &= \frac{1}{3}(\mathbf{e}_{32} + \mathbf{e}_{13} + \mathbf{e}_{12})(\mathbf{e}_{32} + \mathbf{e}_{13} + \mathbf{e}_{12}) \\
&= \frac{1}{3}\left(\mathbf{e}_{32}^2 + \mathbf{e}_{13}^2 + \mathbf{e}_{12}^2 + \mathbf{e}_{3213} + \mathbf{e}_{3212} + \mathbf{e}_{1332} + \mathbf{e}_{1312} + \mathbf{e}_{1232} + \mathbf{e}_{1213}\right) \\
&= \frac{1}{3}\left(-3 + \mathbf{e}_{3321} - \mathbf{e}_{3221} + \mathbf{e}_{12} - \mathbf{e}_{3112} - \mathbf{e}_{1322} - \mathbf{e}_{1123}\right) \\
&= \frac{1}{3}\left(-3 + \mathbf{e}_{21} - \mathbf{e}_{31} + \mathbf{e}_{12} - \mathbf{e}_{32} - \mathbf{e}_{13} - \mathbf{e}_{23}\right) \\
&= \frac{-3}{3} \\
&= -1.
\end{aligned}
\qquad (1.96)
$$

Answer for Exercise 1.12

Here are the basic quaternionic relations

$$\mathbf{ij} = \mathbf{e}_{32}\mathbf{e}_{13} = (-\mathbf{e}_{23})(-\mathbf{e}_{31}) = \mathbf{e}_{21} = \mathbf{k} \tag{1.104}$$

$$\mathbf{jk} = \mathbf{e}_{13}\mathbf{e}_{21} = (-\mathbf{e}_{31})(-\mathbf{e}_{12}) = \mathbf{e}_{32} = \mathbf{i} \tag{1.105}$$

$$\mathbf{ki} = \mathbf{e}_{21}\mathbf{e}_{32} = (-\mathbf{e}_{12})(-\mathbf{e}_{23}) = \mathbf{e}_{13} = \mathbf{j}. \tag{1.106}$$

All these bivectors obviously square to -1, which incidentally shows that $\mathbf{ijk} = \mathbf{k}^2 = -1$, a well known quaternion identity.

Answer for Exercise 1.13

Since $j \ne k$, $\mathbf{e}_j \mathbf{e}_k = -\mathbf{e}_k \mathbf{e}_j$, so for $i = k$

$$\langle \mathbf{e}_i \mathbf{e}_j \mathbf{e}_k \rangle_1 = -\langle \mathbf{e}_i \mathbf{e}_k \mathbf{e}_j \rangle_1 = -\langle \mathbf{e}_k \mathbf{e}_i \mathbf{e}_j \rangle_1 = \langle \mathbf{e}_k \mathbf{e}_j \mathbf{e}_i \rangle_1, \tag{1.131}$$

and for $i = j$

$$\langle \mathbf{e}_i \mathbf{e}_j \mathbf{e}_k \rangle_1 = \langle \mathbf{e}_j \mathbf{e}_i \mathbf{e}_k \rangle_1 = -\langle \mathbf{e}_j \mathbf{e}_k \mathbf{e}_i \rangle_1 = \langle \mathbf{e}_k \mathbf{e}_j \mathbf{e}_i \rangle_1. \tag{1.132}$$

Answer for Exercise 1.14

We can tackle this first looking at the $i = j$ case, where

$$\langle \mathbf{e}_i \mathbf{e}_j \mathbf{e}_k \rangle_1 = \langle (\mathbf{e}_i \cdot \mathbf{e}_j) \mathbf{e}_k \rangle_1 = (\mathbf{e}_i \cdot \mathbf{e}_j) \mathbf{e}_k. \tag{1.134}$$

For the $i = k$ case, we have

$$\langle \mathbf{e}_i \mathbf{e}_j \mathbf{e}_k \rangle_1 = -\langle \mathbf{e}_i \mathbf{e}_k \mathbf{e}_j \rangle_1 = -\langle (\mathbf{e}_i \cdot \mathbf{e}_k) \mathbf{e}_j \rangle_1 = -(\mathbf{e}_i \cdot \mathbf{e}_k) \mathbf{e}_j. \tag{1.135}$$

Combining both possibilities we have

$$\langle \mathbf{e}_i \mathbf{e}_j \mathbf{e}_k \rangle_1 = (\mathbf{e}_i \cdot \mathbf{e}_j) \mathbf{e}_k - (\mathbf{e}_i \cdot \mathbf{e}_k) \mathbf{e}_j. \tag{1.136}$$

Incidentally, note that this only holds when $j \ne k$. More generally

$$\langle \mathbf{e}_i \mathbf{e}_j \mathbf{e}_k \rangle_1 = (\mathbf{e}_i \cdot \mathbf{e}_j) \mathbf{e}_k - (\mathbf{e}_i \cdot \mathbf{e}_k) \mathbf{e}_j + (\mathbf{e}_j \cdot \mathbf{e}_k) \mathbf{e}_i, \tag{1.137}$$

(since there is a term for each permutation of i, j, k and a sign change when that permutuation is not even.)

Answer for Exercise 1.15

Writing the wedge of three vectors as a grade three selection

$$\mathbf{a} \wedge \mathbf{b} \wedge \mathbf{c} = \langle \mathbf{abc} \rangle_3, \tag{1.147}$$

and applying the vector product identity $\mathbf{xy} = -\mathbf{yx} + 2\mathbf{x} \cdot \mathbf{y}$, we have

$$\begin{aligned}
\mathbf{a} \wedge \mathbf{b} \wedge \mathbf{c} &= \langle \mathbf{abc} \rangle_3 \\
&= \langle (-\mathbf{ba} + 2\mathbf{a} \cdot \mathbf{b}) \mathbf{c} \rangle_3 \\
&= -\langle \mathbf{bac} \rangle_3 \\
&= -\mathbf{b} \wedge \mathbf{a} \wedge \mathbf{c}.
\end{aligned} \tag{1.148}$$

Similarly

$$\begin{aligned}
\mathbf{a} \wedge \mathbf{b} \wedge \mathbf{c} &= \langle \mathbf{abc} \rangle_3 \\
&= \langle \mathbf{a} (-\mathbf{cb} + 2\mathbf{b} \cdot \mathbf{c}) \rangle_3 \\
&= -\langle \mathbf{acb} \rangle_3 \\
&= -\mathbf{a} \wedge \mathbf{c} \wedge \mathbf{b}.
\end{aligned} \tag{1.149}$$

We see that any two adjacent wedge products in the wedge of three vectors may be interchanged with a corresponding sign change, a process that can be repeated until all combinations are formed. This includes

$$\begin{aligned}
\mathbf{a} \wedge \mathbf{b} \wedge \mathbf{c} &= -\mathbf{b} \wedge \mathbf{a} \wedge \mathbf{c} \\
&= +\mathbf{b} \wedge \mathbf{c} \wedge \mathbf{a} \\
&= -\mathbf{c} \wedge \mathbf{b} \wedge \mathbf{a} \\
&= \mathbf{c} \wedge \mathbf{a} \wedge \mathbf{b} \\
&= -\mathbf{a} \wedge \mathbf{c} \wedge \mathbf{b}.
\end{aligned} \tag{1.150}$$

Answer for Exercise 1.16

$$(\mathbf{e}_{12} + \mathbf{e}_{34}) \wedge (\mathbf{e}_{12} + \mathbf{e}_{34}) = \mathbf{e}_{1234} + \mathbf{e}_{3412} = 2\mathbf{e}_{1234}. \tag{1.151}$$

A blade is the wedge product of two vectors, or the geometric product of two orthogonal vectors. The grade-2 multivector $\mathbf{e}_{12} + \mathbf{e}_{34}$ is not a blade, since there is no common factor between $\mathbf{e}_{12}$ and $\mathbf{e}_{34}$. It is not possible to factor this multivector into two orthogonal products.

Answer for Exercise 1.17

$$\begin{aligned}(\mathbf{x} - \text{Proj}_{\hat{\mathbf{u}}}(\mathbf{x})) \cdot \hat{\mathbf{u}} &= \mathbf{x} \cdot \hat{\mathbf{u}} - ((\mathbf{x} \cdot \hat{\mathbf{u}}) \hat{\mathbf{u}}) \cdot \hat{\mathbf{u}} \\ &= \mathbf{x} \cdot \hat{\mathbf{u}} - (\mathbf{x} \cdot \hat{\mathbf{u}}) (\hat{\mathbf{u}} \cdot \hat{\mathbf{u}}) \\ &= \mathbf{x} \cdot \hat{\mathbf{u}} - \mathbf{x} \cdot \hat{\mathbf{u}} \\ &= 0.\end{aligned} \quad (1.165)$$

Answer for Exercise 1.18

Solution Part a. Given $\mathbf{x} = a\mathbf{e}_2 + b\mathbf{e}_3$ and $\hat{\mathbf{u}} = (\mathbf{e}_1 + \mathbf{e}_2)/\sqrt{2}$, we found that

$$\mathbf{x} \wedge \hat{\mathbf{u}} = \frac{1}{\sqrt{2}} \left(b \left(\mathbf{e}_{32} + \mathbf{e}_{31} \right) + a \mathbf{e}_{21} \right). \quad (1.166)$$

Multiplying once more by $\hat{\mathbf{u}}$ on the right, we have

$$\begin{aligned}(\mathbf{x} \wedge \hat{\mathbf{u}}) \hat{\mathbf{u}} &= \frac{1}{2} \left(b \left(\mathbf{e}_{32} + \mathbf{e}_{31} \right) + a \mathbf{e}_{21} \right) \left(\mathbf{e}_1 + \mathbf{e}_2 \right) \\ &= \frac{1}{2} \left(b \left(\mathbf{e}_{321} + \mathbf{e}_{311} \right) + a \mathbf{e}_{211} + b \left(\mathbf{e}_{322} + \mathbf{e}_{312} \right) + a \mathbf{e}_{212} \right) \\ &= \frac{1}{2} \left(b \left(\mathbf{e}_{321} + \mathbf{e}_3 \right) + a \mathbf{e}_2 + b \left(\mathbf{e}_3 + \mathbf{e}_{312} \right) - a \mathbf{e}_1 \right) \\ &= \frac{1}{2} \left(2b\mathbf{e}_3 + a \left(\mathbf{e}_2 - \mathbf{e}_1 \right) \right).\end{aligned} \quad (1.167)$$

We are left with $(\mathbf{x} \wedge \hat{\mathbf{u}}) \hat{\mathbf{u}} = (\mathbf{x} \wedge \hat{\mathbf{u}}) \cdot \hat{\mathbf{u}}$, since all the trivector components cancel perfectly.

Solution Part b. Now we can compare the above to $\mathbf{x} - (\mathbf{x} \cdot \hat{\mathbf{u}}) \hat{\mathbf{u}}$. First

$$\mathbf{x} \cdot \hat{\mathbf{u}} = (a\mathbf{e}_2 + b\mathbf{e}_3) \cdot (\mathbf{e}_1 + \mathbf{e}_2) \frac{1}{\sqrt{2}} = \frac{a}{\sqrt{2}}, \quad (1.168)$$

so

$$\begin{aligned}\mathbf{x} - (\mathbf{x} \cdot \hat{\mathbf{u}}) \hat{\mathbf{u}} &= a\mathbf{e}_2 + b\mathbf{e}_3 - \frac{a}{2} (\mathbf{e}_1 + \mathbf{e}_2) \\ &= b\mathbf{e}_3 + \frac{a}{2} (-\mathbf{e}_1 + \mathbf{e}_2),\end{aligned} \quad (1.169)$$

as calculated from $(\mathbf{x} \wedge \hat{\mathbf{u}}) \hat{\mathbf{u}}$.

1.22 PROBLEM SOLUTIONS. 87

Answer for Exercise 1.19

In general, given a bivector $B = a_1 e_{23} + a_2 e_{31} + a_3 e_{12}$, if we pick the coefficient a_i that has the largest absolute magnitude (to avoid numerical instability in case the bivector is ill-conditioned and has a small non-zero component in one direction), and then select one of the two vector factors of the unit blade that is associated with that component, calling this **e**, then we can utilize this vector **e** to find one vector that lies in the plane of B. For example, if the largest absolute magnitude coefficient is a_3 then pick either $\mathbf{e} = \mathbf{e}_1$ or $\mathbf{e} = \mathbf{e}_2$. Now, compute

$$\mathbf{a} = B \cdot \mathbf{e}.$$

This vector lies in the plane that B represents. Specifically, it is the projection of **e** onto B, but rotated 90 degrees, since $(B \cdot \mathbf{e}) \mathbf{e}$ would be the projection itself. If we dot **a** with B then we find another vector that lies in the plane represented by B, but is rotated 90 degrees in the plane, away from **a**. That is:

$$\mathbf{b} = B \cdot \mathbf{a}$$

We've now found two perpendicular vectors that lie in the plane that B represents, so we have

$$B \propto \mathbf{ab} = \mathbf{a} \wedge \mathbf{b}.$$

Let's try these ideas with the bivector of this problem $B = \mathbf{e}_{23} + \mathbf{e}_{31} + \mathbf{e}_{12}$. All components are equally weighted, so let's compute the $B \cdot \mathbf{e}_1$ to start with to find a first factor of B.

$$\begin{aligned} B \cdot \mathbf{e}_1 &= (\mathbf{e}_{23} + \mathbf{e}_{31} + \mathbf{e}_{12}) \cdot \mathbf{e}_1 \\ &= \mathbf{e}_3 - \mathbf{e}_2. \end{aligned} \quad (1.173)$$

Dotting this into B once again will find a second factor

$$\begin{aligned} B \cdot (\mathbf{e}_3 - \mathbf{e}_2) &= (\mathbf{e}_{23} + \mathbf{e}_{31} + \mathbf{e}_{12}) \cdot (\mathbf{e}_3 - \mathbf{e}_2) \\ &= \mathbf{e}_2 - \mathbf{e}_1 + \mathbf{e}_3 - \mathbf{e}_1 \\ &= -2\mathbf{e}_1 + \mathbf{e}_2 + \mathbf{e}_3. \end{aligned} \quad (1.174)$$

Adjusting the scaling appropriately, gives us two orthogonal factors of B

$$\mathbf{e}_{12} + \mathbf{e}_{23} + \mathbf{e}_{31} = \frac{\mathbf{e}_3 - \mathbf{e}_2}{2} (2\mathbf{e}_1 - \mathbf{e}_2 - \mathbf{e}_3). \quad (1.175)$$

Let's see what factors we find by dotting B with $\mathbf{e}_3$ instead. This gives us

$$B \cdot \mathbf{e}_3 = (\mathbf{e}_{12} + \mathbf{e}_{23} + \mathbf{e}_{31}) \cdot \mathbf{e}_3 \qquad (1.176)$$
$$= \mathbf{e}_2 - \mathbf{e}_1.$$

Dotting this into B a second time yields

$$B \cdot (\mathbf{e}_2 - \mathbf{e}_1) = (\mathbf{e}_{12} + \mathbf{e}_{23} + \mathbf{e}_{31}) \cdot (\mathbf{e}_2 - \mathbf{e}_1) \qquad (1.177)$$
$$= \mathbf{e}_1 - \mathbf{e}_3 + \mathbf{e}_2 - \mathbf{e}_3.$$

After rescaling, we find

$$\mathbf{e}_{12} + \mathbf{e}_{23} + \mathbf{e}_{31} = (\mathbf{e}_1 + \mathbf{e}_2 - 2\mathbf{e}_3) \frac{\mathbf{e}_2 - \mathbf{e}_1}{2} \qquad (1.178)$$

Each of the sets of factors of eq. (1.175), eq. (1.178) can be interpreted as the edges of two different rectangular representations of the bivector, for which the total area is fixed. The span of either set of factors describes the plane that the bivector represents.

Answer for Exercise 1.20

The parallelogram area is base times height, that is

$$A = \|\mathbf{a}\| \left\| (\mathbf{b} \wedge \hat{\mathbf{a}}) \hat{\mathbf{a}} \right\| = \left\| (\mathbf{b} \wedge \mathbf{a}) \hat{\mathbf{a}} \right\|, \qquad (1.183)$$

but

$$\mathbf{b} \wedge \mathbf{a} = \begin{vmatrix} b_1 & b_2 \\ a_1 & a_2 \end{vmatrix} \mathbf{e}_{12} = \mathcal{A}i \qquad (1.184)$$

where $\mathcal{A} = \begin{vmatrix} b_1 & b_2 \\ a_1 & a_2 \end{vmatrix}$, and $i = \mathbf{e}_{12}$. Our expression for the area is reduced to

$$A = |\mathcal{A}| \, \|i\hat{\mathbf{a}}\|. \qquad (1.185)$$

Note that

$$\|i\hat{\mathbf{a}}\| = \|\hat{\mathbf{a}}\| = 1, \qquad (1.186)$$

since the multiplicative action of i is to rotate by 90 degrees, not changing the (unit) length at all. That leaves

$$A = \pm \begin{vmatrix} b_1 & b_2 \\ a_1 & a_2 \end{vmatrix}, \qquad (1.187)$$

as expected.

Answer for Exercise 1.21

Solution Part a. We are looking for solutions α, β, γ such that the equality

$$\mathbf{p} + \alpha \mathbf{a} = \mathbf{q} + \beta \mathbf{b} + \gamma \mathbf{c}, \tag{1.214}$$

is satisfied. We have only to wedge with $\mathbf{b} \wedge \mathbf{c}$, to find

$$\mathbf{p} \wedge \mathbf{b} \wedge \mathbf{c} + \alpha \left(\mathbf{a} \wedge \mathbf{b} \wedge \mathbf{c} \right) = \mathbf{q} \wedge \mathbf{b} \wedge \mathbf{c}, \tag{1.215}$$

or

$$\alpha = \frac{(\mathbf{q} - \mathbf{p}) \wedge \mathbf{b} \wedge \mathbf{c}}{\mathbf{a} \wedge \mathbf{b} \wedge \mathbf{c}}. \tag{1.216}$$

Solution Part b. For $\mathbb{R}^3$, a solution exists provided $\mathbf{a} \wedge \mathbf{b} \wedge \mathbf{c} \neq 0$, but for $\mathbb{R}^N$ a solution also requires

$$(\mathbf{q} - \mathbf{p}) \wedge \mathbf{b} \wedge \mathbf{c} \propto \mathbf{a} \wedge \mathbf{b} \wedge \mathbf{c}. \tag{1.217}$$

For instance, there is no solution if $(\mathbf{q} - \mathbf{p}) \wedge \mathbf{b} \wedge \mathbf{c} = \mathbf{e}_{124}$, but $\mathbf{a} \wedge \mathbf{b} \wedge \mathbf{c} = \mathbf{e}_{234}$.

Solution Part c. To solve this equation using coordinates, we seek solutions to

$$\mathbf{p} - \mathbf{q} = -\alpha \mathbf{a} + \beta \mathbf{b} + \gamma \mathbf{c}, \tag{1.218}$$

or

$$(\mathbf{p} - \mathbf{q}) \cdot \mathbf{e}_k = (-\alpha \mathbf{a} + \beta \mathbf{b} + \gamma \mathbf{c}) \cdot \mathbf{e}_k, \tag{1.219}$$

$\forall k \in [1, N]$. In matrix form, this is

$$\begin{bmatrix} p_1 - q_1 \\ p_2 - q_2 \\ \vdots \\ p_N - q_N \end{bmatrix} = \begin{bmatrix} -a_1 & b_1 & c_1 \\ -a_2 & b_2 & c_2 \\ & \vdots & \\ -a_N & b_N & c_N \end{bmatrix} \begin{bmatrix} \alpha \\ \beta \\ \gamma \end{bmatrix}. \tag{1.220}$$

The Cramer's rule solution only applies to the $\mathbb{R}^3$ system, and has the form

$$\begin{bmatrix} \alpha \\ \beta \\ \gamma \end{bmatrix} = \frac{\begin{vmatrix} p_1 - q_1 & b_1 & c_1 \\ p_2 - q_2 & b_2 & c_2 \\ p_3 - q_3 & b_3 & c_3 \end{vmatrix}}{\begin{vmatrix} -a_1 & b_1 & c_1 \\ -a_2 & b_2 & c_2 \\ -a_3 & b_3 & c_3 \end{vmatrix}} = \frac{\begin{vmatrix} q_1 - p_1 & b_1 & c_1 \\ q_2 - p_2 & b_2 & c_2 \\ q_3 - p_3 & b_3 & c_3 \end{vmatrix}}{\begin{vmatrix} a_1 & b_1 & c_1 \\ a_2 & b_2 & c_2 \\ a_3 & b_3 & c_3 \end{vmatrix}}. \tag{1.221}$$

This is obviously equivalent to the GA solution, where the ratio of determinants is found immediately from the coordinate representation of a triple wedge product. We can't solve this system of equations using Cramer's rule for $\mathbb{R}^N$ when $N > 3$ since the system is overspecified in that case. That overspecification is why we require the additional $(\mathbf{q} - \mathbf{p}) \wedge \mathbf{b} \wedge \mathbf{c} \propto \mathbf{a} \wedge \mathbf{b} \wedge \mathbf{c}$ constraint for the GA solution using wedge products. Note that this wedge product solution method is unlikely to be numerically stable for $N > 3$, and we are probably better off solving with SVD, so that we have some estimation of the numerical errors that either rule out or validate the solution.

Answer for Exercise 1.22

To solve for f_r or f_s, first take wedge products with the force direction vectors to eliminate the variable not of interest

$$\begin{aligned} f_r \left(\mathbf{e}_1 e^{i\alpha} \right) \wedge \left(\mathbf{e}_1 e^{i\beta} \right) + mg \mathbf{e}_1 \wedge \left(\mathbf{e}_1 e^{i\beta} \right) &= 0 \\ f_s \left(\mathbf{e}_1 e^{i\beta} \right) \wedge \left(\mathbf{e}_1 e^{i\alpha} \right) + mg \mathbf{e}_1 \wedge \left(\mathbf{e}_1 e^{i\alpha} \right) &= 0. \end{aligned} \tag{1.225}$$

Writing the wedges as grade two selections, and noting that $e^{i\theta} \mathbf{e}_1 = \mathbf{e}_1 e^{-i\theta}$, we have

$$\begin{aligned} f_r &= -mg \frac{\left\langle \mathbf{e}_1^2 e^{i\beta} \right\rangle_2}{\left\langle \mathbf{e}_1^2 e^{-i\alpha} e^{i\beta} \right\rangle_2} = -mg \frac{i \sin \beta}{i \sin (\beta - \alpha)} \\ f_s &= -mg \frac{\left\langle \mathbf{e}_1^2 e^{i\alpha} \right\rangle_2}{\left\langle \mathbf{e}_1^2 e^{-i\beta} e^{i\alpha} \right\rangle_2} = mg \frac{i \sin \alpha}{i \sin (\beta - \alpha)}. \end{aligned} \tag{1.226}$$

The pseudoscalar factors i cancel out. We are left with a ratio of sines, but those fell out of the grade selection directly, without requiring us to pull out our good old tables of trig identities.

Answer for Exercise 1.23

We can reduce that second term, first expanding the bivector inverse explicitly

$$-\frac{1}{\mathbf{a}\wedge\mathbf{b}}\cdot(\mathbf{c}_\perp\wedge\mathbf{b}) = -\frac{\mathbf{a}\wedge\mathbf{b}}{(\mathbf{a}\wedge\mathbf{b})^2}\cdot(\mathbf{c}_\perp\wedge\mathbf{b}). \quad (1.230)$$

We can ignore the scalar $-1/(\mathbf{a}\wedge\mathbf{b})^2$ factor, and expand the bivector dot product, to find

$$\begin{aligned}(\mathbf{a}\wedge\mathbf{b})\cdot(\mathbf{c}_\perp\wedge\mathbf{b}) &= ((\mathbf{a}\wedge\mathbf{b})\cdot\mathbf{c}_\perp)\cdot\mathbf{b} \\ &= (\mathbf{a}\,(\mathbf{b}\cdot\mathbf{c}_\perp) - \mathbf{b}\,(\mathbf{a}\cdot\mathbf{c}_\perp))\cdot\mathbf{b} \\ &= 0.\end{aligned} \quad (1.231)$$

Answer for Exercise 1.24

We follow the usual procedure, by equating all partials to zero

$$\begin{aligned}0 &= \frac{\partial\epsilon}{\partial x} = 2(\mathbf{c}-x\mathbf{a}-y\mathbf{b})\cdot(-\mathbf{a}) \\ 0 &= \frac{\partial\epsilon}{\partial y} = 2(\mathbf{c}-x\mathbf{a}-y\mathbf{b})\cdot(-\mathbf{b}).\end{aligned} \quad (1.233)$$

This is a two equation, two unknown system, which can be expressed in matrix form as

$$\begin{bmatrix}\mathbf{a}^2 & \mathbf{a}\cdot\mathbf{b} \\ \mathbf{a}\cdot\mathbf{b} & \mathbf{b}^2\end{bmatrix}\begin{bmatrix}x \\ y\end{bmatrix} = \begin{bmatrix}\mathbf{a}\cdot\mathbf{c} \\ \mathbf{b}\cdot\mathbf{c}\end{bmatrix}. \quad (1.234)$$

This has solution

$$\begin{bmatrix}x \\ y\end{bmatrix} = \frac{1}{\begin{vmatrix}\mathbf{a}^2 & \mathbf{a}\cdot\mathbf{b} \\ \mathbf{a}\cdot\mathbf{b} & \mathbf{b}^2\end{vmatrix}}\begin{bmatrix}\mathbf{b}^2 & -\mathbf{a}\cdot\mathbf{b} \\ -\mathbf{a}\cdot\mathbf{b} & \mathbf{a}^2\end{bmatrix}\begin{bmatrix}\mathbf{a}\cdot\mathbf{c} \\ \mathbf{b}\cdot\mathbf{c}\end{bmatrix} = \frac{\begin{bmatrix}\mathbf{b}^2(\mathbf{a}\cdot\mathbf{c}) - (\mathbf{a}\cdot\mathbf{b})(\mathbf{b}\cdot\mathbf{c}) \\ \mathbf{a}^2(\mathbf{b}\cdot\mathbf{c}) - (\mathbf{a}\cdot\mathbf{b})(\mathbf{a}\cdot\mathbf{c})\end{bmatrix}}{\mathbf{a}^2\mathbf{b}^2 - (\mathbf{a}\cdot\mathbf{b})^2}. \quad (1.235)$$

All of these differences can be expressed as wedge dot products, using the following expansions in reverse

$$\begin{aligned}(\mathbf{a}\wedge\mathbf{b})\cdot(\mathbf{c}\wedge\mathbf{d}) &= \mathbf{a}\cdot(\mathbf{b}\cdot(\mathbf{c}\wedge\mathbf{d})) \\ &= \mathbf{a}\cdot((\mathbf{b}\cdot\mathbf{c})\mathbf{d} - (\mathbf{b}\cdot\mathbf{d})\mathbf{c}) \\ &= (\mathbf{a}\cdot\mathbf{d})(\mathbf{b}\cdot\mathbf{c}) - (\mathbf{a}\cdot\mathbf{c})(\mathbf{b}\cdot\mathbf{d}).\end{aligned} \quad (1.236)$$

We find

$$
\begin{aligned}
x &= \frac{\mathbf{b}^2 (\mathbf{a} \cdot \mathbf{c}) - (\mathbf{a} \cdot \mathbf{b})(\mathbf{b} \cdot \mathbf{c})}{-(\mathbf{a} \wedge \mathbf{b})^2} \\
&= \frac{(\mathbf{a} \wedge \mathbf{b}) \cdot (\mathbf{b} \wedge \mathbf{c})}{-(\mathbf{a} \wedge \mathbf{b})^2} \\
&= \frac{1}{\mathbf{a} \wedge \mathbf{b}} \cdot (\mathbf{c} \wedge \mathbf{b}),
\end{aligned}
\qquad (1.237)
$$

and

$$
\begin{aligned}
y &= \frac{\mathbf{a}^2 (\mathbf{b} \cdot \mathbf{c}) - (\mathbf{a} \cdot \mathbf{b})(\mathbf{a} \cdot \mathbf{c})}{-(\mathbf{a} \wedge \mathbf{b})^2} \\
&= \frac{-(\mathbf{a} \wedge \mathbf{b}) \cdot (\mathbf{a} \wedge \mathbf{c})}{-(\mathbf{a} \wedge \mathbf{b})^2} \\
&= \frac{1}{\mathbf{a} \wedge \mathbf{b}} \cdot (\mathbf{a} \wedge \mathbf{c}).
\end{aligned}
\qquad (1.238)
$$

Sure enough, we find what was dubbed the least squares solution, which we now know can be written out as a ratio of (dotted) wedge products.

2

MULTIVECTOR CALCULUS.

2.1 RECIPROCAL FRAMES.

2.1.1 *Motivation and definition.*

The end goal of this chapter is to be able to integrate multivector functions along curves and surfaces, known collectively as manifolds. For our purposes, a manifold is defined by a parameterization, such as the vector valued function $\mathbf{x}(a, b)$ where a, b are scalar parameters. With one parameter the vector traces out a curve, with two a surface, three a volume, and so forth. The respective partial derivatives of such a parameterized vector define a local basis for the surface at the point at which the partials are evaluated. The span of such a basis is called the tangent space, and the partials that constitute it are not necessarily orthonormal, or even orthogonal.

Unfortunately, in order to work with the curvilinear non-orthonormal bases that will be encountered in general integration theory, some additional tools are required.

- We introduce a reciprocal frame (basis) which partially generalizes the notion of orthogonality to non-orthonormal bases.

- We will borrow the upper and lower index (tensor) notation from relativistic physics that is useful for the intrinsically non-orthonormal spaces encountered in that study, as this notation works well to define the reciprocal frame.

> **Definition 2.1: Reciprocal frame**
>
> Given a subspace basis $\beta = \{\mathbf{x}_1, \mathbf{x}_2, \cdots \mathbf{x}_m\}$, not necessarily orthonormal, the *reciprocal frame* is the set $\{\mathbf{x}^1, \mathbf{x}^2, \cdots \mathbf{x}^m\} \in \text{span}\,\beta$ satisfying
>
> $$\mathbf{x}_i \cdot \mathbf{x}^j = \delta_i{}^j,$$

> where the vector $\mathbf{x}^j$ is not the j-th power of $\mathbf{x}$, but is a superscript index, the conventional way of denoting a reciprocal frame vector, and $\delta_i{}^j$ is the Kronecker delta.

If a basis $\{\mathbf{x}_i\}$ is orthogonal, then the reciprocal frame vectors are, literally, the reciprocals $\mathbf{x}^i = 1/\mathbf{x}_i$ (exercise 2.1). Any orthonormal basis, where every basis vector is its own inverse, is also its reciprocal basis.

In general, if the original basis is not-orthogonal, every reciprocal basis vector is orthogonal to all but one of the original basis vectors, but may not be orthogonal to any other reciprocal basis vector. Techniques for computation of reciprocal bases will be developed for the non-orthogonal case.

Mixed index variables have been introduced above for the first time in this text, which may be unfamiliar. These are most often used in tensor algebra, where any expression that has pairs of upper and lower indexes implies a sum, and is called the summation (or Einstein) convention. For example:

$$\begin{aligned} a_i b^i &\equiv \sum_i a_i b^i \\ A^i{}_j B_i C^j &\equiv \sum_{i,j} A^i{}_j B_i C^j. \end{aligned} \quad (2.1)$$

Summation convention will not be used explicitly in this text, as it deviates from normal practises in electrical engineering[1].

2.1.1.1 Vector coordinates.

The most important application of a reciprocal frame is for the computation of the coordinates of a vector with respect to a non-orthonormal frame. Let a vector $\mathbf{a}$ have coordinates a^i with respect to a basis $\{\mathbf{x}_i\}$

$$\mathbf{a} = \sum_j a^j \mathbf{x}_j, \quad (2.2)$$

where j in a^j is an index not a power[2].

---

[1] Generally, when summation convention is used, explicit summation is only used explicitly when upper and lower indexes are not perfectly matched, but summation is still implied. Readers of texts that use summation convention can check for proper matching of upper and lower indexes to ensure that the expressions make sense. Such matching is the reason a mixed index Kronecker delta has been used in the definition of the reciprocal frame.

[2] In tensor algebra, any index that is found in matched upper and lower index pairs, is known as a dummy summation index, whereas an index that is unmatched is known as a

Dotting with the reciprocal frame vectors $\mathbf{x}^i$ provides these coordinates a^i

$$\begin{aligned}\mathbf{a} \cdot \mathbf{x}^i &= \left(\sum_j a^j \mathbf{x}_j\right) \cdot \mathbf{x}^i \\ &= \sum_j a^j \delta_j{}^i \\ &= a^i.\end{aligned} \quad (2.3)$$

Alternatively, coordinates can be computed with respect to the reciprocal frame. Let those coordinates be a_i, so that

$$\mathbf{a} = \sum_i a_i \mathbf{x}^i. \quad (2.4)$$

Dotting with the basis vectors $\mathbf{x}_i$ provides the reciprocal frame relative coordinates a_i

$$\begin{aligned}\mathbf{a} \cdot \mathbf{x}_i &= \left(\sum_j a_j \mathbf{x}^j\right) \cdot \mathbf{x}_i \\ &= \sum_j a_j \delta^j{}_i \\ &= a_i.\end{aligned} \quad (2.5)$$

We can summarize eq. (2.3) and eq. (2.5) by stating that a vector can be expressed in terms of coordinates relative to either the original or reciprocal basis as follows

$$\mathbf{a} = \sum_j \left(\mathbf{a} \cdot \mathbf{x}^j\right) \mathbf{x}_j = \sum_j \left(\mathbf{a} \cdot \mathbf{x}_j\right) \mathbf{x}^j. \quad (2.6)$$

In tensor algebra the basis is generally implied[3].

---

free index. For example, in $a^j b_{ij}$ (summation implied) j is a summation index, and i is a free index. We are free to make a change of variables of any summation index, so for the same example we can write $a^k b_{ik}$. These index tracking conventions are obvious when summation symbols are included explicitly, as we will do.

[3] In tensor algebra, a vector, identified by the coordinates a^i is called a contravariant vector. When that vector is identified by the coordinates a_i it is called a covariant vector. These labels relate to how the coordinates transform with respect to norm preserving transformations. We have no need of this nomenclature, since we never transform coordinates in isolation, but will always transform the coordinates along with their associated basis vectors.

An example of a 2D oblique Euclidean basis and a corresponding reciprocal basis is plotted in fig. 2.1. Also plotted are the superposition of the projections required to arrive at point (4, 2) along the $\mathbf{x}_1, \mathbf{x}_2$ directions or the $\mathbf{x}^1, \mathbf{x}^2$ directions. In this plot, neither of the reciprocal frame vectors $\mathbf{x}^i$ are orthogonal to the corresponding basis vectors $\mathbf{x}_i$. When one of $\mathbf{x}_i$ is increased(decreased) in magnitude, there will be a corresponding decrease(increase) in the magnitude of $\mathbf{x}^i$, but if the orientation is remained fixed, the corresponding direction of the reciprocal frame vector stays the same.

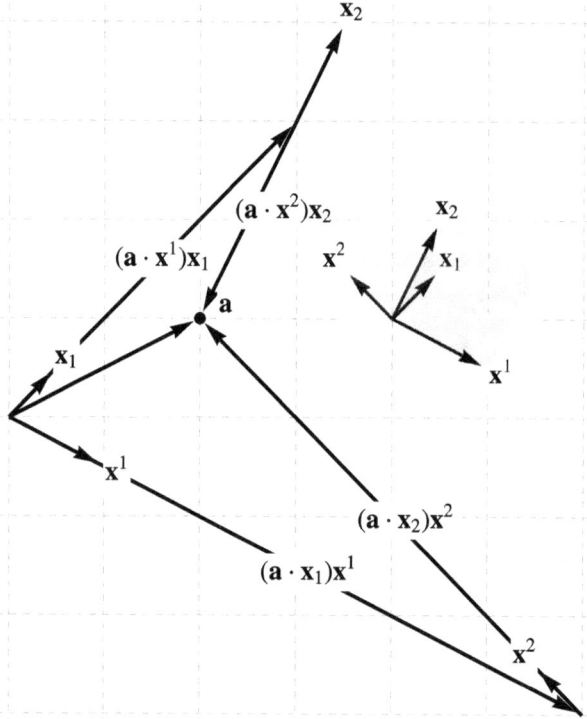

Figure 2.1: Oblique and reciprocal bases.

2.1.1.2 *Bivector coordinates.*

Higher grade multivector objects may also be represented in curvilinear coordinates. Illustrating by example, we will calculate the coordinates of a bivector constrained to a three parameter manifold span $\{\mathbf{x}_1, \mathbf{x}_2, \mathbf{x}_3\}$ which can be represented as

$$B = \frac{1}{2} \sum_{i,j} B^{ij} \mathbf{x}_i \wedge \mathbf{x}_j = \sum_{i<j} B^{ij} \mathbf{x}_i \wedge \mathbf{x}_j. \tag{2.7}$$

The coordinates B^{ij} can be determined by dotting B with $\mathbf{x}^j \wedge \mathbf{x}^i$, where $i \neq j$, yielding

$$\begin{aligned} B \cdot \left(\mathbf{x}^j \wedge \mathbf{x}^i\right) &= \frac{1}{2} \sum_{r,s} B^{rs} \left(\mathbf{x}_r \wedge \mathbf{x}_s\right) \cdot \left(\mathbf{x}^j \wedge \mathbf{x}^i\right) \\ &= \frac{1}{2} \sum_{r,s} B^{rs} \left(\left(\mathbf{x}_r \wedge \mathbf{x}_s\right) \cdot \mathbf{x}^j\right) \cdot \mathbf{x}^i \\ &= \frac{1}{2} \sum_{r,s} B^{rs} \left(\mathbf{x}_r \delta_s{}^j - \mathbf{x}_s \delta_r{}^j\right) \cdot \mathbf{x}^i \\ &= \frac{1}{2} \sum_{r,s} B^{rs} \left(\delta_r{}^i \delta_s{}^j - \delta_s{}^i \delta_r{}^j\right) \\ &= \frac{1}{2} \left(B^{ij} - B^{ji}\right). \end{aligned} \tag{2.8}$$

We see that the coordinates of a bivector, even with respect to a non-orthonormal basis, are antisymmetric, so eq. (2.8) is just B^{ij} as claimed. That is

$$B^{ij} = B \cdot \left(\mathbf{x}^j \wedge \mathbf{x}^i\right). \tag{2.9}$$

Just as the reciprocal frame was instrumental for computation of the coordinates of a vector with respect to an arbitrary (i.e. non-orthonormal frame), we use the reciprocal frame to calculate the coordinates of a bivector, and could do the same for higher grade k-vectors as well.

2.1.2 $\mathbb{R}^2$ *reciprocal frame.*

How are the reciprocal frame vectors computed? While these vectors have a natural GA representation, this is not intrinsically a GA problem, and can be solved with standard linear algebra, using a matrix inversion. For example, given a 2D basis $\{\mathbf{x}_1, \mathbf{x}_2\}$, the reciprocal basis can be assumed to have a coordinate representation in the original basis

$$\begin{aligned} \mathbf{x}^1 &= a\mathbf{x}_1 + b\mathbf{x}_2 \\ \mathbf{x}^2 &= c\mathbf{x}_1 + d\mathbf{x}_2. \end{aligned} \tag{2.10}$$

Imposing the constraints of definition 2.1 leads to a pair of 2x2 linear systems that are easily solved to find

$$\mathbf{x}^1 = \frac{1}{(\mathbf{x}_1)^2(\mathbf{x}_2)^2 - (\mathbf{x}_1 \cdot \mathbf{x}_2)^2} \left((\mathbf{x}_2)^2 \mathbf{x}_1 - (\mathbf{x}_1 \cdot \mathbf{x}_2) \mathbf{x}_2 \right)$$
$$\mathbf{x}^2 = \frac{1}{(\mathbf{x}_1)^2(\mathbf{x}_2)^2 - (\mathbf{x}_1 \cdot \mathbf{x}_2)^2} \left((\mathbf{x}_1)^2 \mathbf{x}_2 - (\mathbf{x}_1 \cdot \mathbf{x}_2) \mathbf{x}_1 \right).$$
(2.11)

The reader may notice that for $\mathbb{R}^3$ the denominator is related to the norm of the cross product $\mathbf{x}_1 \times \mathbf{x}_2$. More generally, this can be expressed as the square of the bivector $\mathbf{x}_1 \wedge \mathbf{x}_2$

$$\begin{aligned} -(\mathbf{x}_1 \wedge \mathbf{x}_2)^2 &= -(\mathbf{x}_1 \wedge \mathbf{x}_2) \cdot (\mathbf{x}_1 \wedge \mathbf{x}_2) \\ &= -((\mathbf{x}_1 \wedge \mathbf{x}_2) \cdot \mathbf{x}_1) \cdot \mathbf{x}_2 \\ &= (\mathbf{x}_1)^2(\mathbf{x}_2)^2 - (\mathbf{x}_1 \cdot \mathbf{x}_2)^2. \end{aligned}$$
(2.12)

Additionally, the numerators are each dot products of $\mathbf{x}_1, \mathbf{x}_2$ with that same bivector

$$\mathbf{x}^1 = \frac{\mathbf{x}_2 \cdot (\mathbf{x}_1 \wedge \mathbf{x}_2)}{(\mathbf{x}_1 \wedge \mathbf{x}_2)^2}$$
$$\mathbf{x}^2 = \frac{\mathbf{x}_1 \cdot (\mathbf{x}_2 \wedge \mathbf{x}_1)}{(\mathbf{x}_1 \wedge \mathbf{x}_2)^2},$$
(2.13)

or

$$\boxed{\begin{aligned} \mathbf{x}^1 &= \mathbf{x}_2 \cdot \frac{1}{\mathbf{x}_1 \wedge \mathbf{x}_2} \\ \mathbf{x}^2 &= \mathbf{x}_1 \cdot \frac{1}{\mathbf{x}_2 \wedge \mathbf{x}_1}. \end{aligned}}$$
(2.14)

Recall that dotting with the unit bivector of a plane (or its inverse) rotates a vector in that plane by $\pi/2$. In a plane subspace, such a rotation is exactly the transformation to ensure that $\mathbf{x}_1 \cdot \mathbf{x}^2 = \mathbf{x}_2 \cdot \mathbf{x}^1 = 0$. This shows that the reciprocal frame for the basis of a two dimensional subspace is found by a duality transformation of each of the curvilinear coordinates, plus a subsequent scaling operation. As $\mathbf{x}_1 \wedge \mathbf{x}_2$, the pseudoscalar for the subspace spanned by $\{\mathbf{x}_1, \mathbf{x}_2\}$, is not generally a unit bivector, the dot product with its inverse also has a scaling effect.

Numerical example: Here is a Mathematica calculation of the reciprocal frame depicted in fig. 2.1.

In[6]:=
```
ClearAll[x1, x2, inverse]
x1 = e[1] + e[2]; x2 = e[1] + 2 e[2];
x12 = OuterProduct[x1, x2];
inverse[a_] := a / GeometricProduct[a, a] ;
x12inverse = inverse[x12];
s1 = InnerProduct[x2, x12inverse];
s2 = InnerProduct[x1, -x12inverse];
s1
s2
dots[a_,b_] := {a , ".", b, " = ",
                InnerProduct[a // ReleaseHold,
    b // ReleaseHold]};
MapThread[dots, {{x1 // HoldForm, x2 // HoldForm,

                     x1 // HoldForm, x2 //
    HoldForm},
                  {s1 // HoldForm, s1 // HoldForm,
```

Out[6]= $2\,e[1] - e[2]$

Out[7]= $-e[1] + e[2]$

Out[8]=
x1 · s1 = 1
x2 · s1 = 0
x1 · s2 = 0

This shows the reciprocal vector calculations using eq. (2.14) and that the defining property $\mathbf{x}_i \cdot \mathbf{x}^j = \delta_i{}^j$ of the reciprocal frame vectors is satisfied.

Example: $\mathbb{R}^2$: Given a pair of arbitrary oriented vectors in $\mathbb{R}^2$, $\mathbf{x}_1 = a_1 \mathbf{e}_1 + a_2 \mathbf{e}_2, \mathbf{x}_2 = b_1 \mathbf{e}_1 + b_2 \mathbf{e}_2$, the pseudoscalar associated with the basis $\{\mathbf{x}_1, \mathbf{x}_2\}$ is

$$\begin{aligned}\mathbf{x}_1 \wedge \mathbf{x}_2 &= (a_1 \mathbf{e}_1 + a_2 \mathbf{e}_2) \wedge (b_1 \mathbf{e}_1 + b_2 \mathbf{e}_2) \\ &= (a_1 b_2 - a_2 b_1)\, \mathbf{e}_{12}.\end{aligned} \quad (2.15)$$

The inverse of this pseudoscalar is

$$\frac{1}{\mathbf{x}_1 \wedge \mathbf{x}_2} = \frac{1}{a_1 b_2 - a_2 b_1} \mathbf{e}_{21}. \quad (2.16)$$

So for this fixed oblique $\mathbb{R}^2$ basis, the reciprocal frame is just

$$\begin{aligned} \mathbf{x}^1 &= \mathbf{x}_2 \frac{\mathbf{e}_{21}}{a_1 b_2 - a_2 b_1} \\ \mathbf{x}^2 &= \mathbf{x}_1 \frac{\mathbf{e}_{12}}{a_1 b_2 - a_2 b_1}. \end{aligned} \quad (2.17)$$

The vector $\mathbf{x}^1$ is obtained by rotating $\mathbf{x}_2$ by $-\pi/2$, and rescaling it by the area of the parallelogram spanned by $\mathbf{x}_1, \mathbf{x}_2$. The vector $\mathbf{x}^2$ is obtained with the same scaling plus a rotation of $\mathbf{x}_1$ by $\pi/2$.

2.1.3 $\mathbb{R}^3$ reciprocal frame.

In this section we generalize eq. (2.14) to $\mathbb{R}^3$ vectors, which will illustrate the general case by example.

Given a subspace spanned by a three vector basis $\{\mathbf{x}_1, \mathbf{x}_2, \mathbf{x}_3\}$ the reciprocal frame vectors can be written as dot products

$$\begin{aligned} \mathbf{x}^1 &= (\mathbf{x}_2 \wedge \mathbf{x}_3) \cdot \left(\mathbf{x}^3 \wedge \mathbf{x}^2 \wedge \mathbf{x}^1\right) \\ \mathbf{x}^2 &= (\mathbf{x}_3 \wedge \mathbf{x}_1) \cdot \left(\mathbf{x}^1 \wedge \mathbf{x}^3 \wedge \mathbf{x}^2\right) \\ \mathbf{x}^3 &= (\mathbf{x}_1 \wedge \mathbf{x}_2) \cdot \left(\mathbf{x}^2 \wedge \mathbf{x}^1 \wedge \mathbf{x}^3\right). \end{aligned} \quad (2.18)$$

Each of those trivector terms equals $-\mathbf{x}^1 \wedge \mathbf{x}^2 \wedge \mathbf{x}^3$ and can be related to the (known) pseudoscalar $\mathbf{x}_1 \wedge \mathbf{x}_2 \wedge \mathbf{x}_3$ by observing that

$$\begin{aligned} \left(\mathbf{x}^1 \wedge \mathbf{x}^2 \wedge \mathbf{x}^3\right) \cdot (\mathbf{x}_3 \wedge \mathbf{x}_2 \wedge \mathbf{x}_1) &= \mathbf{x}^1 \cdot \left(\mathbf{x}^2 \cdot \left(\mathbf{x}^3 \cdot (\mathbf{x}_3 \wedge \mathbf{x}_2 \wedge \mathbf{x}_1)\right)\right) \\ &= \mathbf{x}^1 \cdot \left(\mathbf{x}^2 \cdot (\mathbf{x}_2 \wedge \mathbf{x}_1)\right) \\ &= \mathbf{x}^1 \cdot \mathbf{x}_1 \\ &= 1, \end{aligned} \quad (2.19)$$

which means that

$$\begin{aligned} -\mathbf{x}^1 \wedge \mathbf{x}^2 \wedge \mathbf{x}^3 &= -\frac{1}{\mathbf{x}_3 \wedge \mathbf{x}_2 \wedge \mathbf{x}_1} \\ &= \frac{1}{\mathbf{x}_1 \wedge \mathbf{x}_2 \wedge \mathbf{x}_3}, \end{aligned} \quad (2.20)$$

and

$$\boxed{\begin{aligned} \mathbf{x}^1 &= (\mathbf{x}_2 \wedge \mathbf{x}_3) \cdot \frac{1}{\mathbf{x}_1 \wedge \mathbf{x}_2 \wedge \mathbf{x}_3} \\ \mathbf{x}^2 &= (\mathbf{x}_3 \wedge \mathbf{x}_1) \cdot \frac{1}{\mathbf{x}_1 \wedge \mathbf{x}_2 \wedge \mathbf{x}_3} \\ \mathbf{x}^3 &= (\mathbf{x}_1 \wedge \mathbf{x}_2) \cdot \frac{1}{\mathbf{x}_1 \wedge \mathbf{x}_2 \wedge \mathbf{x}_3} \end{aligned}} \qquad (2.21)$$

Geometrically, dotting with this trivector is a duality transformation within the subspace spanned by the three vectors $\mathbf{x}_1, \mathbf{x}_2, \mathbf{x}_3$, also scaling the result so that the $\mathbf{x}_i \cdot \mathbf{x}^j = \delta_i{}^j$ condition is satisfied. The scaling factor is the volume of the parallelepiped spanned by $\mathbf{x}_1, \mathbf{x}_2, \mathbf{x}_3$.

2.1.4 Problems.

Exercise 2.1 Orthogonal reciprocals.

Given an orthogonal basis $\{\mathbf{x}_i\}$, show that

$$\mathbf{x}^i = \frac{1}{\mathbf{x}_i}.$$

Exercise 2.2 Reciprocal frame for two dimensional subspace.

Prove eq. (2.11).

Hint: Take dot products of eq. (2.10) with $\mathbf{x}_1, \mathbf{x}_2$, group the resulting equations into matrix form (you'll find the same matrix for both sets of unknowns), and then invert the matrix to find the solution.

Exercise 2.3 Two vector reciprocal frame

Calculate the reciprocal frame for the $\mathbb{R}^3$ subspace spanned by $\{\mathbf{x}_1, \mathbf{x}_2\}$ where

$$\begin{aligned} \mathbf{x}_1 &= \mathbf{e}_1 + 2\mathbf{e}_2 \\ \mathbf{x}_2 &= \mathbf{e}_2 - \mathbf{e}_3. \end{aligned} \qquad (2.27)$$

2.2 CURVILINEAR BASES.

2.2.1 *Two parameters.*

Curvilinear coordinates can be defined for any subspace spanned by a parameterized vector into that space. As an example, consider a two parameter planar subspace parameterized by the following continuous vector function

$$\mathbf{x}(u_1, u_2) = u_1 \left(\mathbf{e}_1 \cos u_2 + \frac{1}{2} \mathbf{e}_2 \sin u_2 \right), \tag{2.33}$$

where $u_1 \in [0, 1]$ and $u_2 \in [0, \pi/2]$. This parameterization spans the first quadrant of the ellipse with semi-major axis length 1, and semi-minor axis length $1/2$. A parameterization of an elliptic area may or may not be of much use in electrodynamics, but it happens to provide a non-trivial, yet simple, example of a non-orthonormal parameterization. Contours for this parameterization are plotted in fig. 2.2. The radial contours are for fixed values of u_2 and the elliptical contours fix the value of u_1, and depict a set of elliptic curves with a semi-major/major axis ratio of $1/2$.

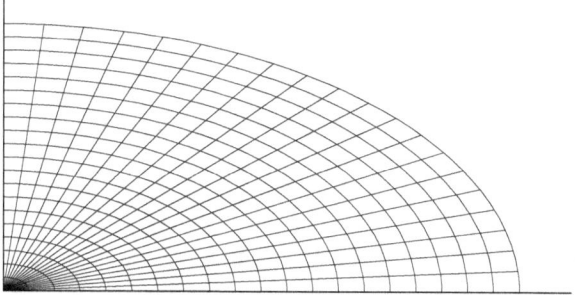

Figure 2.2: Contours for an elliptical region.

We define a curvilinear basis associated with each point in the region by the partials

$$\begin{aligned} \mathbf{x}_1 &= \frac{\partial \mathbf{x}}{\partial u_1} \\ \mathbf{x}_2 &= \frac{\partial \mathbf{x}}{\partial u_2}. \end{aligned} \tag{2.34}$$

2.2 CURVILINEAR BASES.

For eq. (2.33) our curvilinear basis elements are

$$\begin{aligned}\mathbf{x}_1 &= \mathbf{e}_1 \cos u_2 + \frac{1}{2}\mathbf{e}_2 \sin u_2 \\ \mathbf{x}_2 &= u_1\left(-\mathbf{e}_1 \sin u_2 + \frac{1}{2}\mathbf{e}_2 \cos u_2\right),\end{aligned} \qquad (2.35)$$

We form vector valued differentials for each parameter

$$\begin{aligned}d\mathbf{x}_1 &= \mathbf{x}_1 du_1 \\ d\mathbf{x}_2 &= \mathbf{x}_2 du_2.\end{aligned} \qquad (2.36)$$

For eq. (2.33), the values of these differentials $d\mathbf{x}_1, d\mathbf{x}_2$ with $du_1 = du_2 = 0.1$ are plotted in fig. 2.3 for the points

$$(u_1, u_2) = (0.7, 5\pi/20), (0.9, 3\pi/20), (1.0, 5\pi/20) \qquad (2.37)$$

in (dark-thick) red, blue and purple respectively.

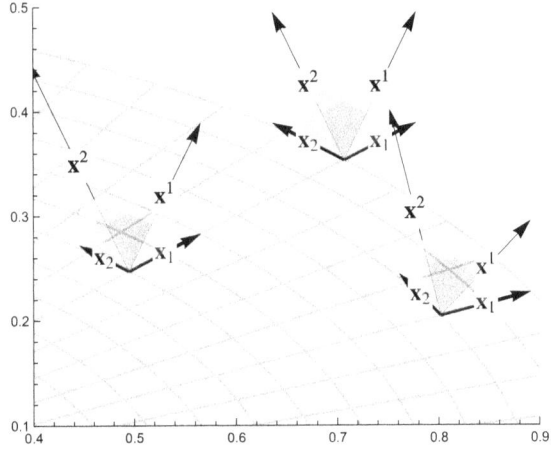

Figure 2.3: Differentials for an elliptical parameterization.

In this case and in general there is no reason to presume that there is any orthonormality constraint on the basis $\{\mathbf{x}_1, \mathbf{x}_2\}$ for a given two parameter subspace.

Should we wish to calculate the reciprocal frame for eq. (2.33), we would find (exercise 2.4) that

$$\begin{aligned}\mathbf{x}^1 &= \mathbf{e}_1 \cos u_2 + 2\mathbf{e}_2 \sin u_2 \\ \mathbf{x}^2 &= \frac{1}{u_1}(-\mathbf{e}_1 \sin u_2 + 2\mathbf{e}_2 \cos u_2)\end{aligned} \qquad (2.38)$$

These are plotted (scaled by $da = 0.1$ so they fit in the image nicely) in fig. 2.3 using thin light arrows.

When evaluating surface integrals, we will form oriented (bivector) area elements from the wedge product of the differentials

$$d^2\mathbf{x} \equiv d\mathbf{x}_1 \wedge d\mathbf{x}_2. \tag{2.39}$$

This absolute value of this area element $\sqrt{-(d^2\mathbf{x})^2}$ is the area of the parallelogram spanned by $d\mathbf{x}_1, d\mathbf{x}_2$. In this example, all such area elements lie in the $x - y$ plane, but that need not be the case.

Also note that we will only perform integrals for those parametrizations for which the area element $d^2\mathbf{x}$ is non-zero.

Exercise 2.4   Elliptic parameterization.

An elliptical area can be parameterized as

$$\mathbf{x}(u_1, u_2) = u_1 \left(\mathbf{e}_1 \cos u_2 + \beta \mathbf{e}_2 \sin u_2 \right), \tag{2.40}$$

where $\beta = \sqrt{1 - \epsilon^2}$, and ϵ is the eccentricity of the ellipse.

a. Compute the curvilinear vectors

$$\begin{aligned}\mathbf{x}_1 &= \partial \mathbf{x}/\partial u_1 \\ \mathbf{x}_2 &= \partial \mathbf{x}/\partial u_2.\end{aligned} \tag{2.41}$$

b. Compute the reciprocal frame vectors

$$\begin{aligned}\mathbf{x}^1 &= \mathbf{x}_2 \cdot \frac{1}{\mathbf{x}_1 \wedge \mathbf{x}_2} \\ \mathbf{x}^2 &= -\mathbf{x}_1 \cdot \frac{1}{\mathbf{x}_1 \wedge \mathbf{x}_2}.\end{aligned} \tag{2.42}$$

c. Verify that $\mathbf{x}_i \cdot \mathbf{x}^j = \delta_i{}^j$.

Exercise 2.5   Hyperbolic identities.

Show that

$$2 \cosh(\mu - i\theta) \sinh(\mu + i\theta) = \sinh(2\mu) + i \sin(2\theta). \tag{2.51}$$

$$2 \cosh(\mu) \sinh(\mu) = \sinh(2\mu). \tag{2.52}$$

$$\cosh(\mu + i\theta) = \cosh\mu \cos\theta + i\sinh\mu \sin\theta. \qquad (2.53)$$

Exercise 2.6 Elliptic curvilinear and reciprocal basis.

a. Show that an ellipse can be parameterized by

$$\mathbf{x} = u_1 \mathbf{e}_1 \cosh(\mu + iu_2), \qquad (2.56)$$

where $i = \mathbf{e}_{12}$, and find the values of the semi-major and semi-minor axes.

b. Determine how μ and the eccentricity $\epsilon = \sqrt{1 - b^2/a^2}$ are related.

c. Compute the curvilinear and reciprocal frame vectors for the parameterization $\mathbf{x}(u_1, u_2)$ above.

d. Check that $\mathbf{x}^i \cdot \mathbf{x}_j = \delta^i_j$.

At the point of evaluation, the span of these differentials is called the tangent space. In this particular case the tangent space at all points in the region is the entire x-y plane. These partials locally span the tangent space at a given point on the surface.

2.2.1.1 Curved two parameter surfaces.

Continuing to illustrate by example, let's now consider a non-planar two parameter surface

$$\mathbf{x}(u_1, u_2) = (u_1 - u_2)^2 \mathbf{e}_1 + (1 - (u_2)^2) \mathbf{e}_2 + u_1 u_2 \mathbf{e}_3. \qquad (2.70)$$

The curvilinear basis elements, and the area element, are

$$\begin{aligned} \mathbf{x}_1 &= 2(u_1 - u_2)\mathbf{e}_1 + u_2 \mathbf{e}_3 \\ \mathbf{x}_2 &= 2(u_2 - u_1)\mathbf{e}_1 - 2u_2 \mathbf{e}_2 + u_1 \mathbf{e}_3 \\ \mathbf{x}_1 \wedge \mathbf{x}_2 &= -4u_2(u_1 - u_2)\mathbf{e}_{12} + 2u_2^2 \mathbf{e}_{23} + 2\left(u_1^2 - u_2^2\right)\mathbf{e}_{13}. \end{aligned} \qquad (2.71)$$

Two examples of these vectors and the associated area element (rescaled to fit) is plotted in fig. 2.4. This plane is called the tangent space at the point in question, and has been evaluated at $(u_1, u_2) = (0.5, 0.5), (0.35, 0.75)$. The results of eq. (2.71) can be calculated easily by hand for this particular parameterization, but also submit to symbolic calculation software. Here's a complete example using CliffordBasic

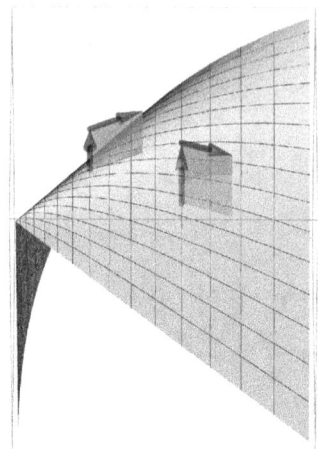

Figure 2.4: Two parameter manifold.

```
In[9]:=   << CliffordBasic`;

In[10]:=  ClearAll[xp, x, x1, x2]
          (* Use dummy parameter values for the
          derivatives,
             and then switch them to function parameter
          values. *)
          xp := (a - b)^2 e[1] + (1 - b^2) e[2] + b a e
             [3];
          x[u_, v_]  := xp /. {a → u, b→v};
          x1[u_, v_] := D[xp, a] /. {a → u, b→v};
          x2[u_, v_] := D[xp, b] /. {a → u, b→v};

          x1[u,v]
          x2[u,v]

Out[10]=  2 (u-v) e[1] + v e[3]

Out[11]=  -2 (u - v) e[1] - 2 v e[2] + u e[3]

Out[12]=  (-4 u v + 4 v^2) e[1,2] + (2 u^2 - 2 v^2) e[1,3] +
          2 v^2 e[2,3]
```

2.2.2 Three (or more) parameters.

We can extend the previous two parameter subspace ideas to higher dimensional (or one dimensional) subspaces associated with a parameterization

2.2 CURVILINEAR BASES. 107

> **Definition 2.2: Curvilinear bases and volume element**
>
> Given a parameterization $\mathbf{x}(u_1, u_2, \cdots, u_k)$ with k degrees of freedom, we define the curvilinear basis elements $\mathbf{x}_i$ by the partials
>
> $$\mathbf{x}_i = \frac{\partial \mathbf{x}}{\partial u_i}.$$
>
> The span of $\{\mathbf{x}_i\}$ at the point of evaluation is called the tangent space. A subspace associated with a parameterization of this sort is also called a manifold. The volume element for the subspace is
>
> $$d^k \mathbf{x} = du_1 du_2 \cdots du_k\, \mathbf{x}_1 \wedge \mathbf{x}_2 \wedge \cdots \wedge \mathbf{x}_k.$$
>
> Such a volume element is a k-vector. The volume of the (hyper-) parallelepiped bounded by $\{\mathbf{x}_i\}$ is $\sqrt{\left|(d^k\mathbf{x})^2\right|}$.

We will assume that the parameterization is non-generate. This means that the volume element $d^k\mathbf{x}$ is non-zero in the region of interest. Note that a zero volume element implies a linear dependency in the curvilinear basis elements $\mathbf{x}_i$.

Given a parameterization $\mathbf{x} = \mathbf{x}(u, v, \cdots, w)$, we may write $\mathbf{x}_u, \mathbf{x}_v, \cdots, \mathbf{x}_w$ for the curvilinear basis elements, and $\mathbf{x}^u, \mathbf{x}^v, \cdots, \mathbf{x}^w$ for the reciprocal frame elements. When doing so, sums over numeric indexes like $\sum_i \mathbf{x}^i \mathbf{x}_i$ should be interpreted as a sum over all the parameter labels, i.e. $\mathbf{x}^u \mathbf{x}_u + \mathbf{x}^v \mathbf{x}_v + \cdots$.

2.2.3 Gradient.

With the introduction of the ideas of reciprocal frame and curvilinear coordinates, we are getting closer to be able to formulate the geometric algebra generalizations of vector calculus.

The next step in the required mathematical preliminaries for geometric calculus is to determine the form of the gradient with respect to curvilinear coordinates and the parameters associated with those coordinates.

Suppose we have a vector parameterization of $\mathbb{R}^N$

$$\mathbf{x} = \mathbf{x}(u_1, u_2, \cdots, u_N). \tag{2.72}$$

We can employ the chain rule to express the gradient in terms of derivatives with respect to u_i

$$\begin{aligned}
\nabla &= \sum_i \mathbf{e}_i \frac{\partial}{\partial x_i} \\
&= \sum_{i,j} \mathbf{e}_i \frac{\partial u_j}{\partial x_i} \frac{\partial}{\partial u_j} \\
&= \sum_j \left(\sum_i \mathbf{e}_i \frac{\partial u_j}{\partial x_i} \right) \frac{\partial}{\partial u_j} \\
&= \sum_j (\nabla u_j) \frac{\partial}{\partial u_j}.
\end{aligned} \quad (2.73)$$

It turns out that the gradients of the parameters are in fact the reciprocal frame vectors

---
Theorem 2.1: Reciprocal frame vectors

Given a curvilinear basis with elements $\mathbf{x}_i = \partial \mathbf{x}/\partial u_i$, the *reciprocal frame* vectors are given by

$$\mathbf{x}^i = \nabla u_i.$$

Proof. This can be proven by direct computation

$$\begin{aligned}
\mathbf{x}^i \cdot \mathbf{x}_j &= (\nabla u_i) \cdot \frac{\partial \mathbf{x}}{\partial u_j} \\
&= \sum_{r,s=1}^n \left(\mathbf{e}_r \frac{\partial u_i}{\partial x_r} \right) \cdot \left(\mathbf{e}_s \frac{\partial x_s}{\partial u_j} \right) \\
&= \sum_{r,s=1}^n (\mathbf{e}_r \cdot \mathbf{e}_s) \frac{\partial u_i}{\partial x_r} \frac{\partial x_s}{\partial u_j} \\
&= \sum_{r,s=1}^n \delta_{rs} \frac{\partial u_i}{\partial x_r} \frac{\partial x_s}{\partial u_j} \\
&= \sum_{r=1}^n \frac{\partial u_i}{\partial x_r} \frac{\partial x_r}{\partial u_j} \\
&= \frac{\partial u_j}{\partial u_i} \\
&= \delta^i{}_j.
\end{aligned} \quad (2.74)$$

This shows that $\mathbf{x}^i = \nabla u_i$ has the properties required of the reciprocal frame, proving the theorem. □

We are now able to define the gradient with respect to an arbitrary set of parameters

> **Theorem 2.2: Curvilinear representation of the gradient**
>
> Given an N-parameter vector parameterization $\mathbf{x} = \mathbf{x}(u_1, u_2, \cdots, u_N)$ of $\mathbb{R}^N$, with curvilinear basis elements $\mathbf{x}_i = \partial \mathbf{x}/\partial u_i$, the *gradient* is
>
> $$\nabla = \sum_i \mathbf{x}^i \frac{\partial}{\partial u_i}.$$
>
> It is convenient to define $\partial_i \equiv \partial/\partial u_i$, so that the gradient can be expressed in mixed index representation
>
> $$\nabla = \sum_i \mathbf{x}^i \partial_i.$$

2.2.4 Vector derivative.

Given curvilinear coordinates defined on a subspace definition 2.2, we don't have enough parameters to define the gradient. For calculus on the k-dimensional subspace, we define the vector derivative

> **Definition 2.3: Vector derivative**
>
> Given a k-parameter vector parameterization $\mathbf{x} = \mathbf{x}(u_1, u_2, \cdots, u_k)$ of $\mathbb{R}^N$ with $k \leq N$, and curvilinear basis elements $\mathbf{x}_i = \partial \mathbf{x}/\partial u_i$, the *vector derivative* ∂ is defined as
>
> $$\partial = \sum_{i=1}^{k} \mathbf{x}^i \partial_i.$$

When the dimension of the subspace (number of parameters) equals the dimension of the underlying vector space, the vector derivative equals the gradient. Otherwise we can write

$$\nabla = \partial + \nabla_\perp, \tag{2.75}$$

and can think of the vector derivative as the projection of the gradient onto the tangent space at the point of evaluation.

Please see [18] for an excellent introduction of the reciprocal frame, the gradient, and the vector derivative, and for details about the connectivity of the manifold ignored here.

2.2.5 Examples.

We've just blasted through a few abstract ideas:

- The curvilinear representation of the gradient.
- The gradient representation of the reciprocal frame.
- The vector derivative.

This completes the mathematical preliminaries required to formulate geometric calculus, the multivector generalization of line, surface, and volume integrals. Before diving into the calculus let's consider some example parameterizations to illustrate how some of the new ideas above fit together.

2.2.5.1 Example parameterization: Polar coordinates.

We will now consider a simple concrete example of a vector parameterization, that of polar coordinates in $\mathbb{R}^2$

$$\mathbf{x}(\rho, \phi) = \rho \mathbf{e}_1 \exp\left(\mathbf{e}_{12}\phi\right), \tag{2.76}$$

as illustrated in fig. 2.5.

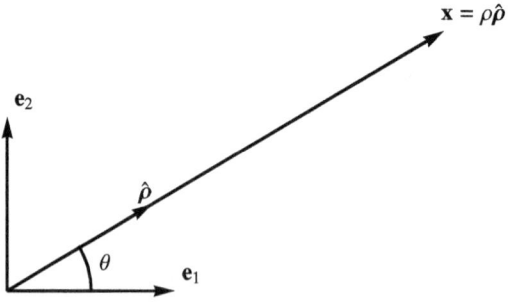

Figure 2.5: Polar coordinates.

Using this example we will

2.2 CURVILINEAR BASES.

- Compute the curvilinear coordinates. We will refer to these as $\mathbf{x}_\rho, \mathbf{x}_\phi$, instead of $\mathbf{x}_1, \mathbf{x}_2$.
- Find the squared length of $\mathbf{x}_\rho, \mathbf{x}_\phi$, and show that they are perpendicular (but not orthonormal.)
- Perform a first bivector valued integral.
- Compute the reciprocal frame vectors with geometric arguments.
- Compute the reciprocal frame explicitly from the gradients of the coordinates.
- Find the polar form of the gradient with respect to this parameterization.

Curvilinear coordinates. The curvilinear coordinate basis can be computed directly

$$\begin{aligned}\mathbf{x}_\rho &= \frac{\partial}{\partial \rho}\left(\rho \mathbf{e}_1 \exp\left(\mathbf{e}_{12}\phi\right)\right) \\ &= \mathbf{e}_1 \exp\left(\mathbf{e}_{12}\phi\right)\end{aligned} \quad (2.77a)$$

$$\begin{aligned}\mathbf{x}_\phi &= \frac{\partial}{\partial \phi}\left(\rho \mathbf{e}_1 \exp\left(\mathbf{e}_{12}\phi\right)\right) \\ &= \rho \mathbf{e}_1 \mathbf{e}_{12} \exp\left(\mathbf{e}_{12}\phi\right) \\ &= \rho \mathbf{e}_2 \exp\left(\mathbf{e}_{12}\phi\right).\end{aligned} \quad (2.77b)$$

For plane configurations, it is often handy to represent the plane pseudoscalar with an imaginary symbol. Here we will use $i = \mathbf{e}_{12}$, allowing for the compact representations $\mathbf{x}_\rho = \mathbf{e}_1 e^{i\phi}$ and $\mathbf{x}_\phi = \rho \mathbf{e}_2 e^{i\phi}$. This also highlights the geometric interpretation of the basis vectors, as we see that the $\{\mathbf{x}_\rho, \mathbf{x}_\phi\}$ basis vectors are constructed by rotating $\{\mathbf{e}_1, \rho \mathbf{e}_2\}$ by ϕ radians in the direction from $\mathbf{e}_1$ to $\mathbf{e}_2$.

Normality. To show that these vectors are perpendicular, we first compute their product, which we will find has no scalar part. From theorem 1.13, property (c), observe that $\mathbf{x}e^{i\phi} = e^{-i\phi}\mathbf{x}$ for any vector $\mathbf{x}$ in the plane, so

$$\begin{aligned}\mathbf{x}_\rho \mathbf{x}_\phi &= \left(\mathbf{e}_1 e^{i\phi}\right)\left(\rho \mathbf{e}_2 e^{i\phi}\right) \\ &= \rho \mathbf{e}_1 \mathbf{e}_2 e^{-i\phi} e^{i\phi} \\ &= \rho \mathbf{e}_{12}.\end{aligned} \quad (2.78)$$

Since this has no scalar part $\mathbf{x}_\rho \cdot \mathbf{x}_\phi = \langle \mathbf{x}_\rho \mathbf{x}_\phi \rangle = 0$.

Length of basis elements. We can use scalar selection to find the (squared) length of the vectors, finding

$$\begin{aligned}
\mathbf{x}_\rho^2 &= \langle \mathbf{e}_1 e^{i\phi} \mathbf{e}_1 e^{i\phi} \rangle \\
&= \langle \mathbf{e}_1 e^{i\phi} e^{-i\phi} \mathbf{e}_1 \rangle \\
&= \langle \mathbf{e}_1^2 \rangle \\
&= 1,
\end{aligned} \qquad (2.79)$$

and

$$\begin{aligned}
\mathbf{x}_\phi^2 &= \langle (\rho \mathbf{e}_2 e^{i\phi})(\rho \mathbf{e}_2 e^{i\phi}) \rangle \\
&= \rho^2 \langle \mathbf{e}_2 e^{i\phi} e^{-i\phi} \mathbf{e}_2 \rangle \\
&= \rho^2 \langle \mathbf{e}_2^2 \rangle \\
&= \rho^2.
\end{aligned} \qquad (2.80)$$

A bivector integral. One of our goals is to understand the multivector generalization of Stokes' theorem and the divergence theorem, but even before that, we can evaluate some simple multivector integrals. In particular, we can calculate the (oriented) area of a circle, given a bivector representation of the area element.

$$\begin{aligned}
\int_{\rho=0}^{r} \int_{\phi=0}^{2\pi} d\mathbf{x}_\rho \wedge d\mathbf{x}_\phi &= \int_{\rho=0}^{r} \int_{\phi=0}^{2\pi} d\rho d\phi\, \mathbf{x}_\rho \wedge \mathbf{x}_\phi \\
&= \int_{\rho=0}^{r} \int_{\phi=0}^{2\pi} \rho d\rho d\phi\, \mathbf{e}_{12} = \pi r^2 \mathbf{e}_{12}.
\end{aligned} \qquad (2.81)$$

Integrating the bivector area over a circular region gives us the area of that region, but weighted by the $\mathbb{R}^2$ pseudoscalar. This is an oriented area.

Reciprocal basis. Because $\mathbf{x}_\rho, \mathbf{x}_\phi$ are mutually perpendicular, we have only to rescale them to determine the reciprocal basis, and can do so by inspection

$$\begin{aligned}
\mathbf{x}^\rho &= \mathbf{e}_1 e^{i\phi} \\
\mathbf{x}^\phi &= \frac{1}{\rho} \mathbf{e}_2 e^{i\phi}.
\end{aligned} \qquad (2.82)$$

According to theorem 2.1 we should be able to find eq. (2.82) by computing the gradients of ρ and ϕ respectively. If we do so using the $\mathbb{R}^2$ standard basis representation of the gradient, we must first solve for $\rho = \rho(x, y), \phi = \phi(x, y)$, inverting

$$x = \rho \cos \phi$$
$$y = \rho \sin \phi. \tag{2.83}$$

An implicit solution to this inversion problem is

$$\rho^2 = x^2 + y^2$$
$$\tan \phi = y/x, \tag{2.84}$$

which we can implicitly differentiate to evaluate the components of the desired gradients

$$2\rho \frac{\partial \rho}{\partial x} = 2x$$
$$2\rho \frac{\partial \rho}{\partial y} = 2y$$
$$\frac{1}{\cos^2 \phi} \frac{\partial \phi}{\partial x} = -\frac{y}{x^2} \tag{2.85}$$
$$\frac{1}{\cos^2 \phi} \frac{\partial \phi}{\partial y} = \frac{1}{x}.$$

So the gradients are

$$\begin{aligned} \nabla \rho &= (\partial \rho/\partial x, \partial \rho/\partial y) \\ &= (x/\rho, y/\rho) \\ &= (\cos \phi, \sin \phi) \\ &= \mathbf{e}_1 e^{\mathbf{e}_{12}\phi} \\ &= \mathbf{x}^\rho \end{aligned} \tag{2.86a}$$

$$\begin{aligned} \nabla \phi &= \cos^2 \phi \left(-\frac{y}{x^2}, \frac{1}{x} \right) \\ &= \frac{1}{\rho}(-\sin \phi, \cos \phi) \\ &= \frac{\mathbf{e}_2}{\rho}(\cos \phi + \mathbf{e}_{12} \sin \phi) \\ &= \frac{\mathbf{e}_2}{\rho} e^{\mathbf{e}_{12}\phi} \\ &= \mathbf{x}^\phi, \end{aligned} \tag{2.86b}$$

which is consistent with eq. (2.82), as expected.

114   MULTIVECTOR CALCULUS.

Gradient. The polar form of the $\mathbb{R}^2$ gradient is

$$\begin{aligned}\nabla &= \mathbf{x}^\rho \frac{\partial}{\partial \rho} + \mathbf{x}^\phi \frac{\partial}{\partial \phi} \\ &= \hat{\boldsymbol{\rho}} \frac{\partial}{\partial \rho} + \frac{1}{\rho} \hat{\boldsymbol{\phi}} \frac{\partial}{\partial \phi},\end{aligned} \qquad (2.87)$$

where

$$\begin{aligned}\hat{\boldsymbol{\rho}} &= \frac{\mathbf{x}_\rho}{\|\mathbf{x}_\rho\|} = \mathbf{x}_\rho = \mathbf{e}_1 e^{i\phi} = \mathbf{x}^\rho \\ \hat{\boldsymbol{\phi}} &= \frac{\mathbf{x}_\phi}{\|\mathbf{x}_\phi\|} = \frac{1}{\rho} \mathbf{x}_\phi = \mathbf{e}_2 e^{i\phi} = \rho \mathbf{x}^\phi.\end{aligned} \qquad (2.88)$$

Should we extend this vector space to $\mathbb{R}^3$, the parameterization of eq. (2.76) covers the subspace of the x-y plane, and for that subspace, the vector derivative is

$$\begin{aligned}\partial &= \mathbf{x}^\rho \frac{\partial}{\partial \rho} + \mathbf{x}^\phi \frac{\partial}{\partial \phi} \\ &= \hat{\boldsymbol{\rho}} \frac{\partial}{\partial \rho} + \frac{1}{\rho} \hat{\boldsymbol{\phi}} \frac{\partial}{\partial \phi}.\end{aligned} \qquad (2.89)$$

2.2.5.2 *Example parameterization: Spherical coordinates.*

We will use the physics and engineering convention for spherical polar angles, as illustrated in fig. 2.6. The conventional way to introduce a spherical polar position vector representation is through coordinates, where inspection of the geometry shows that

$$\mathbf{x} = r \left(\sin\theta \cos\phi, \sin\theta \sin\phi, \cos\theta \right), \qquad (2.90)$$

We can find a compact GA representation from this coordinate representation without too much trouble (exercise 2.10), but we can also examine the geometry of the situation directly.

For the unit bivector for the azimuthal (x-y) plane, let's write

$$i = \mathbf{e}_{12}, \qquad (2.91)$$

We see that the projection of $\mathbf{x}$ onto the azimuthal plane has direction $\text{Proj}_i \mathbf{x} \propto \mathbf{e}_1 e^{i\phi}$. As we rotate from the north pole $\mathbf{e}_3$ down through $\mathbf{x}$ to the azimuthal plane, we are rotating in the plane

$$\begin{aligned}j &= \mathbf{e}_3 \wedge \left(\mathbf{e}_1 e^{i\phi} \right) \\ &= \mathbf{e}_{31} e^{i\phi}.\end{aligned} \qquad (2.92)$$

2.2 CURVILINEAR BASES. 115

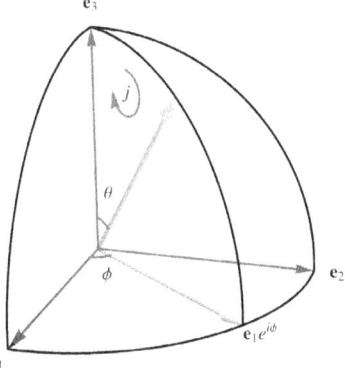

Figure 2.6: Spherical polar conventions.

So, just by inspection, we've found the spherical polar representation of **x**, which is

$$\mathbf{x} = r\mathbf{e}_3 e^{j\theta}. \tag{2.93}$$

Observe that all the ϕ dependency sneakily hides out in the unit bivector $j = j(\phi)$.

Given a parameterized representation of **x**, we may compute the basis elements

$$\mathbf{x}_r = \mathbf{e}_3 e^{j\theta} \tag{2.94a}$$

$$\begin{aligned}\mathbf{x}_\theta &= r\mathbf{e}_3 j e^{j\theta} \\ &= r\mathbf{e}_3 e^{j(\theta+\pi/2)}\end{aligned} \tag{2.94b}$$

$$\begin{aligned}\mathbf{x}_\phi &= \frac{\partial}{\partial \phi} \left(r\mathbf{e}_3 e^{j\theta} \right) \\ &= r\mathbf{e}_3 \frac{\partial}{\partial \phi} \left(\cos\theta + j\sin\theta \right) \\ &= r\mathbf{e}_3 \sin\theta \frac{\partial}{\partial \phi} \left(\mathbf{e}_{31} e^{i\phi} \right) \\ &= r\sin\theta \mathbf{e}_1 \mathbf{e}_{12} e^{i\phi} \\ &= r\sin\theta \mathbf{e}_2 e^{i\phi}.\end{aligned} \tag{2.94c}$$

See exercise 1.7 for why it is necessary to first expand the exponential into its trigonometric constituents. These vectors are all mutually orthogonal (exercise 2.7).

Orthonormalization of the curvilinear basis is now possible by inspection

$$\hat{\mathbf{r}} = \mathbf{x}_r = \mathbf{e}_3 e^{j\theta}$$
$$\hat{\boldsymbol{\theta}} = \frac{1}{r}\mathbf{x}_\theta = \mathbf{e}_3 e^{j(\theta+\pi/2)} = \hat{\mathbf{r}} j \qquad (2.95)$$
$$\hat{\boldsymbol{\phi}} = \frac{1}{r\sin\theta}\mathbf{x}_\phi = \mathbf{e}_2 e^{i\phi},$$

so

$$\mathbf{x}^r = \hat{\mathbf{r}} = \mathbf{e}_3 e^{j\theta}$$
$$\mathbf{x}^\theta = \frac{1}{r}\hat{\boldsymbol{\theta}} = \frac{1}{r}\mathbf{e}_3 e^{j(\theta+\pi/2)} \qquad (2.96)$$
$$\mathbf{x}^\phi = \frac{1}{r\sin\theta}\hat{\boldsymbol{\phi}} = \frac{1}{r\sin\theta}\mathbf{e}_2 e^{i\phi}.$$

The unit vectors $\hat{\mathbf{r}}, \hat{\boldsymbol{\theta}}, \hat{\boldsymbol{\phi}}$ are illustrated in fig. 2.7.

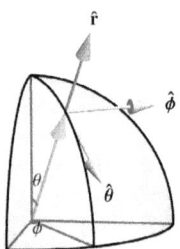

Figure 2.7: Spherical polar unit vectors.

Having computed the reciprocals, we may now form the spherical polar representation of the gradient

$$\begin{aligned}\boldsymbol{\nabla} &= \mathbf{x}^r \frac{\partial}{\partial r} + \mathbf{x}^\theta \frac{\partial}{\partial \theta} + \mathbf{x}^\phi \frac{\partial}{\partial \phi} \\ &= \hat{\mathbf{r}}\frac{\partial}{\partial r} + \frac{1}{r}\hat{\boldsymbol{\theta}}\frac{\partial}{\partial \theta} + \frac{1}{r\sin\theta}\hat{\boldsymbol{\phi}}\frac{\partial}{\partial \phi}.\end{aligned} \qquad (2.97)$$

2.2 CURVILINEAR BASES. 117

It's worth pointing out that computations like finding the curvilinear basis vectors is easily performed in software

In[13]:= ```
ClearAll[i, j, ej, x, xr, xt, xp]
i = e[1, 2];
j[phi_] = GeometricProduct[e[3, 1], Cos[phi] + i
   Sin[phi]];
ej[t_, p_] = Cos[t] + j[p] Sin[t];
x[r_, t_, p_] = r GeometricProduct[e[3], ej[t, p
   ]];
xr[r_, theta_, phi_] = D[x[a, theta, phi], a] /.
   a→r;
xt[r_, theta_, phi_] = D[x[r, t, phi], t] /. t
   →theta;
xp[r_, theta_, phi_] = D[x[r, theta, p], p] /. p
   →phi;
{x[r, θ, φ],
xr[r, θ, φ],
xt[r, θ, φ],
```

Out[13]= r (Cos[θ] e[3] + Cos[φ] e[1] Sin[θ] + e[2] Sin[θ] Sin[φ])
Cos[θ] e[3] + Cos[φ] e[1] Sin[θ] + e[2] Sin[ θ] Sin[φ]
r (Cos[θ] Cos[φ] e[1] - e[3] Sin[θ] + Cos[θ] e[2] Sin[φ])

where it also easy to show that these vectors are mutually perpendicular

In[14]:= ```
ClearAll[x1, x2, x3]
x1 = xr[r, θ, φ];
x2 = xt[r, θ, φ];
x3 = xp[r, θ, φ];
```

Out[14]= {0,0,0}

Unfortunately, if we perform these computations in software, we loose our compact representation.

The spherical (oriented) volume element can also be computed in a compact fashion

$$\frac{d^3\mathbf{x}}{dr d\theta d\phi} = \mathbf{x}_r \wedge \mathbf{x}_\theta \wedge \mathbf{x}_\phi$$
$$= \langle \mathbf{x}_r \mathbf{x}_\theta \mathbf{x}_\phi \rangle_3$$
$$= \langle \hat{\mathbf{r}} r \hat{\mathbf{r}} j r \sin\theta \mathbf{e}_2 e^{i\phi} \rangle_3 \quad (2.98)$$
$$= r^2 \sin\theta \langle \mathbf{e}_{31} e^{i\phi} \mathbf{e}_2 e^{i\phi} \rangle_3$$
$$= r^2 \sin\theta \, \mathbf{e}_{123}.$$

The scalar factor is the Jacobian with respect to the spherical parameterization

$$\frac{dV}{dr d\theta d\phi} = \frac{\partial(x_1, x_2, x_3)}{\partial(r, \theta, \phi)}$$
$$= \begin{vmatrix} \sin\theta\cos\phi & \sin\theta\sin\phi & \cos\theta \\ r\cos\theta\cos\phi & r\cos\theta\sin\phi & -r\sin\theta \\ -r\sin\theta\sin\phi & r\sin\theta\cos\phi & 0 \end{vmatrix} \quad (2.99)$$
$$= r^2 \sin\theta.$$

The final reduction of eq. (2.98), and the expansion of the Jacobian eq. (2.99), are both easily verified with software.

```
In[15]:=  OuterProduct[ xr[r, θ, ϕ],
          xt[r, θ, ϕ],
          xp[r, θ, ϕ]]

          {e1,e2,e3} = IdentityMatrix[3];
          jacobian = {xr[r, θ, ϕ],
          xt[r, θ, ϕ],
          xp[r, θ, ϕ]} /. {e[1] → e1, e[2] → e2, e[3]→e3
          };

Out[15]=  r² e[1,2,3] Sin[θ]

Out[16]=  r² Sin[θ]
```

Performing these calculations manually are left as problems for the student (exercise 2.9, exercise 2.8).

2.2 CURVILINEAR BASES.

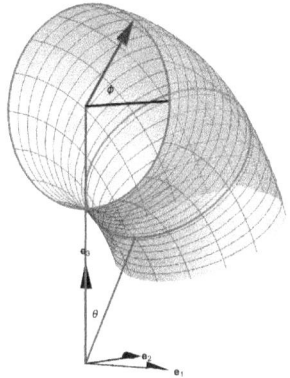

Figure 2.8: Toroidal parameterization.

2.2.5.3 *Example parameterization: Toroidal coordinates.*

Here is a 3D example of a parameterization with a non-orthogonal curvilinear basis, that of a toroidal subspace specified by two angles and a radial distance to the center of the toroid, as illustrated in fig. 2.8.

The position vector on the surface of a toroid of radius ρ within the torus can be stated directly

$$\mathbf{x}(\rho, \theta, \phi) = e^{-j\theta/2} \left(\rho \mathbf{e}_1 e^{i\phi} + R\mathbf{e}_3\right) e^{j\theta/2} \tag{2.100a}$$

$$i = \mathbf{e}_1 \mathbf{e}_3 \tag{2.100b}$$

$$j = \mathbf{e}_3 \mathbf{e}_2 \tag{2.100c}$$

It happens that the unit bivectors i and j used in this construction happen to have the quaternion-ic properties $ij = -ji$, and $i^2 = j^2 = -1$ which can be verified easily.

The curvilinear basis is found (exercise 2.11) to be

$$\mathbf{x}_\rho = \frac{\partial \mathbf{x}}{\partial \rho} = e^{-j\theta/2} \mathbf{e}_1 e^{i\phi} e^{j\theta/2} \tag{2.101a}$$

$$\mathbf{x}_\theta = \frac{\partial \mathbf{x}}{\partial \theta} = e^{-j\theta/2} (R + \rho \sin \phi) \mathbf{e}_2 e^{j\theta/2} \tag{2.101b}$$

$$\mathbf{x}_\phi = \frac{\partial \mathbf{x}}{\partial \phi} = e^{-j\theta/2} \rho \mathbf{e}_3 e^{i\phi} e^{j\theta/2}. \tag{2.101c}$$

The oriented volume element can be computed using a trivector selection operation, which conveniently wipes out a number of the interior exponentials

$$\frac{\partial \mathbf{x}}{\partial \rho} \wedge \frac{\partial \mathbf{x}}{\partial \theta} \wedge \frac{\partial \mathbf{x}}{\partial \phi} = \rho \left(R + \rho \sin \phi \right) \left\langle e^{-j\theta/2} \mathbf{e}_1 e^{i\phi} \mathbf{e}_2 \mathbf{e}_3 e^{i\phi} e^{j\theta/2} \right\rangle_3. \quad (2.102)$$

Note that $\mathbf{e}_1$ commutes with $j = \mathbf{e}_3 \mathbf{e}_2$, so also with $e^{-j\theta/2}$. Also $\mathbf{e}_2 \mathbf{e}_3 = -j$ anticommutes with i, so there is a conjugate commutation effect $e^{i\phi} j = j e^{-i\phi}$. This gives

$$\begin{aligned}
\left\langle e^{-j\theta/2} \mathbf{e}_1 e^{i\phi} \mathbf{e}_2 \mathbf{e}_3 e^{i\phi} e^{j\theta/2} \right\rangle_3 &= -\left\langle \mathbf{e}_1 e^{-j\theta/2} j e^{-i\phi} e^{i\phi} e^{j\theta/2} \right\rangle_3 \\
&= -\left\langle \mathbf{e}_1 e^{-j\theta/2} j e^{j\theta/2} \right\rangle_3 \\
&= -\left\langle \mathbf{e}_1 j \right\rangle_3 \\
&= I.
\end{aligned} \quad (2.103)$$

Together the trivector grade selection reduces almost magically to just

$$\frac{\partial \mathbf{x}}{\partial \rho} \wedge \frac{\partial \mathbf{x}}{\partial \theta} \wedge \frac{\partial \mathbf{x}}{\partial \phi} = \rho \left(R + \rho \sin \phi \right) I. \quad (2.104)$$

Thus the (scalar) volume element is

$$dV = \rho \left(R + \rho \sin \phi \right) d\rho d\theta d\phi. \quad (2.105)$$

As a check, it should be the case that the volume of the complete torus using this volume element has the expected $V = (2\pi R)(\pi r^2)$ value.

That volume is

$$V = \int_{\rho=0}^{r} \int_{\theta=0}^{2\pi} \int_{\phi=0}^{2\pi} \rho \left(R + \rho \sin \phi \right) d\rho d\theta d\phi. \quad (2.106)$$

The sine term conveniently vanishes over the 2π interval, leaving just

$$V = \frac{1}{2} r^2 R (2\pi)(2\pi), \quad (2.107)$$

as expected.

2.2.6 Problems.

Exercise 2.7   Spherical coordinate basis orthogonality.

Using scalar selection, show that the spherical curvilinear basis of eq. (2.94) are all mutually orthogonal.

Exercise 2.8   Spherical volume element pseudoscalar.

Using geometric algebra, perform the reduction of the grade three selection made in the final step of eq. (2.98).

Exercise 2.9   Spherical volume Jacobian.

Without software, expand and simplify the determinant of eq. (2.99).

Exercise 2.10   Spherical polar coordinates.

Starting with

$$\mathbf{x} = r\left(\mathbf{e}_1 \sin\theta \cos\phi + \mathbf{e}_2 \sin\theta \sin\phi + \mathbf{e}_3 \cos\theta\right),$$

first factor out $\mathbf{e}_1$ from the $\sin\theta$ terms, and group the remaining factors into complex exponential form. Then factor our $\mathbf{e}_3$ from both remaining terms to factor out a ϕ dependent unit bivector, and put the entire expression into complex exponential form. Hint: $\mathbf{e}_2 = \mathbf{e}_1 \mathbf{e}_1 \mathbf{e}_2$.

Exercise 2.11   Curvilinear basis for toroidal parameterization.

Prove eq. (2.101).

2.3 INTEGRATION THEORY.

2.3.1 *Line integral.*

In geometric algebra, the integrand of a multivector line integral contains a product of multivector(s) and a single parameter differential.

> **Definition 2.4: Multivector line integral.**
>
> Given a continuous and differentiable curve described by a vector function $\mathbf{x}(a)$, parameterized by single value a with differential
>
> $$d^1\mathbf{x} \equiv d\mathbf{x}_a = \frac{\partial \mathbf{x}}{\partial a} da = \mathbf{x}_a da,$$

and multivector functions F, G, the integral

$$\int F d^1 \mathbf{x} G$$

is called a multivector line integral.

An illustration of a single parameter curve and its differential with respect to that parameter, is given in fig. 2.9. Observe that the differential is tangent to the curve at all points. Possible physical realizations of the parameter describing the curve include time, arclength, or angle.

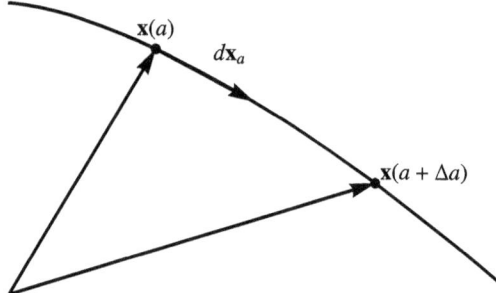

Figure 2.9: One parameter manifold.

Suppose that $\mathbf{f}(\mathbf{x}(a))$ is a vector valued function defined along the curve. The conventional line integral from vector calculus, a dot product of a differential and the function $\mathbf{f}$ may be obtained by the sum of two multivector line integrals one with $F, G = \mathbf{f}/2, 1$, and the other with $F, G = 1, \mathbf{f}/2$

$$\int d\mathbf{x} \frac{\mathbf{f}}{2} + \int \frac{\mathbf{f}}{2} d\mathbf{x} = \int d\mathbf{x} \cdot \mathbf{f}. \tag{2.119}$$

Unlike the conventional dot product line integral, the multivector line integral of a vector function such as $\int d\mathbf{x}\mathbf{f}$ is generally multivector valued, with both a scalar and a bivector component. Let's consider some examples of multivector line integrals.

2.3 INTEGRATION THEORY.

Example: Circular path. Let $f(t) = \mathbf{a}t + \mathbf{b}t^2$, where $\mathbf{a}, \mathbf{b}$ are constant vectors, t is a scalar parameter, and the integration path is circular $\mathbf{x}(t) = \mathbf{e}_1 e^{it}$, where $i = \mathbf{e}_1 \mathbf{e}_2$. The line integral of $\mathbf{f}\, d\mathbf{x}$ is

$$\int \mathbf{f}(t)\, d\mathbf{x} = \mathbf{a}\mathbf{e}_2 \int t e^{it} dt + \mathbf{b}\mathbf{e}_2 \int t^2 e^{it} dt$$
$$= \left(\mathbf{a}\mathbf{e}_2 (1 - it) + \mathbf{b}\mathbf{e}_2 \left(2i + 2t - it^2\right)\right) e^{it} \qquad (2.120)$$
$$= (\mathbf{a} + 2\mathbf{b}t)\, \mathbf{e}_2 e^{it} + \left(\mathbf{a}t - 2\mathbf{b} + \mathbf{b}t^2\right) \mathbf{e}_1 e^{it},$$

and the line integral of $d\mathbf{x}\, \mathbf{f}$ is

$$\int d\mathbf{x}\, \mathbf{f} = \mathbf{e}_2 \int t e^{it} dt\, \mathbf{a} + \mathbf{e}_2 \int t^2 e^{it} dt\, \mathbf{b}$$
$$= \mathbf{e}_2 e^{it} \left((1 - it)\,\mathbf{a} + \left(2i + 2t - it^2\right)\mathbf{b}\right) \qquad (2.121)$$
$$= \mathbf{e}_2 e^{it} (\mathbf{a} + 2\mathbf{b}t) + \mathbf{e}_1 e^{it} \left(\mathbf{a}t - 2\mathbf{b} + \mathbf{b}t^2\right).$$

Unless the vector constants $\mathbf{a}, \mathbf{b}$ have only components along the z-axis, eq. (2.120) and eq. (2.121) are not generally equal.

Example: Circular bivector. Given a bivector valued function $F(t) = \mathbf{e}_2 \wedge \left(\mathbf{e}_3 e^{it}\right)$, where $i = \mathbf{e}_3 \mathbf{e}_1$, and a curve $\mathbf{x}(t) = \mathbf{e}_3 + \mathbf{e}_2 t + \mathbf{e}_1 t^2/2$, we can compute the line integral with the differential on the right

$$\int F\, d\mathbf{x} = \mathbf{e}_{23} \int e^{it} (\mathbf{e}_2 + \mathbf{e}_1 t)\, dt$$
$$= -\mathbf{e}_3 \int e^{it} dt + \mathbf{e}_{123} \int t e^{-it} dt \qquad (2.122)$$
$$= \mathbf{e}_1 e^{it} + \mathbf{e}_{123} (1 + it)\, e^{-it}$$
$$= \mathbf{e}_1 e^{it} - \mathbf{e}_2 t e^{-it} + \mathbf{e}_{123} e^{-it},$$

or the line integral with the differential on the left

$$\int d\mathbf{x}\, F = \int (\mathbf{e}_2 + \mathbf{e}_1 t)\, \mathbf{e}_{23} e^{it} dt$$
$$= \mathbf{e}_3 \int e^{it} dt + \mathbf{e}_{123} \int t e^{it} \qquad (2.123)$$
$$= -\mathbf{e}_1 e^{it} + \mathbf{e}_{123} (1 - it)\, e^{it}$$
$$= -\mathbf{e}_1 e^{it} + \mathbf{e}_{123} e^{it} + \mathbf{e}_2 t e^{it}.$$

In both eq. (2.122) and eq. (2.123) the end result has both vector and trivector grades. While both integrals are equal (zero) when the angular velocity parameter t is a multiple of 2π, this shows that the order of the products in the integrand makes a difference once again.

Example: Function with only scalar and pseudoscalar grades. In $\mathbb{R}^3$, given any function with only scalar and pseudoscalar grades, say $F(t) = f(t) + Ig(t)$, where f, g are both scalar functions, then the order of the products in a line integrand do not matter. For any such function we have

$$\int F d\mathbf{x} = \int d\mathbf{x} F, \tag{2.124}$$

since both the scalar and pseudoscalar grades commute with any vector differential.

2.3.2 Surface integral.

Definition 2.5: Multivector surface integral.

Given a continuous and differentiable surface described by a vector function $\mathbf{x}(a, b)$, parameterized by two scalars a, b with differential

$$d^2\mathbf{x} \equiv d\mathbf{x}_a \wedge d\mathbf{x}_b = \frac{\partial \mathbf{x}}{\partial a} \wedge \frac{\partial \mathbf{x}}{\partial b} da\, db = \mathbf{x}_a \wedge \mathbf{x}_b da\, db,$$

and multivector functions F, G, the integral

$$\int F d^2\mathbf{x} G$$

is called a multivector surface integral.

An example of a two parameter surface, and the corresponding differentials with respect to those parameters, is illustrated in fig. 2.10.

In $\mathbb{R}^3$ it will often be convenient to utilize a dual representation of the area element $d^2\mathbf{x} = I\hat{\mathbf{n}} dA$, where dA is a scalar area element, and $\hat{\mathbf{n}}$ is a normal vector to the surface. With such an area element representation we will call $I \int dA\, F\hat{\mathbf{n}} G$ a surface integral.

Example: Spherical surface integral. From eq. (2.98), we know that

$$\mathbf{x}_r \mathbf{x}_\theta \mathbf{x}_\phi = Ir^2 \sin\theta, \tag{2.125}$$

so

$$\begin{aligned}\mathbf{x}_\theta \wedge \mathbf{x}_\phi &= \mathbf{x}_\theta \mathbf{x}_\phi \\ &= \mathbf{x}_r Ir^2 \sin\theta,\end{aligned} \tag{2.126}$$

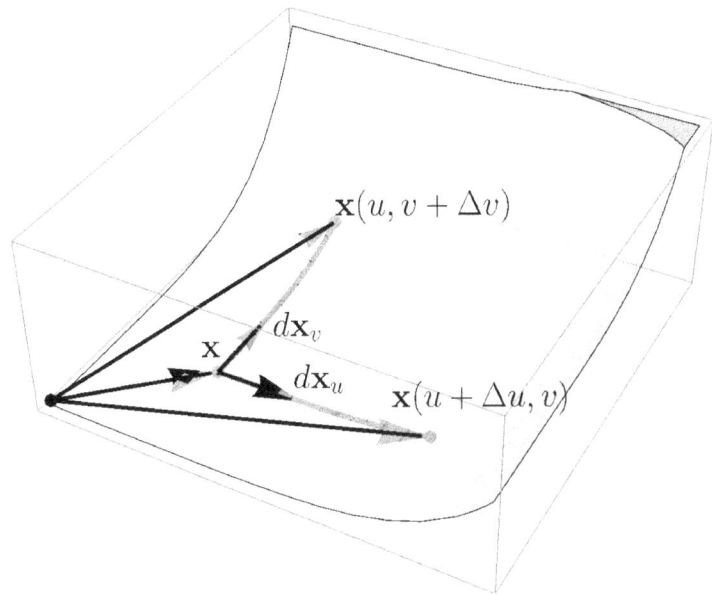

Figure 2.10: Two parameter manifold differentials.

so the (bivector-valued) area element for a spherical surface is

$$d^2\mathbf{x} = I\mathbf{x}_r r^2 \sin\theta d\theta d\phi. \tag{2.127}$$

Suppose we integrate a vector valued function $F(\theta,\phi) = \alpha\mathbf{x}^r + \beta\mathbf{x}^\theta + \gamma\mathbf{x}^\phi$, where α,β,γ are constants, over the surface of a sphere of radius r, then the surface integral (with the area element on the right) is

$$\int F d^2\mathbf{x} = \alpha I r^2 \int \mathbf{x}^r \mathbf{x}_r \sin\theta d\theta d\phi + \beta I r^2 \int \mathbf{x}^\theta \mathbf{x}_r \sin\theta d\theta d\phi \\ + \gamma I r^2 \int \mathbf{x}^\phi \mathbf{x}_r \sin\theta d\theta d\phi. \tag{2.128}$$

This can be simplified using $\hat{\mathbf{r}}\hat{\theta}\hat{\phi} = I$, and eq. (2.96), to find

$$\mathbf{x}^r \mathbf{x}_r = 1$$
$$I\mathbf{x}^\theta \mathbf{x}_r = \frac{1}{r} I \hat{\theta}\hat{\mathbf{r}} = \frac{1}{r}\hat{\phi} \tag{2.129}$$
$$I\mathbf{x}^\phi \mathbf{x}_r = \frac{1}{r\sin\theta} I \hat{\phi}\hat{\mathbf{r}} = -\frac{1}{r\sin\theta}\hat{\theta},$$

so

$$\int F d^2\mathbf{x} = \alpha I 4\pi r^2 + \beta r \int \hat{\phi} \sin\theta d\theta d\phi - \gamma r \int \hat{\theta} d\theta d\phi \\ = \alpha I 4\pi r^2, \tag{2.130}$$

where the integrands containing $\hat{\theta}, \hat{\phi}$ are killed by the integral over $\phi \in [0, 2\pi]$. If integrated over a subset of the spherical surface, where such perfect cancellation does not occur, this surface integral may have both vector and trivector components.

Example: Bivector function. Given a bivector valued function $F(a,b) = (a+b)\mathbf{e}_2\mathbf{e}_1 + 2(a\mathbf{e}_1 - b\mathbf{e}_2)\mathbf{e}_3$ defined over the unit square $a, b \in [0, 1]$, and a surface $\mathbf{x}(a,b) = a\mathbf{e}_1 + b\mathbf{e}_2$, the multivector surface integral (with the area element on the right) is

$$\begin{aligned} \int F d^2\mathbf{x} &= \int_0^1 \int_0^1 (a+b)\, da\, db + 2\int_0^1 \int_0^1 (a\mathbf{e}_1 - b\mathbf{e}_2)\mathbf{e}_3\mathbf{e}_1\mathbf{e}_2\, da\, db \\ &= 1 + I \int_0^1 a^2\Big|_0^1 \mathbf{e}_1\, db - I\int_0^1 b^2\Big|_0^1 \mathbf{e}_2\, da \\ &= 1 + I(\mathbf{e}_1 - \mathbf{e}_2) \\ &= 1 + (\mathbf{e}_1 + \mathbf{e}_2)\mathbf{e}_3. \end{aligned}$$

(2.131)

In this example, the integral of a bivector valued function over a (bivector-valued) surface area element results in a multivector with a scalar and bivector grade. In higher dimensional spaces, such an integral may also have grade-4 components.

2.3.3 Volume integral.

Definition 2.6: Multivector volume integral.

Given a continuous and differentiable volume described by a vector function $\mathbf{x}(a, b, c)$, parameterized by scalars a, b, c with volume element

$$d^3\mathbf{x} \equiv d\mathbf{x}_a \wedge d\mathbf{x}_b \wedge d\mathbf{x}_c = \frac{\partial \mathbf{x}}{\partial a} \wedge \frac{\partial \mathbf{x}}{\partial b} \wedge \frac{\partial \mathbf{x}}{\partial c}\, da\, db\, dc = \mathbf{x}_a \mathbf{x}_b \mathbf{x}_c\, da\, db\, dc,$$

and multivector functions F, G, the integral

$$\int F d^3\mathbf{x}\, G$$

is called a multivector volume integral.

In $\mathbb{R}^3$ the volume element is always a pseudoscalar, which commutes with all grades, so we are free to write $\int F d^3\mathbf{x} G = \int d^3\mathbf{x} F G$ for any multivectors F, G. It will often be useful to make the pseudoscalar nature of the volume element explicit, writing $d^3\mathbf{x} = I dV$, where dV is a scalar volume element.

As an example, let $F(\mathbf{x}) = r(\mathbf{x}) + \mathbf{s}(\mathbf{x}) + I\mathbf{t}(\mathbf{x}) + Iu(\mathbf{x})$ be an arbitrary multivector function in $\mathbb{R}^3$, where r, u are scalar functions and $\mathbf{s}, \mathbf{t}$ are vector functions. Integrating over a unit cube in rectangular coordinates $d^3\mathbf{x} = I dx dy dz = I dV$, the volume integral of such a multivector function is

$$\begin{aligned} \int F d^3\mathbf{x} &= \int \left(r(\mathbf{x}) + \mathbf{s}(\mathbf{x}) + I\mathbf{t}(\mathbf{x}) + Iu(\mathbf{x}) \right) I dV \\ &= \int \left(Ir(\mathbf{x}) + I\mathbf{s}(\mathbf{x}) - \mathbf{t}(\mathbf{x}) - u(\mathbf{x}) \right) dV. \end{aligned} \quad (2.132)$$

The result still has all grades, but each of the original grade components is mapped onto its dual space.

2.3.4 Bidirectional derivative operators.

Having generalized line, surface, and volume integrals to multivector functions, we wish to state the form of the integrand that is perfectly integrable. That statement requires bidirectional integration operators, denoted using left, right, or left-right overarrows, as follows.

Definition 2.7: Bidirectional vector derivative operators.

Given a hypervolume parameterized by k parameters, k-volume volume element $d^k\mathbf{x}$, and multivector functions F, G, let

$$\overleftrightarrow{L} = \sum_i \mathbf{x}^i \partial_i,$$

designate a linear differential operator (i.e. the gradient or vector derivative), where the partials act on multivector functions to the left or right (but not the reciprocal frame vectors $\mathbf{x}^i$).

> To express unidirection action of the operator only to the left or right, we use arrows to designate the scope of the derivatives, writing respectively
>
> $$\int_V F d^k \mathbf{x} \overleftarrow{L} G = \sum_i \int_V (\partial_i F) d^k \mathbf{x} \mathbf{x}^i G$$
>
> $$\int_V F d^k \mathbf{x} \overrightarrow{L} G = \sum_i \int_V F d^k \mathbf{x} \mathbf{x}^i (\partial_i G),$$
>
> and designate bidirectional action as
>
> $$\int_V F d^k \mathbf{x} \overleftrightarrow{L} G \equiv \int_V \left(F d^k \mathbf{x} \overleftarrow{L} \right) G + \int_V F d^k \mathbf{x} \left(\overrightarrow{L} G \right).$$
>
> In all such cases L operates on F and G, but not the volume element $d^k \mathbf{x}$, which may also be a function of the implied parameterization.

The vector derivative may not commute with F, G nor the volume element $d^k \mathbf{x}$, so we are forced to use some notation to indicate what the vector derivative (or gradient) acts on. In conventional right acting cases, where there is no ambiguity, arrows will usually be omitted, but braces may also be used to indicate the scope of derivative operators. This bidirectional notation will also be used for the gradient, especially for volume integrals in $\mathbb{R}^3$ where the vector derivative is identical to the gradient.

Some authors use the Hestenes dot notation, with overdots or primes to indicating the exact scope of multivector derivative operators, as in

$$\dot{F} d^k \mathbf{x} \dot{\partial} \dot{G} = \dot{F} d^k \mathbf{x} \dot{\partial} G + F d^k \mathbf{x} \dot{\partial} \dot{G}. \tag{2.133}$$

The dot notation has the advantage of emphasizing that the action of the vector derivative (or gradient) is on the functions F, G, and not on the hypervolume element $d^k \mathbf{x}$. However, in this book, where primed operators such as ∇' are used to indicate that derivatives are taken with respect to primed $\mathbf{x}'$ variables, a mix of dots and ticks would have been confusing.

2.3.5 Fundamental theorem.

The fundamental theorem of geometric calculus is a generalization of many conventional scalar and vector integral theorems, and relates a hypervolume integral to its boundary. This is a powerful theorem, which we will use with Green's functions to solve Maxwell's equation, but also to derive the geometric algebra form of Stokes' theorem, from which most of the familiar integral calculus results follow.

> **Theorem 2.3: Fundamental theorem of geometric calculus**
>
> Given multivectors F, G, a parameterization $\mathbf{x} = \mathbf{x}(u_1, u_2, \cdots, u_k)$, with hypervolume element $d^k\mathbf{x} = d^k u I_k$, where $I_k = \mathbf{x}_1 \wedge \mathbf{x}_2 \wedge \cdots \wedge \mathbf{x}_k$, the hypervolume integral is related to the boundary integral by
>
> $$\int_V F d^k\mathbf{x}\, \overleftrightarrow{\partial}\, G = \int_{\partial V} F d^{k-1}\mathbf{x} G,$$
>
> where ∂V represents the boundary of the volume, and $d^{k-1}\mathbf{x}$ is the hypersurface element. This is called the *Fundamental theorem of geometric calculus*.
>
> The hypersurface element and boundary integral is defined for $k > 1$ as
>
> $$\int_{\partial V} F d^{k-1}\mathbf{x} G \equiv \sum_{i=1}^{k} \int d^{k-1} u_i \left(F \left(I_k \cdot \mathbf{x}^i \right) G \right) \Big|_{\Delta u_i},$$
>
> where $d^{k-1}u_i$ is the product of all du_j except for du_i. For $k = 1$ the hypersurface element and associated boundary "integral" is really just convenient general shorthand, and should be taken to mean the evaluation of the FG multivector product over the range of the parameter
>
> $$\int_{\partial V} F d^0 \mathbf{x} G \equiv FG|_{\Delta u_1}.$$

The geometry of the hypersurface element $d^{k-1}\mathbf{x}$ will be made more clear when we consider the specific cases of $k = 1, 2, 3$, representing generalized line, surface, and volume integrals respectively. Instead of terrorizing the reader with a general proof theorem 2.3, which requires some unpleasant index gymnastics, this book will separately state and prove the fundamental theorem of calculus for each of the $k = 1, 2, 3$ cases that are of interest for problems in $\mathbb{R}^2$ and $\mathbb{R}^3$. For the interested reader, a sketch of the general proof of theorem 2.3 is available in chapter B.

Before moving on to the line, surface, and volume integral cases, we will state and prove the general Stokes' theorem in its geometric algebra form.

2.3.6 Stokes' theorem.

An important consequence of the fundamental theorem of geometric calculus is the geometric algebra generalization of Stokes' theorem. The Stokes' theorem that we know from conventional vector calculus relates $\mathbb{R}^3$ surface integrals to the line integral around a bounding surface. The geometric algebra form of Stokes' theorem is equivalent to Stokes' theorem from the theory of differential forms, which relates hypervolume integrals of blades[4] to the integrals over their hypersurface boundaries, a much more general result.

> **Theorem 2.4: Stokes' theorem**
>
> *Stokes' theorem* relates the dot product of a k volume element $d^k\mathbf{x}$ with the wedge product "curl" of an s-blade F, $s < k$ as follows
>
> $$\int_V d^k\mathbf{x} \cdot (\partial \wedge F) = \int_{\partial V} d^{k-1}\mathbf{x} \cdot F.$$

We will see that most of the well known scalar and vector integral theorems can easily be derived as direct consequences of theorem 2.4, itself a special case of theorem 2.3.

Proof. We can prove Stokes' theorem from theorem 2.3 by setting $F = 1$, and requiring that G is an s-blade, with grade $s < k$. The proof follows by selecting the $k - (s + 1)$ grade, the lowest grade of $d^k\mathbf{x}(\partial \wedge G)$, from both sides of theorem 2.3.

For the grade selection of the hypervolume integral we have

$$\left\langle \int_V d^k\mathbf{x}\partial G \right\rangle_{k-(s+1)} = \left\langle \int_V d^k\mathbf{x}(\partial \cdot G) + \int_V d^k\mathbf{x}(\partial \wedge G) \right\rangle_{k-(s-1)}, \quad (2.134)$$

however, the lowest grade of $d^k\mathbf{x}(\partial \cdot G)$ is $k - (s - 1) = k - s + 1 > k - (s + 1)$, so the divergence integral is zero. As $d^{k-1}\mathbf{x}$ is a $k - 1$ blade

$$\int_V d^k\mathbf{x} \cdot (\partial \wedge G) = \int_{\partial V} \left\langle d^{k-1}\mathbf{x}G \right\rangle_{k-(s+1)}$$
$$= \int_{\partial V} d^{k-1}\mathbf{x} \cdot G. \quad (2.135)$$

□

---

[4] Blades are isomorphic to the k-forms found in the theory of differential forms.

2.3.7 Fundamental theorem for Line integral.

The line integral specialization of theorem 2.3 is

> **Theorem 2.5: Fundamental theorem for line integrals.**
>
> Given a continuous curve $C = \mathbf{x}(u)$ parameterized by $u \in [u_0, u_1]$, and multivector functions $F(\mathbf{x}), G(\mathbf{x})$ that are differentable over C, then
> $$\int_C F d\mathbf{x} \overleftrightarrow{\partial} G = FG|_{\Delta u} = F(\mathbf{x}(u_1))G(\mathbf{x}(u_1)) - F(\mathbf{x}(u_0))G(\mathbf{x}(u_0)).$$

The differential form $d\mathbf{x} = d^1\mathbf{x} = du\,\mathbf{x}_u = du\,\partial \mathbf{x}/\partial u$ varies over the curve, and the vector derivative is just $\partial = \mathbf{x}^u \partial_u$ (no sum).

Proof. The proof follows by expansion. For

$$\begin{aligned}
\int_C F d\mathbf{x} \overleftrightarrow{\partial} G &= \int_C \left(F d\mathbf{x}\, \overleftarrow{\partial} \right) G + \int_C F d\mathbf{x} \left(\overrightarrow{\partial}\, G \right) \\
&= \int_C \frac{\partial F}{\partial u} du\, \mathbf{x}_u \mathbf{x}^u G + \int_C F du\, \mathbf{x}_u \mathbf{x}^u \frac{\partial G}{\partial u} \\
&= \int_C du\, \frac{\partial F}{\partial u} G + \int_C du\, F \frac{\partial G}{\partial u} \qquad (2.136) \\
&= \int_C du\, \frac{\partial}{\partial u}(FG) \\
&= F(u_1)G(u_1) - F(u_0)G(u_0),
\end{aligned}$$

We have a perfect cancellation of the reciprocal frame $\mathbf{x}^u$ with the vector $\mathbf{x}_u$ that lies along the curve, since $\mathbf{x}^u \mathbf{x}_u = 1$. This leaves a perfect derivative of the product of FG, which can be integrated over the length of the curve, yielding the difference of the product with respect to the parameterization of the end points of the curve. $\square$

For a single parameter subspace the reciprocal frame vector $\mathbf{x}^u$ is trivial to calculate, as it is just the inverse of $\mathbf{x}_u$, that is $\mathbf{x}^u = \mathbf{x}_u / \|\mathbf{x}_u\|^2$. Observe that we did not actually have to calculate it, but instead only require that the vector is invertible.

An important (and familiar) special case of theorem 2.5 is the fundamental theorem of calculus for line integrals, which can be obtained by using a single scalar function f

> **Theorem 2.6: Stokes' theorem for scalar functions.**
>
> Given a continuous curve $C = \mathbf{x}(u)$ parameterized by parameter $u \in [u_0, u_1]$, and a scalar function $f(\mathbf{x})$ differentable over C, then
>
> $$\int_C d\mathbf{x} \cdot \nabla f = \int_C d\mathbf{x} \cdot \partial f = f|_{\Delta u}.$$

Proof. Theorem 2.6 is no doubt familiar in its gradient form. Our proof starts with theorem 2.5 setting $F = 1, G = f(\mathbf{x}(u))$

$$\int_C d\mathbf{x}\, \partial f = f|_{\Delta u}, \tag{2.137}$$

which is a multivector equation with scalar and bivector grades on the left hand side, but only scalar grades on the right. Equating grades yields two equations

$$\int_C d\mathbf{x} \cdot \partial f = f|_{\Delta u} \tag{2.138a}$$

$$\int_C d\mathbf{x} \wedge \partial f = 0. \tag{2.138b}$$

Equation (2.138a), the scalar grade of eq. (2.137), proves part of theorem 2.6. To complete the proof, consider the specific case of $\mathbb{R}^3$ which is representitive. Suppose, that we have an $\mathbb{R}^3$ volume parameterization $\mathbf{x}(u, v, w)$ sharing an edge with the curve $C = \mathbf{x}(u, 0, 0)$. The curvilinear representation of the $\mathbb{R}^3$ gradient is

$$\nabla = \mathbf{x}^u \partial_u + \mathbf{x}^v \partial_v + \mathbf{x}^w \partial_w = \partial + \mathbf{x}^v \partial_v + \mathbf{x}^w \partial_w, \tag{2.139}$$

Over the curve C

$$d\mathbf{x} \cdot \nabla = du\, \mathbf{x}_u \cdot (\partial + \mathbf{x}^v \partial_v + \mathbf{x}^w \partial_w), \tag{2.140}$$

but $\mathbf{x}_u \cdot \mathbf{x}^v = \mathbf{x}_u \cdot \mathbf{x}^w = 0$, so $d\mathbf{x} \cdot \nabla = d\mathbf{x} \cdot \partial$ over the curve. $\square$

2.3.8 Fundamental theorem for Surface integral.

The surface integral specialization of theorem 2.3 is

2.3 INTEGRATION THEORY.

> **Theorem 2.7: Fundamental theorem for surface integrals.**
>
> Given a continuous and connected surface $S = \mathbf{x}(u, v)$ parameterized by $u \in [u_0, u_1]$, $v \in [v_0, v_1]$, multivector functions $F(\mathbf{x}), G(\mathbf{x})$ that are differentable over S, and an (bivector-valued) area element $d^2\mathbf{x} = d\mathbf{x}_1 \wedge d\mathbf{x}_2 = dudv\, \mathbf{x}_u \wedge \mathbf{x}_v$
>
> $$\int_S F d^2\mathbf{x}\, \overleftrightarrow{\partial}\, G = \oint_{\partial S} F d\mathbf{x} G,$$
>
> where ∂S is the boundary of the surface S.

Proof. To prove theorem 2.7 we start by expanding the multivector product $d^2\mathbf{x}\,\partial$ in curvilinear coordinates, where we discover that this product has only a vector grade. The vector derivative, the projection of the gradient onto the surface at the point of integration (also called the tangent space), now has two components

$$\begin{aligned}\partial &= \sum_i \mathbf{x}^i (\mathbf{x}_i \cdot \boldsymbol{\nabla}) \\ &= \mathbf{x}^u \frac{\partial}{\partial u} + \mathbf{x}^v \frac{\partial}{\partial v} \equiv \mathbf{x}^u \partial_u + \mathbf{x}^v \partial_v.\end{aligned} \qquad (2.141)$$

To see why the product of the area elements and the vector derivative

$$d^2\mathbf{x}\,\partial = dudv\, (\mathbf{x}_u \wedge \mathbf{x}_v)(\mathbf{x}^u \partial_u + \mathbf{x}^v \partial_v), \qquad (2.142)$$

has only a vector grade, observe that $\mathbf{x}^u \in \text{span}\{\mathbf{x}_u, \mathbf{x}_v\}$, so

$$\begin{aligned}(\mathbf{x}_u \wedge \mathbf{x}_v) \mathbf{x}^u &= (\mathbf{x}_u \wedge \mathbf{x}_v) \cdot \mathbf{x}^u + \cancel{(\mathbf{x}_u \wedge \mathbf{x}_v) \wedge \mathbf{x}^u} \\ &= (\mathbf{x}_u \wedge \mathbf{x}_v) \cdot \mathbf{x}^u \\ &= \mathbf{x}_u (\mathbf{x}_v \cdot \mathbf{x}^u) - \mathbf{x}_v (\mathbf{x}_u \cdot \mathbf{x}^u) \\ &= -\mathbf{x}_v.\end{aligned} \qquad (2.143)$$

Similarly

$$\begin{aligned}(\mathbf{x}_u \wedge \mathbf{x}_v) \mathbf{x}^v &= (\mathbf{x}_u \wedge \mathbf{x}_v) \cdot \mathbf{x}^v + \cancel{(\mathbf{x}_u \wedge \mathbf{x}_v) \wedge \mathbf{x}^v} \\ &= (\mathbf{x}_u \wedge \mathbf{x}_v) \cdot \mathbf{x}^v \\ &= \mathbf{x}_u (\mathbf{x}_v \cdot \mathbf{x}^v) - \mathbf{x}_v (\mathbf{x}_u \cdot \mathbf{x}^v) \\ &= \mathbf{x}_u.\end{aligned} \qquad (2.144)$$

Not only does eq. (2.142) have only a vector grade, that product reduces to just

$$d^2\mathbf{x}\,\partial = \mathbf{x}_u\partial_v - \mathbf{x}_v\partial_u. \qquad (2.145)$$

Inserting eq. (2.145) into the surface integral, we find

$$\begin{aligned}\int_S F d^2\mathbf{x}\,\partial G &= \int_S \left(F d^2\mathbf{x}\,\overleftarrow{\partial}\right)G + \int_S F d^2\mathbf{x}\left(\overrightarrow{\partial}\,G\right) \\ &= \int_S dudv\,(\partial_v F \mathbf{x}_u - \partial_u F \mathbf{x}_v)G + \int_S dudv\,F\,(\mathbf{x}_u\partial_v G - \mathbf{x}_v\partial_u G) \\ &= \int_S dudv\left(\frac{\partial F}{\partial v}\frac{\partial \mathbf{x}}{\partial u} - \frac{\partial F}{\partial u}\frac{\partial \mathbf{x}}{\partial v}\right)G + \int_S dudv\,F\left(\frac{\partial \mathbf{x}}{\partial u}\frac{\partial G}{\partial v} - \frac{\partial \mathbf{x}}{\partial v}\frac{\partial G}{\partial u}\right) \\ &= \int_S dudv\,\frac{\partial}{\partial v}\left(F\frac{\partial \mathbf{x}}{\partial u}G\right) - \int_S dudv\,\frac{\partial}{\partial u}\left(F\frac{\partial \mathbf{x}}{\partial v}G\right) \\ &\quad - \int_S dudv\,F\left(\frac{\partial}{\partial v}\frac{\partial \mathbf{x}}{\partial u} - \frac{\partial}{\partial u}\frac{\partial \mathbf{x}}{\partial v}\right)G \\ &= \int_S dudv\,\frac{\partial}{\partial v}\left(F\frac{\partial \mathbf{x}}{\partial u}G\right) - \int_S dudv\,\frac{\partial}{\partial u}\left(F\frac{\partial \mathbf{x}}{\partial v}G\right).\end{aligned} \qquad (2.146)$$

This leaves two perfect differentials, which can both be integrated separately

$$\begin{aligned}\int_S F d^2\mathbf{x}\,\partial G &= \int_{\Delta u} du\left(F\frac{\partial \mathbf{x}}{\partial u}G\right)\bigg|_{\Delta v} - \int_{\Delta v} dv\left(F\frac{\partial \mathbf{x}}{\partial v}G\right)\bigg|_{\Delta u} \\ &= \int_{\Delta u}(F d\mathbf{x}_u G)\big|_{\Delta v} - \int_{\Delta v}(F d\mathbf{x}_v G)\big|_{\Delta u}.\end{aligned} \qquad (2.147)$$

Equation (2.147) is an explicit algebraic expression of the boundary integral of theorem 2.7. To complete the proof, we are left with the task of geometrically interpretting this integrand. Suppose we are integrating over the unit parameter volume space $[u, v] \in [0, 1] \otimes [0, 1]$ as illustrated in fig. 2.11. Comparing to the figure we see that we've ended up with a clockwise line integral around the boundary of the surface. For a given subset of the surface, the bivector area element can be chosen small enough that it lies in the tangent space to the surface at the point of integration. In that case, a larger bounding loop can be conceptualized as the sum of a number of smaller ones, as sketched in fig. 2.12, in which case the contributions of the interior loop paths (red and blue) cancel out, leaving only the exterior loop contributions (green.) When that subdivision is made small

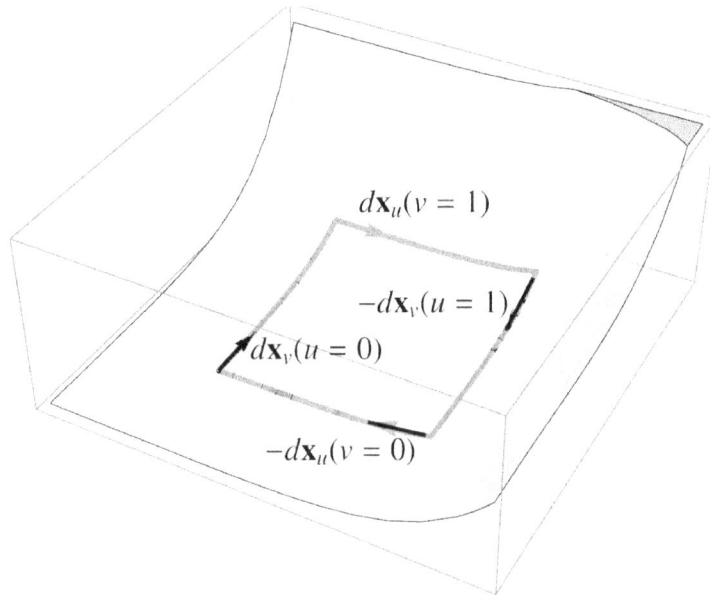

Figure 2.11: Contour for two parameter surface boundary.

enough (assuming that the surface is continuous and differentiable along each of the parameterization paths) then each area element approximates the tangent space at the point of evaluation. □

2.3.8.1 *Two parameter Stokes' theorem.*

Two special cases of theorem 2.7, both variations of Stokes' theorem, result by considering scalar and vector functions. For the scalar case we have

---

Theorem 2.8: Surface integral of scalar function (Stokes').

Given a scalar function $f(\mathbf{x})$ its *surface integral* is given by

$$\int_S d^2\mathbf{x} \cdot \boldsymbol{\partial} f = \int_S d^2\mathbf{x} \cdot \boldsymbol{\nabla} f = \oint_{\partial S} d\mathbf{x} f.$$

In $\mathbb{R}^3$, this can be written as

$$\int_S dA\, \hat{\mathbf{n}} \times \boldsymbol{\nabla} f = \oint_{\partial S} d\mathbf{x}\, f,$$

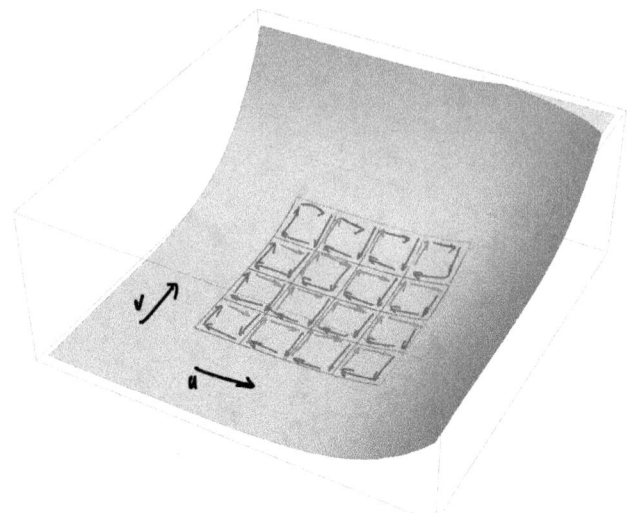

Figure 2.12: Sum of infinitesimal loops.

where $\hat{\mathbf{n}}$ is the normal specified by $d^2\mathbf{x} = I\hat{\mathbf{n}}dA$.

Proof. To show the first part, we can split the (multivector) surface integral into vector and trivector grades

$$\int_S d^2\mathbf{x}\,\partial f = \int_S d^2\mathbf{x} \cdot \partial f + \int_S d^2\mathbf{x} \wedge \partial f. \tag{2.148}$$

Since $\mathbf{x}^u, \mathbf{x}^v$ both lie in the span of $\{\mathbf{x}_u, \mathbf{x}_v\}$, $d^2\mathbf{x} \wedge \partial = 0$, killing the second integral in eq. (2.148). If the gradient is decomposed into its projection along the tangent space (the vector derivative) and its perpendicular components, only the vector derivative components of the gradient contribute to its dot product with the area element. That is

$$\begin{aligned}d^2\mathbf{x} \cdot \nabla &= d^2\mathbf{x} \cdot \left(\mathbf{x}^u \partial_u + \mathbf{x}^v \partial_v + \cdots\right) \\ &= d^2\mathbf{x} \cdot \left(\mathbf{x}^u \partial_u + \mathbf{x}^v \partial_v\right) \\ &= d^2\mathbf{x} \cdot \partial.\end{aligned} \tag{2.149}$$

This means that for a scalar function

$$\int_S d^2\mathbf{x}\,\partial f = \int_S d^2\mathbf{x} \cdot \nabla f. \tag{2.150}$$

The second part of the theorem follows by grade selection, and application of a duality transformation for the area element

$$\begin{aligned}
d^2\mathbf{x} \cdot \boldsymbol{\nabla} f &= \langle d^2\mathbf{x}\boldsymbol{\nabla} f \rangle_1 \\
&= dA \langle I\hat{\mathbf{n}}\boldsymbol{\nabla} f \rangle_1 \\
&= dA \langle I \left(\hat{\mathbf{n}} \cdot \boldsymbol{\nabla} f + I\hat{\mathbf{n}} \times \boldsymbol{\nabla} f \right) \rangle_1 \\
&= -dA\, \hat{\mathbf{n}} \times \boldsymbol{\nabla} f.
\end{aligned} \qquad (2.151)$$

back substitution of eq. (2.151) completes the proof. □

For vector functions we have

Theorem 2.9: Surface integral of a vector function (Stokes').

Given a vector function $\mathbf{f}(\mathbf{x})$, the *surface integral* is given by

$$\int_S d^2\mathbf{x} \cdot (\boldsymbol{\nabla} \wedge \mathbf{f}) = \oint_{\partial S} d\mathbf{x} \cdot \mathbf{f}.$$

In $\mathbb{R}^3$, this can be written as

$$\int_S dA\, \hat{\mathbf{n}} \cdot (\boldsymbol{\nabla} \times \mathbf{f}) = \oint_{\partial S} d\mathbf{x} \cdot \mathbf{f},$$

where $\hat{\mathbf{n}}$ is the normal specified by $d^2\mathbf{x} = I\hat{\mathbf{n}}dA$.

2.3.8.2 Green's theorem.

Theorem 2.9, when stated in terms of coordinates, is another well known result.

Theorem 2.10: Green's theorem.

Given a vector $\mathbf{f} = \sum_i f_i \mathbf{x}^i$ in $\mathbb{R}^N$, and a surface parameterized by $\mathbf{x} = \mathbf{x}(u_1, u_2)$, *Green's theorem* states

$$\int_S du_1 du_2 \left(\frac{\partial f_1}{\partial u_2} - \frac{\partial f_2}{\partial u_1} \right) = \oint_{\partial S} du_1 f_1 + du_2 f_2.$$

This is often stated for vectors $\mathbf{f} = P\mathbf{e}_1 + Q\mathbf{e}_2 \in \mathbb{R}^2$ with a Cartesian x, y parameterization as

$$\int_S dxdy \left(\frac{\partial P}{\partial y} - \frac{\partial Q}{\partial x} \right) = \oint_{\partial S} Pdx + Qdy.$$

Proof. The first equality in theorem 2.10 holds in $\mathbb{R}^N$ for vectors expressed in terms of an arbitrary curvilinear basis. Only the (curvilinear) coordinates of the vector $\mathbf{f}$ contribute to this integral, and only those that lie in the tangent space. The reciprocal basis vectors $\mathbf{x}^i$ are also nowhere to be seen. This is because they are either obliterated in dot products with $\mathbf{x}_j$, or cancel due to mixed partial equality.

To see how this occurs let's look at the area integrand of theorem 2.9

$$\begin{aligned}
d^2\mathbf{x} \cdot (\boldsymbol{\nabla} \wedge \mathbf{f}) &= du_1 du_2 \, (\mathbf{x}_1 \wedge \mathbf{x}_2) \cdot \left[\sum_{ij} \left(\mathbf{x}^i \partial_i \right) \wedge \left(f_j \mathbf{x}^j \right) \right] \\
&= du_1 du_2 \sum_{ij} \left((\mathbf{x}_1 \wedge \mathbf{x}_2) \cdot \mathbf{x}^i \right) \cdot \left(\partial_i (f_j \mathbf{x}^j) \right) \\
&= du_1 du_2 \sum_{ij} \left((\mathbf{x}_1 \wedge \mathbf{x}_2) \cdot \mathbf{x}^i \right) \cdot \mathbf{x}^j \partial_i f_j \\
&\quad + du_1 du_2 \sum_{ij} f_j \left((\mathbf{x}_1 \wedge \mathbf{x}_2) \cdot \mathbf{x}^i \right) \cdot (\partial_i \mathbf{x}^j).
\end{aligned} \quad (2.152)$$

With a bit of trouble, we will see that the second integrand is zero. On the other hand, the first integrand simplifies without too much trouble

$$\begin{aligned}
\sum_{ij} \left((\mathbf{x}_1 \wedge \mathbf{x}_2) \cdot \mathbf{x}^i \right) \cdot \mathbf{x}^j \partial_i f_j &= \sum_{ij} (\mathbf{x}_1 \delta_{2i} - \mathbf{x}_2 \delta_{1i}) \cdot \mathbf{x}^j \partial_i f_j \\
&= \sum_j \mathbf{x}_1 \cdot \mathbf{x}^j \partial_2 f_j - \mathbf{x}_2 \cdot \mathbf{x}^j \partial_1 f_j \quad (2.153) \\
&= \partial_2 f_1 - \partial_1 f_2.
\end{aligned}$$

For the second integrand, we have

$$\begin{aligned}
\sum_{ij} f_j \left((\mathbf{x}_1 \wedge \mathbf{x}_2) \cdot \mathbf{x}^i \right) \cdot (\partial_i \mathbf{x}^j) & \\
= \sum_j f_j \sum_i (\mathbf{x}_1 \delta_{2i} - \mathbf{x}_2 \delta_{1i}) \cdot (\partial_i \mathbf{x}_j) & \quad (2.154) \\
= \sum_j f_j \left(\mathbf{x}_1 \cdot (\partial_2 \mathbf{x}^j) - \mathbf{x}_2 \cdot (\partial_1 \mathbf{x}^j) \right). &
\end{aligned}$$

We can apply the chain rule (backwards) to the portion in brackets to find

$$\begin{aligned}
\mathbf{x}_1 \cdot (\partial_2 \mathbf{x}^j) - \mathbf{x}_2 \cdot (\partial_1 \mathbf{x}^j) &= \partial_2 \overline{(\mathbf{x}_1 \cdot \mathbf{x}^j)} - (\partial_2 \mathbf{x}_1) \cdot \mathbf{x}^j - \partial_1 \overline{(\mathbf{x}_2 \cdot \mathbf{x}^j)} + (\partial_1 \mathbf{x}_2) \cdot \mathbf{x}^j \\
&= \mathbf{x}_j \cdot (\partial_1 \mathbf{x}_2 - \partial_2 \mathbf{x}_1) \\
&= \mathbf{x}_j \cdot \left(\frac{\partial}{\partial u_1} \frac{\partial \mathbf{x}}{\partial u_2} - \frac{\partial}{\partial u_2} \frac{\partial \mathbf{x}}{\partial u_1} \right) \\
&= 0.
\end{aligned}$$

(2.155)

In this reduction the derivatives of $\mathbf{x}_i \cdot \mathbf{x}^j = \delta_i{}^j$ were killed since those are constants (either zero or one). The final step relies on the fact that we assume our vector parameterization is well behaved enough that the mixed partials are zero.

Substituting these results into theorem 2.9 we find

$$\begin{aligned}
\oint_{\partial S} d\mathbf{x} \cdot \mathbf{f} &= \oint_{\partial S} (du_1 \mathbf{x}_1 + du_2 \mathbf{x}_2) \cdot \left(\sum_i f_i \mathbf{x}^i \right) \\
&= \oint_{\partial S} du_1 \, f_1 + du_2 \, f_2 \\
&= \int_S du_1 du_2 \, (\partial_2 f_1 - \partial_1 f_2).
\end{aligned}$$

(2.156)

$\square$

2.3.9 Fundamental theorem for Volume integral.

The volume integral specialization of theorem 2.3 follows.

Theorem 2.11: Fundamental theorem for volume integrals.

Given a continuous and connected volume $V = \mathbf{x}(u, v, w)$ parameterized by $u \in [u_0, u_1], v \in [v_0, v_1], w \in [w_0, w_1]$, multivector functions $F(\mathbf{x}), G(\mathbf{x})$ that are differentable over V, and an (trivector-valued) volume element $d^3\mathbf{x} = d\mathbf{x}_1 \wedge d\mathbf{x}_2 \wedge d\mathbf{x}_3 = du\,dv\,dw\, \mathbf{x}_u \wedge \mathbf{x}_v \wedge \mathbf{x}_w$

$$\int_V F d^3\mathbf{x} \stackrel{\leftrightarrow}{\partial} G = \oint_{\partial V} F d^2\mathbf{x} G,$$

where ∂V is the boundary of the volume V, and $d^2\mathbf{x}$ is the counter-clockwise oriented area element on the boundary of the volume, that is

$$\oint_{\partial V} F d^2\mathbf{x} G = \int (F d\mathbf{x}_1 \wedge d\mathbf{x}_2 G)\big|_{\Delta w}$$
$$+ \int (F d\mathbf{x}_2 \wedge d\mathbf{x}_3 G)\big|_{\Delta u}$$
$$+ \int (F d\mathbf{x}_3 \wedge d\mathbf{x}_1 G)\big|_{\Delta v}.$$

In $\mathbb{R}^3$ with $d^3\mathbf{x} = I dV$, $d^2\mathbf{x} = I\hat{n} dA$, this integral can be written using a scalar volume element, as

$$\int_V dV\, F \overset{\leftrightarrow}{\partial} G = \int_{\partial V} dA\, F \hat{n} G.$$

Before diving into the proof of theorem 2.11, let's consider the geometry of the volume element briefly.

For uniformity, let $u = u_1, v = u_2, w = u_3$, so that the differentials along each of the parameterization directions are

$$d\mathbf{x}_1 = \frac{\partial \mathbf{x}}{\partial u_1} du_1 = \mathbf{x}_1 du_1$$
$$d\mathbf{x}_2 = \frac{\partial \mathbf{x}}{\partial u_2} du_2 = \mathbf{x}_2 du_2 \qquad (2.157)$$
$$d\mathbf{x}_3 = \frac{\partial \mathbf{x}}{\partial u_3} du_3 = \mathbf{x}_3 du_3.$$

The trivector valued volume element for this parameterization is

$$d^3\mathbf{x} = d\mathbf{x}_1 \wedge d\mathbf{x}_2 \wedge d\mathbf{x}_3 = d^3 u\, (\mathbf{x}_1 \wedge \mathbf{x}_2 \wedge \mathbf{x}_3), \qquad (2.158)$$

where $d^3 u = du_1 du_2 du_3$. A volume and its corresponding differentials with respect to three parameters is sketched in fig. 2.13.

In $\mathbb{R}^3$ the vector derivative for a volume parameterization and the gradient are identical. In higher dimensional spaces the projection of the gradient onto the volume at the point of integration (also called the tangent space), has three components

$$\partial = \sum_i \mathbf{x}^i (\mathbf{x}_i \cdot \boldsymbol{\nabla})$$
$$= \mathbf{x}^1 \frac{\partial}{\partial u_1} + \mathbf{x}^2 \frac{\partial}{\partial u_2} + \mathbf{x}^3 \frac{\partial}{\partial u_3} \qquad (2.159)$$
$$\equiv \mathbf{x}^1 \partial_1 + \mathbf{x}^2 \partial_2 + \mathbf{x}^3 \partial_3.$$

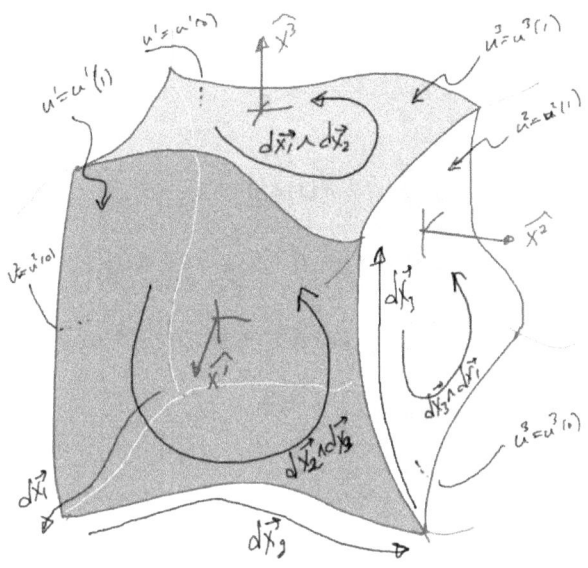

Figure 2.13: Three parameter volume element.

With the volume element and the vector derivative spelled out explicitly, we can proceed with a proof of theorem 2.11.

Proof. The first step, is the reduction of the product of the volume element and the vector derivative, which we will see is a bivector.

$$d^3\mathbf{x}\partial = d^3u \left(\mathbf{x}_1 \wedge \mathbf{x}_2 \wedge \mathbf{x}_3\right)\left(\mathbf{x}^1\partial_1 + \mathbf{x}^2\partial_2 + \mathbf{x}^3\partial_3\right). \quad (2.160)$$

Since all $\mathbf{x}^i$ lie within span $\{\mathbf{x}_1, \mathbf{x}_2, \mathbf{x}_3\}$, this multivector product has only a vector grade. That is

$$\left(\mathbf{x}_1 \wedge \mathbf{x}_2 \wedge \mathbf{x}_3\right)\mathbf{x}^i = \left(\mathbf{x}_1 \wedge \mathbf{x}_2 \wedge \mathbf{x}_3\right) \cdot \mathbf{x}^i + \cancel{\left(\mathbf{x}_1 \wedge \mathbf{x}_2 \wedge \mathbf{x}_3\right) \wedge \mathbf{x}^i}, \quad (2.161)$$

for all $\mathbf{x}^i$. These products reduces to

$$\begin{aligned}
\left(\mathbf{x}_2 \wedge \mathbf{x}_3 \wedge \mathbf{x}_1\right)\mathbf{x}^1 &= \mathbf{x}_2 \wedge \mathbf{x}_3 \\
\left(\mathbf{x}_3 \wedge \mathbf{x}_1 \wedge \mathbf{x}_2\right)\mathbf{x}^2 &= \mathbf{x}_3 \wedge \mathbf{x}_1 \\
\left(\mathbf{x}_1 \wedge \mathbf{x}_2 \wedge \mathbf{x}_3\right)\mathbf{x}^3 &= \mathbf{x}_1 \wedge \mathbf{x}_2.
\end{aligned} \quad (2.162)$$

Inserting eq. (2.162) into the volume integral, we find

$$\int_V F d^3\mathbf{x} \partial G$$
$$= \int_V \left(F d^3\mathbf{x} \overleftarrow{\partial}\right) G + \int_V F d^3\mathbf{x} \left(\overrightarrow{\partial} G\right)$$
$$= \int_V d^3u \, ((\partial_1 F)\mathbf{x}_2 \wedge \mathbf{x}_3 G + (\partial_2 F)\mathbf{x}_3 \wedge \mathbf{x}_1 G + (\partial_3 F)\mathbf{x}_1 \wedge \mathbf{x}_2 G)$$
$$+ \int_V d^3u \, (F\mathbf{x}_2 \wedge \mathbf{x}_3(\partial_1 G) + F\mathbf{x}_3 \wedge \mathbf{x}_1(\partial_2 G) + F\mathbf{x}_1 \wedge \mathbf{x}_2(\partial_3 G))$$
$$= \int_V d^3u \, (\partial_1(F\mathbf{x}_2 \wedge \mathbf{x}_3 G) + \partial_2(F\mathbf{x}_3 \wedge \mathbf{x}_1 G) + \partial_3(F\mathbf{x}_1 \wedge \mathbf{x}_2 G))$$
$$- \int_V d^3u \, (F(\partial_1(\mathbf{x}_2 \wedge \mathbf{x}_3))G + F(\partial_2(\mathbf{x}_3 \wedge \mathbf{x}_1))G + F(\partial_3(\mathbf{x}_1 \wedge \mathbf{x}_2))G)$$
$$= \int_V d^3u \, (\partial_1(F\mathbf{x}_2 \wedge \mathbf{x}_3 G) + \partial_2(F\mathbf{x}_3 \wedge \mathbf{x}_1 G) + \partial_3(F\mathbf{x}_1 \wedge \mathbf{x}_2 G))$$
$$- \int_V d^3u \, F \left(\partial_1(\mathbf{x}_2 \wedge \mathbf{x}_3) + \partial_2(\mathbf{x}_3 \wedge \mathbf{x}_1) + \partial_3(\mathbf{x}_1 \wedge \mathbf{x}_2)\right) G.$$

(2.163)

The sum within the second integral is

$$\sum_{i=1}^{3} \partial_i \left(I_k \cdot \mathbf{x}^i\right)$$
$$= \partial_3 \left((\mathbf{x}_1 \wedge \mathbf{x}_2 \wedge \mathbf{x}_3) \cdot \mathbf{x}^3\right)$$
$$+ \partial_1 \left((\mathbf{x}_2 \wedge \mathbf{x}_3 \wedge \mathbf{x}_1) \cdot \mathbf{x}^1\right)$$
$$+ \partial_2 \left((\mathbf{x}_3 \wedge \mathbf{x}_1 \wedge \mathbf{x}_2) \cdot \mathbf{x}^2\right)$$
$$= \partial_3 (\mathbf{x}_1 \wedge \mathbf{x}_2) + \partial_1 (\mathbf{x}_2 \wedge \mathbf{x}_3) + \partial_2 (\mathbf{x}_3 \wedge \mathbf{x}_1)$$
$$= (\partial_3 \mathbf{x}_1) \wedge \mathbf{x}_2 + \mathbf{x}_1 \wedge (\partial_3 \mathbf{x}_2)$$
$$+ (\partial_1 \mathbf{x}_2) \wedge \mathbf{x}_3 + \mathbf{x}_2 \wedge (\partial_1 \mathbf{x}_3)$$
$$+ (\partial_2 \mathbf{x}_3) \wedge \mathbf{x}_1 + \mathbf{x}_3 \wedge (\partial_2 \mathbf{x}_1)$$
$$= \mathbf{x}_2 \wedge (-\partial_3 \mathbf{x}_1 + \partial_1 \mathbf{x}_3) + \mathbf{x}_3 \wedge (-\partial_1 \mathbf{x}_2 + \partial_2 \mathbf{x}_1) + \mathbf{x}_1 \wedge (-\partial_2 \mathbf{x}_3 + \partial_3 \mathbf{x}_2)$$
$$= \mathbf{x}_2 \wedge \left(-\frac{\partial^2 \mathbf{x}}{\partial_3 \partial_1} + \frac{\partial^2 \mathbf{x}}{\partial_1 \partial_3}\right) + \mathbf{x}_3 \wedge \left(-\frac{\partial^2 \mathbf{x}}{\partial_1 \partial_2} + \frac{\partial^2 \mathbf{x}}{\partial_2 \partial_1}\right) + \mathbf{x}_1 \wedge \left(-\frac{\partial^2 \mathbf{x}}{\partial_2 \partial_3} + \frac{\partial^2 \mathbf{x}}{\partial_3 \partial_2}\right),$$

(2.164)

which is zero by equality of mixed partials. This leaves three perfect differentials, which can integrated separately, giving

$$\int_V F d^3\mathbf{x} \partial G = \int du_2 du_3 \left(F\mathbf{x}_2 \wedge \mathbf{x}_3 G \right)\big|_{\Delta u_1}$$
$$+ \int du_3 du_1 \left(F\mathbf{x}_3 \wedge \mathbf{x}_1 G \right)\big|_{\Delta u_2} + \int du_1 du_2 \left(F\mathbf{x}_1 \wedge \mathbf{x}_2 G \right)\big|_{\Delta u_3}$$
$$= \int \left(F d\mathbf{x}_2 \wedge d\mathbf{x}_3 G \right)\big|_{\Delta u_1}$$
$$+ \int \left(F d\mathbf{x}_3 \wedge d\mathbf{x}_1 G \right)\big|_{\Delta u_2} + \int \left(F d\mathbf{x}_1 \wedge d\mathbf{x}_2 G \right)\big|_{\Delta u_3}.$$
(2.165)

This proves the theorem from an algebraic point of view. □

With the aid of a geometrical model, such as that of fig. 2.14, if assuming that $d\mathbf{x}_1, d\mathbf{x}_2, d\mathbf{x}_3$ is a right handed triple). it is possible to convince oneself that the two parameter integrands describe an integral over a counterclockwise oriented surface.

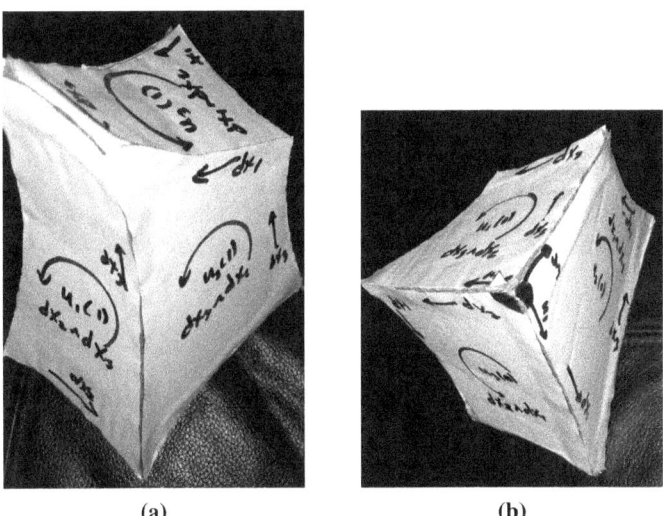

(a) (b)

Figure 2.14: Differential surface of a volume.

We obtain the RHS of theorem 2.11 if we introduce a mnemonic for the bounding oriented surface of the volume

$$d^2\mathbf{x} \equiv d\mathbf{x}_1 \wedge d\mathbf{x}_2 + d\mathbf{x}_2 \wedge d\mathbf{x}_3 + d\mathbf{x}_3 \wedge d\mathbf{x}_1,$$
(2.166)

where it is implied that each component of this area element and anything that it is multiplied with is evaluated on the boundaries of the integration volume (for the parameter omitted) as detailed explicitly in eq. (2.165).

2.3.9.1 Three parameter Stokes' theorem.

Three special cases of theorem 2.11 can be obtained by integrating scalar, vector or bivector functions over the volume, as follows

> **Theorem 2.12: Volume integral of scalar function (Stokes').**
>
> Given a scalar function $f(\mathbf{x})$ its volume integral is given by
>
> $$\int_V d^3\mathbf{x} \cdot \partial f = \int_V d^3\mathbf{x} \cdot \nabla f = \oint_{\partial V} d^2\mathbf{x} f.$$
>
> In $\mathbb{R}^3$, this can be written as
>
> $$\int_V dV \nabla f = \int_{\partial V} dA\, \hat{\mathbf{n}} f$$
>
> where $\hat{\mathbf{n}}$ is the outwards normal specified by $d^2\mathbf{x} = I\hat{\mathbf{n}} dA$, and $d^3\mathbf{x} = I dV$.

> **Theorem 2.13: Volume integral of vector function (Stokes').**
>
> The specialization of *Stokes' theorem* for a volume integral of the (bivector) curl of a vector function $\mathbf{f}(\mathbf{x})$, relates the volume integral to a surface area over the boundary as follows
>
> $$\int_V d^3\mathbf{x} \cdot (\partial \wedge \mathbf{f}) = \int_V d^3\mathbf{x} \cdot (\nabla \wedge \mathbf{f}) = \oint_{\partial V} d^2\mathbf{x} \cdot \mathbf{f}.$$
>
> In $\mathbb{R}^3$, this can be written as
>
> $$\int_V dV \nabla \times \mathbf{f} = \int_{\partial V} dA\, \hat{\mathbf{n}} \times \mathbf{f},$$
>
> or with a duality transformation $\mathbf{f} = IB$, where B is a bivector
>
> $$\int_V dV \nabla \cdot B = \int_{\partial V} dA\, \hat{\mathbf{n}} \cdot B,$$

where $\hat{\mathbf{n}}$ is the normal specified by $d^2\mathbf{x} = I\hat{\mathbf{n}}dA$, and $d^3\mathbf{x} = IdV$.

Theorem 2.14: Volume integral of bivector function (Stokes').

Given a bivector function $B(\mathbf{x})$, the volume integral of the (trivector) curl is related to a surface integral by

$$\int_V d^3\mathbf{x} \cdot (\boldsymbol{\partial} \wedge B) = \int_V d^3\mathbf{x} \cdot (\boldsymbol{\nabla} \wedge B) = \oint_{\partial V} d^2\mathbf{x} \cdot B.$$

In $\mathbb{R}^3$, this can be written as

$$\int_V dV \, \boldsymbol{\nabla} \wedge B = \int_{\partial V} dA \, \hat{\mathbf{n}} \wedge B,$$

which yields the *divergence theorem* after making a duality transformation $B(\mathbf{x}) = I\mathbf{f}(\mathbf{x})$, where $\mathbf{f}$ is a vector, by

$$\int_V dV \, \boldsymbol{\nabla} \cdot \mathbf{f} = \int_{\partial V} dA \, \hat{\mathbf{n}} \cdot \mathbf{f},$$

where $\hat{\mathbf{n}}$ is the normal specified by $d^2\mathbf{x} = I\hat{\mathbf{n}}dA$, and $d^3\mathbf{x} = IdV$.

2.3.9.2 Divergence theorem.

Observe that for $\mathbb{R}^3$ we there are dot product relations in each of theorem 2.12, theorem 2.13 and theorem 2.14 which can be summarized as

---

Theorem 2.15: Divergence theorem.

The *divergence theorem* may be generalized in $\mathbb{R}^3$ to multivectors M containing grades 0,1, or 2, but no grade 3 components

$$\int_V dV \, \boldsymbol{\nabla} \cdot M = \int_{\partial V} dA \, \hat{\mathbf{n}} \cdot M,$$

where $\hat{\mathbf{n}}$ is the normal to the surface bounding V given by $d^2\mathbf{x} = I\hat{\mathbf{n}}dA$.

2.4 VECTOR CALCULUS IDENTITIES.

2.4.1 Curl.

> **Definition 2.8: Curl of a k-blade.**
>
> Let A_k be a k-blade. We define the curl of a k-blade as the wedge product of the gradient with that k-blade, designated
>
> $$\boldsymbol{\nabla} \wedge A_k.$$

From the definition of the multivector wedge product definition 1.29, it is worth noting that the curl of a scalar function f, is in fact just the gradient of that function

$$\boldsymbol{\nabla} \wedge f = \langle \boldsymbol{\nabla} f \rangle_{1+0} = \boldsymbol{\nabla} f. \tag{2.167}$$

Recall that the conventional curl of an $\mathbb{R}^3$ vector, is written in terms of the cross product, as $\boldsymbol{\nabla} \times \mathbf{v}$. The cross product curl can be thought of a measure of how much a vector field rotates, and is proportional to the rotational axis. Our wedge product curl, when applied to a vector, is also a measure of the rotational nature of the vector field, but is a bivector that describes the rotational plane. Our use of a wedge product based curl risks some ambiguity, compared to the conventional $\mathbb{R}^3$ cross product based curl, but this ambiguity is worthwhile, since a wedge product based curl is much more useful in a geometric algebra context.

Let's consider some examples of curls of vector fields, starting with

$$\mathbf{v} = x\mathbf{e}_2, \tag{2.168}$$

as plotted in fig. 2.15. The curl is

$$(\mathbf{e}_1 \partial_x + \mathbf{e}_2 \partial_y + \mathbf{e}_3 \partial_z) \wedge (x\mathbf{e}_2) = \mathbf{e}_{12}, \tag{2.169}$$

describing a positive oriented rotation in the x-y plane. The conventional cross product curl of this field is

$$\begin{aligned} \boldsymbol{\nabla} \times \mathbf{v} &= -I(\boldsymbol{\nabla} \wedge \mathbf{v}) \\ &= -\mathbf{e}_{12312} \\ &= \mathbf{e}_3, \end{aligned} \tag{2.170}$$

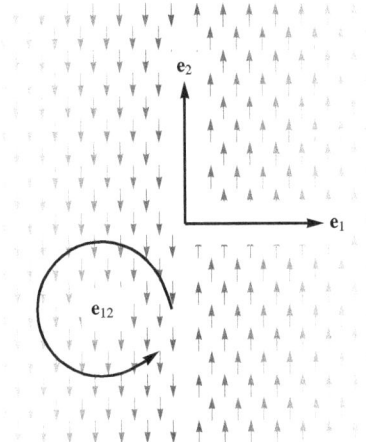

Figure 2.15: Curl example 1.

also describes a positive rotation in the x-y plane (i.e.: around the z-axis.)
As a second example, consider the purely rotational vector field

$$\mathbf{v} = -y\mathbf{e}_1 + x\mathbf{e}_2, \tag{2.171}$$

as plotted in fig. 2.16. That curl, computed in this Cartesian representation

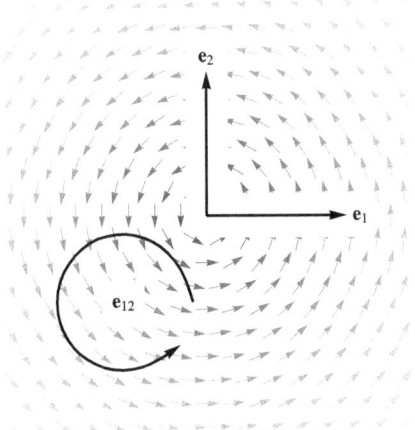

Figure 2.16: Curl example 2.

is

$$\begin{aligned}\boldsymbol{\nabla} \wedge \mathbf{v} &= \left(\mathbf{e}_1 \partial_x + \mathbf{e}_2 \partial_y + \mathbf{e}_3 \partial_z\right) \wedge \left(-y\mathbf{e}_1 + x\mathbf{e}_2\right) \\ &= \mathbf{e}_1 \wedge \mathbf{e}_2 - \mathbf{e}_2 \wedge \mathbf{e}_1 \\ &= 2\mathbf{e}_{12}.\end{aligned} \tag{2.172}$$

This vector field also describes a positive rotation in the x-y plane. As this field is intrinsically rotational, this curl can also be computed in polar coordinates (exercise 2.12.)

As a third example, consider a non-planar vector field

$$\mathbf{v} = y\mathbf{e}_1 + z\mathbf{e}_2 + x\mathbf{e}_3, \tag{2.173}$$

for which the curl is

$$\begin{aligned}\mathbf{\nabla} \wedge \mathbf{v} &= \left(\mathbf{e}_1 \partial_x + \mathbf{e}_2 \partial_y + \mathbf{e}_3 \partial_z\right) \wedge \left(y\mathbf{e}_1 + z\mathbf{e}_2 + x\mathbf{e}_3\right)\\ &= \mathbf{e}_{13} + \mathbf{e}_{21} + \mathbf{e}_{32} \\ &= \left(\mathbf{e}_2 - \mathbf{e}_3\right) \wedge \left(\mathbf{e}_1 - \mathbf{e}_3\right).\end{aligned} \tag{2.174}$$

In the last step, an arbitrary wedge product factorization was selected to illustrate the orientation of the plane, as in fig. 2.17.

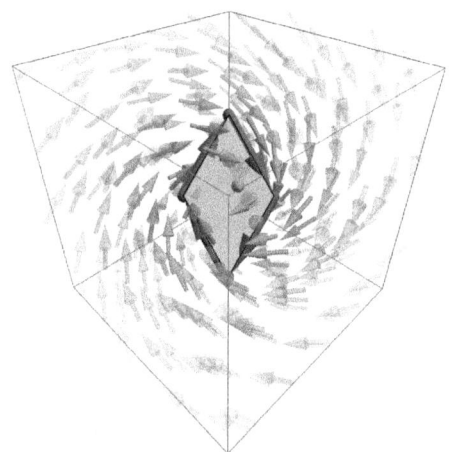

Figure 2.17: Curl of 3D vector field.

These three examples were all simple vector fields that had constant (bivector-valued) curl, but that need not be the case in general.

In $\mathbb{R}^3$, the curl of a bivector, is related to the vector divergence, since we can write $B = I\mathbf{v}$ for an $\mathbb{R}^3$ bivector, so

$$\mathbf{\nabla} \wedge B = \langle \mathbf{\nabla} I \mathbf{v} \rangle_3 = I\left(\mathbf{\nabla} \cdot \mathbf{v}\right), \tag{2.175}$$

so $\mathbb{R}^3$ bivectors dual to divergence free vectors will have zero curl. See exercise 2.13 for some examples of bivector curl.

2.4 VECTOR CALCULUS IDENTITIES. 149

> **Lemma 2.1: Repeated curl identities.**
>
> Let A be a smooth k-blade, then
>
> $$\nabla \wedge (\nabla \wedge A) = 0.$$
>
> For $\mathbb{R}^3$, this result, for a scalar function f, and a vector function $\mathbf{f}$, in terms of the cross product, as
>
> $$\begin{aligned} \nabla \times (\nabla f) &= 0 \\ \nabla \cdot (\nabla \times \mathbf{f}) &= 0. \end{aligned} \quad (2.176)$$

Proof. First consider the 0-blade case

$$\begin{aligned} \nabla \wedge (\nabla \wedge A) &= \nabla \wedge (\nabla A) \\ &= \sum_{ij} \mathbf{e}_i \wedge \mathbf{e}_j \frac{\partial^2 A}{\partial x_i \partial x_j} \\ &= 0. \end{aligned} \quad (2.177)$$

The smooth criteria of for the function A is assumed to imply that we have equality of mixed partials, and since this is a sum of an antisymmetric term with respect to indexes i, j (the wedge) and a symmetric term in indexes i, j (the partials), we have zero overall.

Now consider a k-blade $A, k > 0$. Expanding the gradients, we have

$$\nabla \wedge (\nabla \wedge A) = \sum_{ij} \mathbf{e}_i \wedge \mathbf{e}_j \wedge \frac{\partial^2 A}{\partial x_i \partial x_j}. \quad (2.178)$$

It may be obvious that this is zero for the same reasons as above (sum of product of symmetric and antisymmetric entities). We can, however, make it more obvious, at the cost of some hellish indexing, by expressing A in coordinate form. Let

$$A = \sum_{i_1, i_2, \cdots, i_k} A_{i_1, i_2, \cdots, i_k} \mathbf{e}_{i_1} \wedge \mathbf{e}_{i_2} \wedge \cdots \wedge \mathbf{e}_{i_k}, \quad (2.179)$$

then

$$\begin{aligned} \nabla \wedge (\nabla \wedge A) &= \sum_{i,j,i_1,i_2,\cdots,i_k} \mathbf{e}_i \wedge \mathbf{e}_j \wedge \mathbf{e}_{i_1} \wedge \mathbf{e}_{i_2} \wedge \cdots \wedge \mathbf{e}_{i_k} \frac{\partial^2}{\partial x_i \partial x_j} A_{i_1,i_2,\cdots,i_k} \\ &= 0. \end{aligned}$$

$$\tag{2.180}$$

Now we clearly have a sum of an antisymmetric term (the wedges), and a symmetric term (assuming smooth A means that we have equality of mixed partials), so the sum is zero.

Finally, for the $\mathbb{R}^3$ identities, we have

$$\begin{aligned}\boldsymbol{\nabla} \times (\boldsymbol{\nabla} f) &= -I\left(\boldsymbol{\nabla} \wedge (\boldsymbol{\nabla} f)\right) \\ &= 0,\end{aligned} \tag{2.181}$$

since $\boldsymbol{\nabla} \wedge (\boldsymbol{\nabla} f) = 0$. For a vector $\mathbf{f}$, we have

$$\begin{aligned}\boldsymbol{\nabla} \cdot (\boldsymbol{\nabla} \times \mathbf{f}) &= \langle \boldsymbol{\nabla} (\boldsymbol{\nabla} \times \mathbf{f}) \rangle \\ &= \langle \boldsymbol{\nabla}(-I)(\boldsymbol{\nabla} \wedge \mathbf{f}) \rangle \\ &= -\langle I\boldsymbol{\nabla} (\boldsymbol{\nabla} \wedge \mathbf{f}) \rangle \\ &= -I\boldsymbol{\nabla} \wedge (\boldsymbol{\nabla} \wedge \mathbf{f}) \\ &= 0,\end{aligned} \tag{2.182}$$

again, because $\boldsymbol{\nabla} \wedge (\boldsymbol{\nabla} \wedge \mathbf{f}) = 0$. □

2.4.2 Chain rule identities.

Lemma 2.2: Chain rule identities.

Let f be a scalar function and A be a k-blade, then

$$\boldsymbol{\nabla}(fA) = (\boldsymbol{\nabla} f)A + f(\boldsymbol{\nabla} A).$$

For A with grade $k > 0$, the grade $k - 1$ and $k + 1$ components of this product are

$$\begin{aligned}\boldsymbol{\nabla} \cdot (fA) &= (\boldsymbol{\nabla} f) \cdot A + f(\boldsymbol{\nabla} \cdot A) \\ \boldsymbol{\nabla} \wedge (fA) &= (\boldsymbol{\nabla} f) \wedge A + f(\boldsymbol{\nabla} \wedge A).\end{aligned}$$

For $\mathbb{R}^3$, and vector $A = \mathbf{A}$, the wedge product relation above can be written in dual form as

$$\boldsymbol{\nabla} \times (f\mathbf{A}) = (\boldsymbol{\nabla} f) \times \mathbf{A} + f(\boldsymbol{\nabla} \times \mathbf{A}).$$

Proving this is left to the reader.

Next up is another chain rule identity

2.4 VECTOR CALCULUS IDENTITIES.

> **Lemma 2.3: Gradient of dot product.**
>
> If $\mathbf{a}, \mathbf{b}$ are vectors, then
>
> $$\nabla\left(\mathbf{a}\cdot\mathbf{b}\right) = \left(\mathbf{a}\cdot\nabla\right)\mathbf{b} + \left(\mathbf{b}\cdot\nabla\right)\mathbf{a} + \left(\nabla\wedge\mathbf{b}\right)\cdot\mathbf{a} + \left(\nabla\wedge\mathbf{a}\right)\cdot\mathbf{b}$$
>
> For $\mathbb{R}^3$, this can be written as
>
> $$\nabla\left(\mathbf{a}\cdot\mathbf{b}\right) = \left(\mathbf{a}\cdot\nabla\right)\mathbf{b} + \left(\mathbf{b}\cdot\nabla\right)\mathbf{a} + \mathbf{a}\times\left(\nabla\times\mathbf{b}\right) + \mathbf{b}\times\left(\nabla\times\mathbf{a}\right)$$

Proof. We will use $\vec{\nabla}$ to indicate that the gradient operates on everything to the right, $\overset{\leftrightarrow}{\nabla}$ to indicate that the gradient operates bidirectionally, and $\nabla' A B'$ to indicate that the gradient's scope is limited to the ticked entity (just on B in this case.)

$$\begin{aligned}
\vec{\nabla}\left(\mathbf{a}\cdot\mathbf{b}\right) &= \left\langle\vec{\nabla}\left(\mathbf{ab}-\mathbf{a}\wedge\mathbf{b}\right)\right\rangle_1 \\
&= \left\langle\nabla'\mathbf{a}'\mathbf{b}+\nabla'\mathbf{ab}'\right\rangle_1 - \vec{\nabla}\cdot\left(\mathbf{a}\wedge\mathbf{b}\right) \\
&= \left(\nabla\cdot\mathbf{a}\right)\mathbf{b} + \left(\nabla\wedge\mathbf{a}\right)\cdot\mathbf{b} + \left\langle-\mathbf{a}\nabla\mathbf{b}+2\left(\mathbf{a}\cdot\nabla\right)\mathbf{b}\right\rangle_1 \quad (2.183) \\
&\quad - \nabla'\cdot\left(\mathbf{a}'\wedge\mathbf{b}\right) - \nabla'\cdot\left(\mathbf{a}\wedge\mathbf{b}'\right) \\
&= \left(\nabla\cdot\mathbf{a}\right)\mathbf{b} + \left(\nabla\wedge\mathbf{a}\right)\cdot\mathbf{b} - \mathbf{a}\left(\nabla\cdot\mathbf{b}\right) - \mathbf{a}\cdot\left(\nabla\wedge\mathbf{b}\right) \\
&\quad + 2\left(\mathbf{a}\cdot\nabla\right)\mathbf{b} - \nabla'\cdot\left(\mathbf{a}'\wedge\mathbf{b}\right) - \nabla'\cdot\left(\mathbf{a}\wedge\mathbf{b}'\right).
\end{aligned}$$

We are running out of room, and have not had any cancellation yet, so let's expand those last two terms separately

$$\begin{aligned}
-\nabla'\cdot&\left(\mathbf{a}'\wedge\mathbf{b}\right) - \nabla'\cdot\left(\mathbf{a}\wedge\mathbf{b}'\right) \\
&= -\left(\nabla'\cdot\mathbf{a}'\right)\mathbf{b} + \left(\nabla'\cdot\mathbf{b}\right)\mathbf{a}' - \left(\nabla'\cdot\mathbf{a}\right)\mathbf{b}' + \left(\nabla'\cdot\mathbf{b}'\right)\mathbf{a} \quad (2.184) \\
&= -\left(\nabla\cdot\mathbf{a}\right)\mathbf{b} + \left(\mathbf{b}\cdot\nabla\right)\mathbf{a} - \left(\mathbf{a}\cdot\nabla\right)\mathbf{b} + \left(\nabla\cdot\mathbf{b}\right)\mathbf{a}.
\end{aligned}$$

Now we can cancel some terms, leaving

$$\vec{\nabla}\left(\mathbf{a}\cdot\mathbf{b}\right) = \left(\nabla\wedge\mathbf{a}\right)\cdot\mathbf{b} - \mathbf{a}\cdot\left(\nabla\wedge\mathbf{b}\right) + \left(\mathbf{a}\cdot\nabla\right)\mathbf{b} + \left(\mathbf{b}\cdot\nabla\right)\mathbf{a}. \quad (2.185)$$

After adjustment of the order and sign of the second term, we see that this is the result we wanted. To show the $\mathbb{R}^3$ formulation, we have only apply theorem 1.7. □

Lemma 2.4: Divergence of a bivector.

Let $\mathbf{a}, \mathbf{b} \in \mathbb{R}^N$ be vectors. The divergence of their wedge can be written

$$\nabla \cdot (\mathbf{a} \wedge \mathbf{b}) = \mathbf{b}(\nabla \cdot \mathbf{a}) - \mathbf{a}(\nabla \cdot \mathbf{b}) - (\mathbf{b} \cdot \nabla)\mathbf{a} + (\mathbf{a} \cdot \nabla)\mathbf{b}.$$

For $\mathbb{R}^3$, this can also be written in triple cross product form

$$\nabla \cdot (\mathbf{a} \wedge \mathbf{b}) = -\nabla \times (\mathbf{a} \times \mathbf{b}).$$

Proof.

$$\begin{aligned}
\vec{\nabla} \cdot (\mathbf{a} \wedge \mathbf{b}) &= \nabla' \cdot (\mathbf{a}' \wedge \mathbf{b}) + \nabla' \cdot (\mathbf{a} \wedge \mathbf{b}') \\
&= (\nabla' \cdot \mathbf{a}')\mathbf{b} - (\nabla' \cdot \mathbf{b})\mathbf{a}' + (\nabla' \cdot \mathbf{a})\mathbf{b}' - (\nabla' \cdot \mathbf{b}')\mathbf{a} \quad (2.186)\\
&= (\nabla \cdot \mathbf{a})\mathbf{b} - (\mathbf{b} \cdot \nabla)\mathbf{a} + (\mathbf{a} \cdot \nabla)\mathbf{b} - (\nabla \cdot \mathbf{b})\mathbf{a}.
\end{aligned}$$

For the $\mathbb{R}^3$ part of the story, we have

$$\begin{aligned}
\nabla \times (\mathbf{a} \times \mathbf{b}) &= \langle -I(\nabla \wedge (\mathbf{a} \times \mathbf{b})) \rangle_1 \\
&= \langle -I\nabla (\mathbf{a} \times \mathbf{b}) \rangle_1 \\
&= \langle (-I)^2 \nabla (\mathbf{a} \wedge \mathbf{b}) \rangle_1 \\
&= -\nabla \cdot (\mathbf{a} \wedge \mathbf{b})
\end{aligned} \quad (2.187)$$

$\square$

Lemma 2.5: Curl of a wedge of gradients.

Let f, g, h be smooth functions with smooth derivatives. Then

$$\nabla \wedge (f(\nabla g \wedge \nabla h)) = \nabla f \wedge \nabla g \wedge \nabla h.$$

For $\mathbb{R}^3$ this can be written as

$$\nabla \cdot (f(\nabla g \times \nabla h)) = \nabla f \cdot (\nabla g \times \nabla h).$$

Proof. The GA identity follows by chain rule and application of lemma 2.1.

$$\begin{aligned}
\nabla \wedge (f(\nabla g \wedge \nabla h)) &= \nabla f \wedge (\nabla g \wedge \nabla h) + f \nabla \wedge (\nabla g \wedge \nabla h) \\
&= \nabla f \wedge \nabla g \wedge \nabla h.
\end{aligned} \quad (2.188)$$

2.4 VECTOR CALCULUS IDENTITIES.

The $\mathbb{R}^3$ part of the lemma can be shown using eq. (1.145), or we can compute it directly

$$\begin{aligned}\boldsymbol{\nabla}\cdot(f(\boldsymbol{\nabla} g\times\boldsymbol{\nabla} h)) &= \langle\boldsymbol{\nabla}(f(\boldsymbol{\nabla} g\times\boldsymbol{\nabla} h))\rangle \\ &= \boldsymbol{\nabla} f\cdot((\boldsymbol{\nabla} g\times\boldsymbol{\nabla} h)) + f\langle -I\boldsymbol{\nabla}(\boldsymbol{\nabla} g\wedge\boldsymbol{\nabla} h)\rangle \quad (2.189) \\ &= \boldsymbol{\nabla} f\cdot((\boldsymbol{\nabla} g\times\boldsymbol{\nabla} h)) - fI(\boldsymbol{\nabla}\wedge(\boldsymbol{\nabla} g\wedge\boldsymbol{\nabla} h)).\end{aligned}$$

The last term is clearly zero, since after our chain rule application, we end up with a $\boldsymbol{\nabla}\wedge\boldsymbol{\nabla}$ term on either branch of the chain rule expansion. □

Lemma 2.6: Curl of a bivector.

Let $\mathbf{a}, \mathbf{b}$ be vectors. The curl of their wedge is

$$\boldsymbol{\nabla}\wedge(\mathbf{a}\wedge\mathbf{b}) = \mathbf{b}\wedge(\boldsymbol{\nabla}\wedge\mathbf{a}) - \mathbf{a}\wedge(\boldsymbol{\nabla}\wedge\mathbf{b})$$

For $\mathbb{R}^3$, this can be expressed as the divergence of a cross product

$$\boldsymbol{\nabla}\cdot(\mathbf{a}\times\mathbf{b}) = \mathbf{b}\cdot(\boldsymbol{\nabla}\times\mathbf{a}) - \mathbf{a}\cdot(\boldsymbol{\nabla}\times\mathbf{b})$$

Proof. The GA case is a trivial chain rule application

$$\begin{aligned}\vec{\boldsymbol{\nabla}}\wedge(\mathbf{a}\wedge\mathbf{b}) &= (\boldsymbol{\nabla}'\wedge\mathbf{a}')\wedge\mathbf{b} + (\boldsymbol{\nabla}'\wedge\mathbf{a})\wedge\mathbf{b}' \\ &= \mathbf{b}\wedge(\boldsymbol{\nabla}\wedge\mathbf{a}) - \mathbf{a}\wedge(\boldsymbol{\nabla}\wedge\mathbf{b}).\end{aligned} \quad (2.190)$$

The $\mathbb{R}^3$ case, is less obvious by inspection, but follows from eq. (1.145). □

2.4.3 Problems.

Exercise 2.12 Curl example in polar coordinates.

Find the polar form of eq. (2.171) and then compute the curl, using the polar representation of the gradient found in eq. (2.87).

Exercise 2.13 Some bivector curls.

For each B compute the curl $\boldsymbol{\nabla}\wedge B$.
  a. $B = x\mathbf{e}_{23} + y\mathbf{e}_{31} + z\mathbf{e}_{12}$,
  b. $B = x\mathbf{e}_{31}$,
  c. $B = xyz\mathbf{e}_{23}$.

2.5 MULTIVECTOR FOURIER TRANSFORM AND PHASORS.

It will often be convenient to utilize time harmonic (frequency domain) representations. This can be achieved by utilizing Fourier transform pairs or with a phasor representation.

We may define Fourier transform pairs of multivector fields and sources in the conventional fashion

> **Definition 2.9: Multivector Fourier transform pair.**
>
> The *Fourier transform pair* for a multivector function $F(\mathbf{x}, t)$ will be written as
>
> $$F(\mathbf{x}, t) = \int F_\omega(\mathbf{x}) e^{j\omega t} d\omega$$
>
> $$F_\omega(\mathbf{x}) = \frac{1}{2\pi} \int F(\mathbf{x}, t) e^{-j\omega t} dt,$$
>
> where j is an arbitrary scalar imaginary that commutes with all multivectors.

In these transform pairs, the imaginary j need not be represented by any geometrical imaginary such as $\mathbf{e}_{12}$. In particular, we need not assume that the representation of j is the $\mathbb{R}^3$ pseudoscalar I, despite the fact that I does commute with all $\mathbb{R}^3$ multivectors. We wish to have the freedom to assume that non-geometric real and imaginary operations can be performed without picking or leaving out any specific grade pseudoscalar components of the multivector fields or sources, so we won't impose any a-priori restrictions on the representations of j. In particular, this provides the freedom to utilize phasor (fixed frequency) representations of our multivector functions. We will use the engineering convention for our phasor representations, where assuming a complex exponential time dependence of the following form is assumed

> **Definition 2.10: Multivector phasor representation.**
>
> The *phasor representation* $F(\mathbf{x})$ of a multivector valued (real) function $F(\mathbf{x}, t)$ is defined implicitly as
>
> $$F(\mathbf{x}, t) = \operatorname{Re}\left(F(\mathbf{x}) e^{j\omega t}\right),$$

where j is an arbitrary scalar imaginary that commutes with all multivectors.

The complex valued multivector $f(\mathbf{x})$ is still generated from the real Euclidean basis for $\mathbb{R}^3$, so there will be no reason to introduce complex inner products spaces into the mix.

The reader must take care when reading any literature that utilizes Fourier transforms or phasor representation, since the conventions vary. In particular the physics representation of a phasor typically uses the opposite sign convention $F(\mathbf{x}, t) = \text{Re}\left(F(\mathbf{x})e^{-i\omega t}\right)$, which toggles the sign of all the imaginaries in derived results.

2.6 GREEN'S FUNCTIONS.

2.6.1 *Motivation.*

Every engineer's toolbox includes Laplace and Fourier transform tricks for transforming differential equations to the frequency domain. Here the space and time domain equivalent of the frequency and time domain linear system response function, called the Green's function, is introduced.

Everybody's favorite differential equation, the harmonic oscillator, can be used as an illustrative example

$$x'' + 2kx' + (\omega_0)^2 x = f(t). \tag{2.196}$$

Here derivatives are with respect to time, ω_0 is the intrinsic angular frequency of the oscillator, k is a damping factor, and $f(t)$ is a forcing function. If the oscillator represents a child on a swing at the park (a pendulum system), then k represents the friction in the swing pivot and retardation due to wind, and the forcing function represents the father pushing the swing. The forcing function $f(t)$ could include an initial impulse to get the child up to speed, or could have a periodic aspect, such as the father running underdogs[5] as the child gleefully shouts "Again, again, again!"

The full solution of this problem is $x(t) = x_s(t) + x_0(t)$, where $x_s(t)$ is a solution of eq. (2.196) and x_0 is any solution of the homogeneous

---

5 The underdog is a non-passive swing pushing technique, where you run behind and under the swing and child, giving a push as you go. Before my kids learned to "pump their legs", and even afterwards, this was their favorite way of being pushed on the swing. With two kids the Dad-forcing-function tires quickly, as it is applied repeatedly to both oscillating children.

equation $x_0'' + 2kx_0' + (\omega_0)^2 x_0 = 0$, picked to satisfy the boundary value constraints of the problem.

Let's attack this problem initially with Fourier transforms (definition 2.9)

We can assume zero position and velocity for the non-homogeneous problem, since we can adjust the boundary conditions with the homogeneous solution $x_0(t)$. With zero boundary conditions on x, x', the transform of eq. (2.196) is

$$((j\omega)^2 + 2j\omega k + (\omega_0)^2)X(\omega) = F(\omega), \tag{2.197}$$

so the system is solved in the frequency domain by the system response function $G(\omega)$

$$X(\omega) = G(\omega)F(\omega), \tag{2.198}$$

where

$$G(\omega) = \frac{-1}{\omega^2 - 2j\omega k - (\omega_0)^2}. \tag{2.199}$$

We can apply the inverse transformation to find the time domain solution for the forced oscillator problem.

$$\begin{aligned} x(t) &= \int d\omega G(\omega) F(\omega) e^{j\omega t} \\ &= \int d\omega G(\omega) \left(\frac{1}{2\pi} \int dt' f(t') e^{-j\omega t'} dt' \right) e^{j\omega t} \\ &= \int dt' f(t') \left(\frac{1}{2\pi} \int d\omega G(\omega) e^{j\omega(t-t')} \right). \end{aligned} \tag{2.200}$$

The frequency domain integral is the Green's function. We'll write this as

$$G(t, t') = \frac{1}{2\pi} \int d\omega G(\omega) e^{j\omega(t-t')}. \tag{2.201}$$

If we can evaluate this integral (exercise 2.14), then the system can be considered solved, where a solution is given by the convolution integral

$$x(t) = \int dt' f(t') G(t, t') + x_0(t). \tag{2.202}$$

The Green's function is the weighting factor that determines how much of $f(t')$ for each time t' contributes to the motion of the system that is

2.6 GREEN'S FUNCTIONS.

explicitly due to the forcing function. Green's functions for physical problems are causal, so only forcing events in the past contribute to the current state of the system (i.e. if you were to depend on only future pushes of the swing, you would have a very bored child.)

An alternate way of viewing a linear systems problem is to assume that a convolution solution of the form eq. (2.202) must exist. Since the equation is a linear, it is reasonable to assume that a linear weighted sum of all the forcing function values must contribute to the solution. If such a solution is assumed, then we can operate on that solution with the differential operator for the problem. For our harmonic oscillator problem that operator is

$$\mathcal{L} = \frac{\partial^2}{\partial t^2} + 2k\frac{\partial}{\partial t} + (\omega_0)^2. \tag{2.203}$$

We find

$$\begin{aligned} f(t) &= \left(\frac{\partial^2}{\partial t^2} + 2k\frac{\partial}{\partial t} + (\omega_0)^2\right) x(t) \\ &= \left(\frac{\partial^2}{\partial t^2} + 2k\frac{\partial}{\partial t} + (\omega_0)^2\right) \int dt' f(t') G(t, t') \\ &= \int dt' f(t') \left(\frac{\partial^2}{\partial t^2} + 2k\frac{\partial}{\partial t} + (\omega_0)^2\right) G(t, t'), \end{aligned} \tag{2.204}$$

and see that the Green's function, when acted on by the differential operator, must have the characteristics of a delta function

$$\left(\frac{\partial^2}{\partial t^2} + 2k\frac{\partial}{\partial t} + (\omega_0)^2\right) G(t, t') = \delta(t - t'). \tag{2.205}$$

The problem of determining the Green's function, implicitly determining the solution of any forced system, can be viewed as seeking the solution of distribution equations of the form

$$\boxed{\mathcal{L}G(t, t') = \delta(t - t').} \tag{2.206}$$

Framing the problem this way is independent of whatever techniques (transform or other) that we may choose to use to determine the structure of the Green's function itself. Observe that the Green's function itself is not unique. In particular, we may add any solution of the homogeneous problem $\mathcal{L}G_0(t, t') = 0$ to the Green's function, just as we can do so for the forced system itself.

We will see that Green's functions provide a general method of solving many of the linear differential equations that will be encountered in electromagnetism.

Exercise 2.14   Harmonic oscillator Green's function.

Evaluate the integral

$$G(\tau) = \frac{-1}{2\pi} \int_{-\infty}^{\infty} \frac{1}{\omega^2 - 2j\omega k - \omega_0^2} e^{j\omega\tau} d\omega, \qquad (2.207)$$

using the semicircular infinite contours depicted in fig. 2.18.

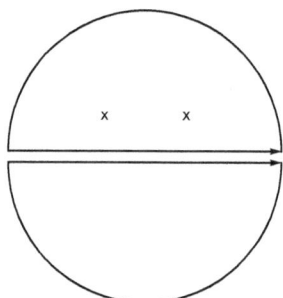

Figure 2.18: Contours for harmonic oscillator Green's function.

2.6.1.1 *Time domain problems in electromagnetism*

Examples of the PDEs that we can apply Green's function techniques to include

$$\left(\nabla + \frac{1}{c}\frac{\partial}{\partial t}\right) F(\mathbf{x}, t) = J(\mathbf{x}, t) \qquad (2.213a)$$

$$\left(\nabla^2 - \frac{1}{c^2}\frac{\partial^2}{\partial t^2}\right) F(\mathbf{x}, t) = \left(\nabla - \frac{1}{c}\frac{\partial}{\partial t}\right)\left(\nabla + \frac{1}{c}\frac{\partial}{\partial t}\right) F(\mathbf{x}, t) = B(\mathbf{x}, t). \quad (2.213b)$$

The reader is no doubt familiar with the wave equation (eq. (2.213b)), where F is the waving function, and B is the forcing function. Scalar and vector valued wave equations are encountered in scalar and vector forms in conventional electromagnetism. We will see multivector variations of the wave equation, so it should be assumed that F and B are multivector valued.

Equation (2.213a) is actually the geometric algebra form of Maxwell's equation (singular), where F is a multivector with grades 1 and 2, and J is a multivector containing all the charge and current density contributions. We will call the operator in eq. (2.213a) the spacetime gradient[6].

Armed with Fourier transform or phasor representations, the frequency domain representations of eq. (2.213) are found to be

$$(\boldsymbol{\nabla} + jk) F(\mathbf{x}) = J(\mathbf{x}) \tag{2.214a}$$

$$\left(\boldsymbol{\nabla}^2 + k^2\right) F(\mathbf{x}) = (\boldsymbol{\nabla} - jk)(\boldsymbol{\nabla} + jk) F(\mathbf{x}) = B(\mathbf{x}), \tag{2.214b}$$

where $k = \omega/c$, and any explicit frequency dependence in our transform pairs has been suppressed. We will call these equations the first and second order Helmholtz equations respectively. The first order equation applies a multivector differential operator to a multivector field, which must equal the multivector forcing function (the sources).

For statics problems ($k = 0$), we may work with real fields and sources, dispensing with any need to take real parts.

2.6.2 Green's function solutions.

2.6.2.1 Unbounded.

The operators in eq. (2.213), and eq. (2.214) all have a similar linear structure. Abstracting that structure, all these problems have the form

$$\mathcal{L} F(\mathbf{x}) = J(\mathbf{x}), \tag{2.215}$$

where $\mathcal{L}$ is an operator formed from a linear combination of linear operators $1, \boldsymbol{\nabla}, \boldsymbol{\nabla}^2, \partial_t, \partial_{tt}$.

Given the linear structure of the PDE that we wish to solve, it makes sense to assume that the solutions also have a linear structure. The most general such solution we can assume has the form

$$F(\mathbf{x}, t) = \int G(\mathbf{x}, \mathbf{x}'; t, t') J(\mathbf{x}', t') dV' dt' + F_0(\mathbf{x}, t), \tag{2.216}$$

---

6 A slightly different operator is also called the spacetime gradient in STA (Space Time Algebra) [6], which employs a non-Euclidean basis to generate a four dimensional relativistic geometric algebra. Our spacetime gradient is related to the STA spacetime gradient by a constant factor.

where $F_0(\mathbf{x}, t)$ is any solution to the equivalent homogeneous equation $\mathcal{L} F_0 = 0$, and $G(\mathbf{x}, \mathbf{x}'; t, t')$ is the Green's function (to be determined) associated with eq. (2.215). Operating on the presumed solution eq. (2.216) with $\mathcal{L}$ yields

$$J(\mathbf{x}, t) = \mathcal{L} F(\mathbf{x}, t) = \mathcal{L} \left(\int G(\mathbf{x}, \mathbf{x}'; t, t') J(\mathbf{x}', t') dV' dt' + F_0(\mathbf{x}, t) \right) \\ = \int \left(\mathcal{L} G(\mathbf{x}, \mathbf{x}'; t, t') \right) J(\mathbf{x}', t') dV' dt', \quad (2.217)$$

which shows that we require the Green's function to have delta function semantics satisfying

$$\mathcal{L} G(\mathbf{x}, \mathbf{x}'; t, t') = \delta(\mathbf{x} - \mathbf{x}') \delta(t - t'). \quad (2.218)$$

The scalar valued Green's functions for the Laplacian and the (2nd order) Helmholtz equations are well known. The Green's functions for the spacetime gradient and the 1st order Helmholtz equation (which is just the gradient when $k = 0$) are multivector valued and will be derived here.

2.6.2.2 Green's theorem.

When the presumed solution is a superposition of only states in a bounded region then life gets a bit more interesting. For instance, consider a problem for which the differential operator is a function of space only, with a presumed solution such as

$$F(\mathbf{x}) = \int_V dV' B(\mathbf{x}') G(\mathbf{x}, \mathbf{x}') + F_0(\mathbf{x}), \quad (2.219)$$

then life gets a bit more interesting. This sort of problem requires different treatment for operators that are first and second order in the gradient.

For the second order problems, we require Green's theorem, which must be generalized slightly for use with multivector fields.

The basic idea is that we can relate the Laplacian of the Green's function and the field $F(\mathbf{x}') \left((\nabla')^2 G(\mathbf{x}, \mathbf{x}') \right) = G(\mathbf{x}, \mathbf{x}') \left((\nabla')^2 F(\mathbf{x}') \right) + \cdots$. That relationship can be expressed as the integral of an antisymmetric sandwich of the two functions

Theorem 2.16: Green's theorem

2.6 GREEN'S FUNCTIONS. 161

Given a multivector function F and a scalar function G

$$\int_V \left(F\nabla^2 G - G\nabla^2 F \right) dV = \int_{\partial V} \left(F\hat{\mathbf{n}} \cdot \nabla G - G\hat{\mathbf{n}} \cdot \nabla F \right),$$

where ∂V is the boundary of the volume V.

Proof. A straightforward, but perhaps inelegant way of proving this theorem is to expand the antisymmetric product in coordinates

$$\begin{aligned} F\nabla^2 G - G\nabla^2 F &= \sum_k F\partial_k \partial_k G - G\partial_k \partial_k F \\ &= \sum_k \partial_k \left(F\partial_k G - G\partial_k F \right) - (\partial_k F)(\partial_k G) + (\partial_k G)(\partial_k F). \end{aligned}$$
(2.220)

Since G is a scalar, the last two terms cancel, and we can integrate

$$\int_V \left(F\nabla^2 G - G\nabla^2 F \right) dV = \sum_k \int_V \partial_k \left(F\partial_k G - G\partial_k F \right). \qquad (2.221)$$

Each integral above involves one component of the gradient. From theorem 2.3 we know that

$$\int_V \nabla Q \, dV = \int_{\partial V} \hat{\mathbf{n}} Q \, dA, \qquad (2.222)$$

for any multivector Q. Equating components gives

$$\int_V \partial_k Q \, dV = \int_{\partial V} \hat{\mathbf{n}} \cdot \mathbf{e}_k Q \, dA, \qquad (2.223)$$

which can be substituted into eq. (2.221) to find

$$\begin{aligned} \int_V \left(F\nabla^2 G - G\nabla^2 F \right) dV &= \sum_k \int_{\partial V} \hat{\mathbf{n}} \cdot \mathbf{e}_k \left(F\partial_k G - G\partial_k F \right) dA \\ &= \int_{\partial V} \left(F(\hat{\mathbf{n}} \cdot \nabla) G - G(\hat{\mathbf{n}} \cdot \nabla) F \right) dA. \end{aligned}$$
(2.224)

$\square$

2.6.2.3 Bounded solutions to first order problems.

For first order problems we will need an intermediate result similar to Green's theorem.

Lemma 2.7: Normal relations for a gradient sandwich.

Given multivector functions $F(\mathbf{x}'), G(\mathbf{x}, \mathbf{x}')$, and a gradient $\boldsymbol{\nabla}'$ acting bidirectionally on functions of $\mathbf{x}'$, we have

$$-\int_V \left(G(\mathbf{x}, \mathbf{x}') \overleftarrow{\boldsymbol{\nabla}}' \right) F(\mathbf{x}')dV' = \int_V G(\mathbf{x}, \mathbf{x}') \left(\overrightarrow{\boldsymbol{\nabla}}' F(\mathbf{x}') \right) dV'$$
$$-\int_{\partial V} G(\mathbf{x}, \mathbf{x}') \hat{\mathbf{n}}' F(\mathbf{x}') dA'.$$

Proof. This follows directly from theorem 2.3

$$\int_{\partial V} G(\mathbf{x}, \mathbf{x}') \hat{\mathbf{n}}' F(\mathbf{x}') dA' = \int_V G(\mathbf{x}, \mathbf{x}') \overleftrightarrow{\boldsymbol{\nabla}}' F(\mathbf{x}') dV'$$
$$= \int_V \left(G(\mathbf{x}, \mathbf{x}') \overleftarrow{\boldsymbol{\nabla}}' \right) F(\mathbf{x}')dV' + \int_V G(\mathbf{x}, \mathbf{x}') \left(\overrightarrow{\boldsymbol{\nabla}}' F(\mathbf{x}') \right) dV',$$
(2.225)

which can be rearranged to prove lemma 2.7. □

2.6.3 *Helmholtz equation.*

2.6.3.1 *Unbounded superposition solutions for the Helmholtz equation.*

The specialization of eq. (2.218) to the Helmholtz equation eq. (2.214b) is

$$\left(\boldsymbol{\nabla}^2 + k^2\right) G(\mathbf{x}, \mathbf{x}') = \delta(\mathbf{x} - \mathbf{x}'). \tag{2.226}$$

While it is possible [21] to derive the Green's function using Fourier transform techniques, we will state the result instead, which is well known

Theorem 2.17: Green's function for the Helmholtz operator.

The advancing (causal), and the receding (acausal) Green's functions satisfying eq. (2.226) are respectively

$$G_{\text{adv}}(\mathbf{x}, \mathbf{x}') = -\frac{e^{-jk\|\mathbf{x}-\mathbf{x}'\|}}{4\pi \|\mathbf{x} - \mathbf{x}'\|}$$

$$G_{\text{rec}}(\mathbf{x}, \mathbf{x}') = -\frac{e^{jk\|\mathbf{x}-\mathbf{x}'\|}}{4\pi \|\mathbf{x} - \mathbf{x}'\|}.$$

We will use the advancing (causal) Green's function, and refer to this function as $G(\mathbf{x}, \mathbf{x}')$ without any subscript. Because it may not be obvious that these Green's function representations are valid in a multivector context, a demonstration of this fact can be found in section C.1.

Observe that as a special case, the Helmholtz Green's function reduces to the Green's function for the Laplacian when $k = 0$

$$G(\mathbf{x}, \mathbf{x}') = -\frac{1}{4\pi \|\mathbf{x} - \mathbf{x}'\|}. \tag{2.227}$$

2.6.3.2 Bounded superposition solutions for the Helmholtz equation.

For our application of theorem 2.17 to the Helmholtz problem, we are actually interested in a antisymmetric sandwich of the Helmholtz operator by the function F and the scalar (Green's) function G, but that reduces to an asymmetric sandwich of our functions around the Laplacian

$$\begin{aligned} F\left(\nabla^2 + k^2\right)G - G\left(\nabla^2 + k^2\right)F \\ = F\nabla^2 G + \cancel{Fk^2 G} - G\nabla^2 F - \cancel{Gk^2 F} \\ = F\nabla^2 G - G\nabla^2 F, \end{aligned} \tag{2.228}$$

so

$$\begin{aligned} \int_V F(\mathbf{x}') \left((\nabla')^2 + k^2\right) G(\mathbf{x}, \mathbf{x}') \\ = \int_V G(\mathbf{x}, \mathbf{x}') \left((\nabla')^2 + k^2\right) F(\mathbf{x}') dV' \\ + \int_{\partial V} \left(F(\mathbf{x}')(\hat{\mathbf{n}}' \cdot \nabla')G(\mathbf{x}, \mathbf{x}') - G(\mathbf{x}, \mathbf{x}')(\hat{\mathbf{n}}' \cdot \nabla')F(\mathbf{x}')\right) dA'. \end{aligned} \tag{2.229}$$

This shows that if we assume the Green's function satisfies the delta function condition eq. (2.226), then the general solution of eq. (2.214b) is formed from a bounded superposition of sources is

$$\begin{aligned} F(\mathbf{x}) = &\int_V G(\mathbf{x}, \mathbf{x}') B(\mathbf{x}') dV' \\ &+ \int_{\partial V} \left(G(\mathbf{x}, \mathbf{x}')(\hat{\mathbf{n}}' \cdot \boldsymbol{\nabla}') F(\mathbf{x}') - F(\mathbf{x}')(\hat{\mathbf{n}}' \cdot \boldsymbol{\nabla}') G(\mathbf{x}, \mathbf{x}') \right) dA'. \end{aligned}$$

(2.230)

We are also free to add in any specific solution $F_0(\mathbf{x})$ that satisfies the homogeneous Helmholtz equation. There is also freedom to add any solution of the homogeneous Helmholtz equation to the Green's function itself, so it is not unique. For a bounded superposition we generally desire that the solution F and its normal derivative, or the Green's function G (and it's normal derivative) or an appropriate combination of the two are zero on the boundary, so that the surface integral is killed.

2.6.4 First order Helmholtz equation.

The specialization of eq. (2.218) to the first order Helmholtz equation eq. (2.214a) is

$$(\boldsymbol{\nabla} + jk) G(\mathbf{x}, \mathbf{x}') = \delta(\mathbf{x} - \mathbf{x}'). \tag{2.231}$$

This Green's function is multivector valued

Theorem 2.18: Green's function for 1st order Helmholtz operator.

The *Green's function for the first order Helmholtz operator* $\boldsymbol{\nabla} + jk$ satisfies

$$\left(\overrightarrow{\boldsymbol{\nabla}} + jk\right) G(\mathbf{x}, \mathbf{x}') = G(\mathbf{x}, \mathbf{x}') \left(-\overleftarrow{\boldsymbol{\nabla}}' + jk\right) = \delta(\mathbf{x} - \mathbf{x}'),$$

and has the value

$$G(\mathbf{x}, \mathbf{x}') = \frac{e^{-jkr}}{4\pi r} \left(jk(1 + \hat{\mathbf{r}}) + \frac{\hat{\mathbf{r}}}{r} \right),$$

2.6 GREEN'S FUNCTIONS.

where $\mathbf{r} = \mathbf{x} - \mathbf{x}'$, $r = \|\mathbf{r}\|$ and $\hat{\mathbf{r}} = \mathbf{r}/r$, and $\boldsymbol{\nabla}'$ denotes differentiation with respect to $\mathbf{x}'$.

A special but important case is the $k = 0$ condition, which provides the Green's function for the gradient, which is vector valued

$$G(\mathbf{x}, \mathbf{x}'; k = 0) = \frac{1}{4\pi} \frac{\hat{\mathbf{r}}}{r^2}. \tag{2.232}$$

Proof. If we denote the (advanced) Green's function for the 2nd order Helmholtz operator theorem 2.17 as $\phi(\mathbf{x}, \mathbf{x}')$, we must have

$$\left(\vec{\boldsymbol{\nabla}} + jk\right) G(\mathbf{x}, \mathbf{x}') = \delta(\mathbf{x} - \mathbf{x}') = \left(\vec{\boldsymbol{\nabla}} + jk\right)\left(\vec{\boldsymbol{\nabla}} - jk\right)\phi(\mathbf{x}, \mathbf{x}'), \tag{2.233}$$

we see that the Green's function is given by

$$G(\mathbf{x}, \mathbf{x}') = \left(\vec{\boldsymbol{\nabla}} - jk\right)\phi(\mathbf{x}, \mathbf{x}'). \tag{2.234}$$

This can be computed directly

$$\begin{aligned}
G(\mathbf{x}, \mathbf{x}') &= \left(\vec{\boldsymbol{\nabla}} - jk\right)\left(-\frac{e^{-jkr}}{4\pi r}\right) \\
&= \left(\hat{\mathbf{r}}\frac{\partial}{\partial r} - jk\right)\left(-\frac{e^{-jkr}}{4\pi r}\right) \\
&= \frac{-e^{-jkr}}{4\pi}\left(\hat{\mathbf{r}}\left(-\frac{jk}{r} - \frac{1}{r^2}\right) - \frac{jk}{r}\right) \\
&= \frac{e^{-jkr}}{4\pi}\left(jk\left(1 + \hat{\mathbf{r}}\right) + \frac{\hat{\mathbf{r}}}{r}\right),
\end{aligned} \tag{2.235}$$

as claimed. $\square$

Observe that since ϕ is scalar valued, we can also rewrite eq. (2.234) in terms of a right acting operator

$$\begin{aligned}
G(\mathbf{x}, \mathbf{x}') &= \phi(\mathbf{x}, \mathbf{x}')\left(\overleftarrow{\boldsymbol{\nabla}} - jk\right) \\
&= \phi(\mathbf{x}, \mathbf{x}')\left(-\overleftarrow{\boldsymbol{\nabla}}' - jk\right),
\end{aligned} \tag{2.236}$$

so

$$G(\mathbf{x}, \mathbf{x}')\left(-\overleftarrow{\boldsymbol{\nabla}}' + jk\right) = \phi(\mathbf{x}, \mathbf{x}')\left((\overleftarrow{\boldsymbol{\nabla}}')^2 + k^2\right) = \delta(\mathbf{x} - \mathbf{x}'). \tag{2.237}$$

This is relevant for bounded superposition states, which we will discuss next now that the proof of theorem 2.18 is complete. In particular addition of $\int_V G(\mathbf{x}, \mathbf{x}') jk F(\mathbf{x}') dV'$ to both sides of lemma 2.7 gives

$$\int_V \left(G(\mathbf{x}, \mathbf{x}') \left(-\overleftarrow{\nabla}' + jk \right) \right) F(\mathbf{x}') dV' = \int_V G(\mathbf{x}, \mathbf{x}') \left(\left(\overrightarrow{\nabla}' + jk \right) F(\mathbf{x}') \right) dV' - \int_{\partial V} G(\mathbf{x}, \mathbf{x}') \hat{\mathbf{n}}' F(\mathbf{x}') dA'. \quad (2.238)$$

Utilizing theorem 2.18, and substituting $J(\mathbf{x}')$ from eq. (2.214a), we find that one solution to the first order Helmholtz equation is

$$F(\mathbf{x}) = \int_V G(\mathbf{x}, \mathbf{x}') J(\mathbf{x}') dV' - \int_{\partial V} G(\mathbf{x}, \mathbf{x}') \hat{\mathbf{n}}' F(\mathbf{x}') dA'. \quad (2.239)$$

We are free to add any specific solution F_0 that satisfies the homogeneous equation $(\nabla + jk) F_0 = 0$.

2.6.5 Spacetime gradient.

We want to find the Green's function that solves spacetime gradient equations of the form eq. (2.213a). For the wave equation operator, it is helpful to introduce a d'Alembertian operator, defined as follows.

Definition 2.11: d'Alembertian (wave equation) operator.

In this book, the symbol $\square$ is used to represent the *d'Alembertian (wave equation) operator*, with a positive sign on the Laplacian term

$$\square = \left(\nabla - \frac{1}{c} \frac{\partial}{\partial t} \right) \left(\nabla + \frac{1}{c} \frac{\partial}{\partial t} \right) = \nabla^2 - \frac{1}{c^2} \frac{\partial^2}{\partial t^2}.$$

We will be able to derive the Green's function for the spacetime gradient from the Green's function for the d'Alembertian. The Green's function for the spacetime gradient is multivector valued and given by the following.

Theorem 2.19: Green's function for the spacetime gradient.

2.6 GREEN'S FUNCTIONS.

The *Green's function for the spacetime gradient* $\nabla + (1/c)\partial_t$ satisfies

$$\left(\nabla + \frac{1}{c}\frac{\partial}{\partial t}\right) G(\mathbf{x} - \mathbf{x}', t - t') = \delta(\mathbf{x} - \mathbf{x}')\delta(t - t'),$$

and has the value

$$G(\mathbf{x} - \mathbf{x}', t - t') = \frac{1}{4\pi}\left(-\frac{\hat{\mathbf{r}}}{r^2}\frac{\partial}{\partial r} + \frac{\hat{\mathbf{r}}}{r^2} + \frac{1}{cr}\frac{\partial}{\partial t}\right)\delta(-r/c + t - t'),$$

where $\mathbf{r} = \mathbf{x} - \mathbf{x}'$, $r = \|\mathbf{r}\|$ and $\hat{\mathbf{r}} = \mathbf{r}/r$.

With the help of eq. (C.18) it is possible to further evaluate the delta function derivatives, however, we will defer doing so until we are ready to apply this Green's function in a convolution integral to solve Maxwell's equation.

Proof. To prove this result, let $\phi(\mathbf{x} - \mathbf{x}', t - t')$ be the retarded time (causal) Green's function for the wave equation, satisfying

$$\Box \phi(\mathbf{x} - \mathbf{x}', t - t') = \left(\nabla + \frac{1}{c}\frac{\partial}{\partial t}\right)\left(\nabla - \frac{1}{c}\frac{\partial}{\partial t}\right)\phi(\mathbf{x} - \mathbf{x}', t - t') \quad (2.240)$$
$$= \delta(\mathbf{x} - \mathbf{x}')\delta(t - t').$$

This function has the value

$$\phi(\mathbf{r}, t - t') = -\frac{1}{4\pi r}\delta(-r/c + t - t'), \quad (2.241)$$

where $\mathbf{r} = \mathbf{x} - \mathbf{x}'$, $r = \|\mathbf{r}\|$. Derivations of this Green's function, and it's acausal advanced time friend, can be found in [21], [15], and use the usual Fourier transform and contour integration tricks.

Comparing eq. (2.240) to the defining statement of theorem 2.19, we see that the spacetime gradient Green's function is given by

$$G(\mathbf{x} - \mathbf{x}', t - t') = \left(\nabla - \frac{1}{c}\frac{\partial}{\partial t}\right)\phi(\mathbf{r}, t - t')$$
$$= \left(\hat{\mathbf{r}}\frac{\partial}{\partial r} - \frac{1}{c}\frac{\partial}{\partial t}\right)\phi(\mathbf{r}, t - t'), \quad (2.242)$$

where $\hat{\mathbf{r}} = \mathbf{r}/r$. Evaluating the derivatives gives

$$G(\mathbf{r}, t - t') = -\frac{1}{4\pi}\left(\hat{\mathbf{r}}\frac{\partial}{\partial r} - \frac{1}{c}\frac{\partial}{\partial t}\right)\frac{\delta(-r/c + t - t')}{r}$$
$$= -\frac{1}{4\pi}\left(\frac{\hat{\mathbf{r}}}{r}\frac{\partial}{\partial r}\delta(-r/c + t - t') - \frac{\hat{\mathbf{r}}}{r^2}\delta(-r/c + t - t') - \frac{1}{cr}\frac{\partial}{\partial t}\delta(-r/c + t - t')\right),$$

(2.243)

which completes the proof after some sign cancellation and minor rearrangement. □

2.7 HELMHOLTZ THEOREM.

In conventional electromagnetism Maxwell's equations are posed in terms of separate divergence and curl equations. It is therefore desirable to show that the divergence and curl of a function and it's normal characteristics on the boundary of an integration volume determine that function uniquely. This is known as the Helmholtz theorem

---

Theorem 2.20: Helmholtz first theorem.

A vector $\mathbf{M}$ is uniquely determined by its divergence

$$\nabla \cdot \mathbf{M} = s,$$

and curl

$$\nabla \times \mathbf{M} = \mathbf{C},$$

and its value over the boundary.

The conventional proof of Helmholtz's theorem uses the Green's function for the (second order) Helmholtz operator. Armed with a vector valued Green's function for the gradient, a first order proof is also possible. As illustrations of the geometric integration theory developed in this chapter, both strategies will be applied here to this problem.

In either case, we start by forming an even grade multivector (gradient) equation containing both the dot and cross product contributions

$$\nabla \mathbf{M} = \nabla \cdot \mathbf{M} + I \nabla \times \mathbf{M} = s + I \mathbf{C}. \tag{2.244}$$

First order proof. For the first order case, we perform a grade one selection of lemma 2.7, setting $F = \mathbf{M}$ where G is the Green's function for the gradient given by eq. (2.232). The proof follows directly

2.7 HELMHOLTZ THEOREM.

$$\begin{aligned} M(\mathbf{x}) &= -\int_V \left(G(\mathbf{x},\mathbf{x}') \overleftarrow{\nabla}' \right) \mathbf{M}(\mathbf{x}') dV' \\ &= \int_V \left\langle G(\mathbf{x},\mathbf{x}') \left(\overrightarrow{\nabla}' \mathbf{M}(\mathbf{x}') \right) \right\rangle_1 dV' - \int_{\partial V} \left\langle G(\mathbf{x},\mathbf{x}') \hat{\mathbf{n}}' \mathbf{M}(\mathbf{x}') \right\rangle_1 dA' \\ &= \int_V \frac{1}{4\pi \|\mathbf{x}-\mathbf{x}'\|^3} \left\langle (\mathbf{x}-\mathbf{x}')(s(\mathbf{x}') + I\mathbf{C}(\mathbf{x}')) \right\rangle_1 dV' \\ &\quad - \int_{\partial V} \frac{1}{4\pi \|\mathbf{x}-\mathbf{x}'\|^3} \left\langle (\mathbf{x}-\mathbf{x}')\hat{\mathbf{n}}' \mathbf{M}(\mathbf{x}') \right\rangle_1 dA' \\ &= \int_V \frac{1}{4\pi \|\mathbf{x}-\mathbf{x}'\|^3} \left((\mathbf{x}-\mathbf{x}')s(\mathbf{x}') - (\mathbf{x}-\mathbf{x}') \times \mathbf{C}(\mathbf{x}') \right) dV' \\ &\quad - \int_{\partial V} \frac{1}{4\pi \|\mathbf{x}-\mathbf{x}'\|^3} \left\langle (\mathbf{x}-\mathbf{x}')\hat{\mathbf{n}}' \mathbf{M}(\mathbf{x}') \right\rangle_1 dA'. \end{aligned}$$
(2.245)

If **M** is well behaved enough that the boundary integral vanishes on an infinite surface, we see that **M** is completely specified by the divergence and the curl. In general, the divergence and the curl, must also be supplemented by the value of vector valued function on the boundary.

Observe that the boundary integral has a particularly simple form for a spherical surface or radius R centered on $\mathbf{x}'$. Switching to spherical coordinates $\mathbf{r} = \mathbf{x}' - \mathbf{x} = R\hat{\mathbf{r}}(\theta,\phi)$ where $\hat{\mathbf{r}} = (\mathbf{x}'-\mathbf{x})/\|\mathbf{x}'-\mathbf{x}\|$ is the outwards normal, we have

$$\begin{aligned} &- \int_{\partial V} \frac{1}{4\pi \|\mathbf{x}-\mathbf{x}'\|^3} \left\langle (\mathbf{x}-\mathbf{x}')\hat{\mathbf{n}}' \mathbf{M}(\mathbf{x}') \right\rangle_1 dA' \\ &= \int_{\partial V} \frac{\mathbf{M}(\mathbf{x}')}{4\pi \|\mathbf{x}-\mathbf{x}'\|^2} dA' \\ &= \frac{1}{4\pi} \int_{\theta=0}^{\pi} \int_{\phi=0}^{2\pi} \mathbf{M}(R,\theta,\phi) \sin\theta d\theta d\phi. \end{aligned}$$
(2.246)

This is an average of **M** over the surface of the radius-R sphere surrounding the point **x** where the field **M** is evaluated.

Second order proof. Again, we use eq. (2.244) to discover the relation between the vector **M** and its divergence and curl. The vector **M** can be expressed at the point of interest as a convolution with the delta function at all other points in space

$$\mathbf{M}(\mathbf{x}) = \int_V dV' \delta(\mathbf{x}-\mathbf{x}')\mathbf{M}(\mathbf{x}').$$
(2.247)

The Laplacian representation of the delta function in $\mathbb{R}^3$ is

$$\delta(\mathbf{x} - \mathbf{x}') = -\frac{1}{4\pi} \nabla^2 \frac{1}{\|\mathbf{x} - \mathbf{x}'\|}, \qquad (2.248)$$

so $\mathbf{M}$ can be represented as the following convolution

$$\mathbf{M}(\mathbf{x}) = -\frac{1}{4\pi} \int_V dV' \, \nabla^2 \frac{1}{\|\mathbf{x} - \mathbf{x}'\|} \mathbf{M}(\mathbf{x}'). \qquad (2.249)$$

Using this relation and proceeding with a few applications of the chain rule, plus the fact that $\nabla 1/\|\mathbf{x} - \mathbf{x}'\| = -\nabla' 1/\|\mathbf{x} - \mathbf{x}'\|$, we find

$$-4\pi \mathbf{M}(\mathbf{x})$$

$$= \int_V dV' \, \nabla^2 \frac{1}{\|\mathbf{x} - \mathbf{x}'\|} \mathbf{M}(\mathbf{x}')$$

$$= \left\langle \int_V dV' \, \nabla^2 \frac{1}{\|\mathbf{x} - \mathbf{x}'\|} \mathbf{M}(\mathbf{x}') \right\rangle_1$$

$$= -\left\langle \int_V dV' \, \nabla \left(\nabla' \frac{1}{\|\mathbf{x} - \mathbf{x}'\|} \right) \mathbf{M}(\mathbf{x}') \right\rangle_1$$

$$= -\left\langle \nabla \int_V dV' \left(\nabla' \frac{\mathbf{M}(\mathbf{x}')}{\|\mathbf{x} - \mathbf{x}'\|} - \frac{\nabla' \mathbf{M}(\mathbf{x}')}{\|\mathbf{x} - \mathbf{x}'\|} \right) \right\rangle_1$$

$$= -\left\langle \nabla \int_{\partial V} dA' \, \hat{\mathbf{n}} \frac{\mathbf{M}(\mathbf{x}')}{\|\mathbf{x} - \mathbf{x}'\|} \right\rangle_1 + \left\langle \nabla \int_V dV' \, \frac{s(\mathbf{x}') + I\mathbf{C}(\mathbf{x}')}{\|\mathbf{x} - \mathbf{x}'\|} \right\rangle_1$$

$$= -\left\langle \nabla \int_{\partial V} dA' \, \hat{\mathbf{n}} \frac{\mathbf{M}(\mathbf{x}')}{\|\mathbf{x} - \mathbf{x}'\|} \right\rangle_1 + \nabla \int_V dV' \, \frac{s(\mathbf{x}')}{\|\mathbf{x} - \mathbf{x}'\|} + \nabla \cdot \int_V dV' \, \frac{I\mathbf{C}(\mathbf{x}')}{\|\mathbf{x} - \mathbf{x}'\|}. \qquad (2.250)$$

By inserting a no-op grade selection operation in the second step, the trivector terms that would show up in subsequent steps are automatically filtered out. This leaves us with a boundary term dependent on the surface and the normal and tangential components of $\mathbf{M}$. Added to that is a pair of volume integrals that provide the unique dependence of $\mathbf{M}$ on its divergence and curl. When the surface is taken to infinity, which requires $\|\mathbf{M}\|/\|\mathbf{x} - \mathbf{x}'\| \to 0$, then the dependence of $\mathbf{M}$ on its divergence and curl is unique.

In order to express final result in traditional vector algebra form, a couple transformations are required. The first is that

$$\langle \mathbf{a}I\mathbf{b} \rangle_1 = I^2 \mathbf{a} \times \mathbf{b} = -\mathbf{a} \times \mathbf{b}. \qquad (2.251)$$

For the grade selection in the boundary integral, note that

$$\begin{aligned}\langle \nabla \hat{\mathbf{n}} \mathbf{X} \rangle_1 &= \langle \nabla (\hat{\mathbf{n}} \cdot \mathbf{X}) \rangle_1 + \langle \nabla (\hat{\mathbf{n}} \wedge \mathbf{X}) \rangle_1 \\ &= \nabla (\hat{\mathbf{n}} \cdot \mathbf{X}) + \langle \nabla I (\hat{\mathbf{n}} \times \mathbf{X}) \rangle_1 \\ &= \nabla (\hat{\mathbf{n}} \cdot \mathbf{X}) - \nabla \times (\hat{\mathbf{n}} \times \mathbf{X}). \end{aligned} \quad (2.252)$$

These give

$$\boxed{\begin{aligned} \mathbf{M}(\mathbf{x}) = {}& \nabla \frac{1}{4\pi} \int_{\partial V} dA' \, \hat{\mathbf{n}} \cdot \frac{\mathbf{M}(\mathbf{x}')}{\|\mathbf{x} - \mathbf{x}'\|} - \nabla \times \frac{1}{4\pi} \int_{\partial V} dA' \, \hat{\mathbf{n}} \times \frac{\mathbf{M}(\mathbf{x}')}{\|\mathbf{x} - \mathbf{x}'\|} \\ & - \nabla \frac{1}{4\pi} \int_V dV' \frac{s(\mathbf{x}')}{\|\mathbf{x} - \mathbf{x}'\|} + \nabla \times \frac{1}{4\pi} \int_V dV' \frac{\mathbf{C}(\mathbf{x}')}{\|\mathbf{x} - \mathbf{x}'\|}. \end{aligned}}$$

(2.253)

2.8 PROBLEM SOLUTIONS.

Answer for Exercise 2.1

Proof. Since each reciprocal vector must each satisfy $\mathbf{x}^i \cdot \mathbf{x}_i = 1$, let $\mathbf{x}^i = \alpha \mathbf{x}_i$, then

$$\begin{aligned} 1 &= \mathbf{x}^i \cdot \mathbf{x}_i \\ &= (\alpha \mathbf{x}_i) \cdot \mathbf{x}_i \\ &= \alpha (\mathbf{x}_i \cdot \mathbf{x}_i), \end{aligned} \quad (2.22)$$

or

$$\mathbf{x}^i = \frac{1}{\mathbf{x}_i \cdot \mathbf{x}_i} \mathbf{x}_i = \frac{1}{\mathbf{x}_i}. \quad (2.23)$$

□

Answer for Exercise 2.2

Assuming the representation of eq. (2.10), the dot products are

$$1 = \mathbf{x}_1 \cdot \mathbf{x}^1 = a\mathbf{x}_1^2 + b\mathbf{x}_1 \cdot \mathbf{x}_2$$
$$0 = \mathbf{x}_2 \cdot \mathbf{x}^1 = a\mathbf{x}_2 \cdot \mathbf{x}_1 + b\mathbf{x}_2^2$$
$$0 = \mathbf{x}_1 \cdot \mathbf{x}^2 = c\mathbf{x}_1^2 + d\mathbf{x}_1 \cdot \mathbf{x}_2 \quad (2.24)$$
$$1 = \mathbf{x}_2 \cdot \mathbf{x}^2 = c\mathbf{x}_2 \cdot \mathbf{x}_1 + d\mathbf{x}_2^2.$$

This can be written out as a pair of matrix equations

$$\begin{bmatrix} 1 \\ 0 \end{bmatrix} = \begin{bmatrix} \mathbf{x}_1^2 & \mathbf{x}_1 \cdot \mathbf{x}_2 \\ \mathbf{x}_2 \cdot \mathbf{x}_1 & \mathbf{x}_2^2 \end{bmatrix} \begin{bmatrix} a \\ b \end{bmatrix}$$
$$\begin{bmatrix} 0 \\ 1 \end{bmatrix} = \begin{bmatrix} \mathbf{x}_1^2 & \mathbf{x}_1 \cdot \mathbf{x}_2 \\ \mathbf{x}_2 \cdot \mathbf{x}_1 & \mathbf{x}_2^2 \end{bmatrix} \begin{bmatrix} c \\ d \end{bmatrix}. \quad (2.25)$$

The matrix inverse is

$$\begin{bmatrix} \mathbf{x}_1^2 & \mathbf{x}_1 \cdot \mathbf{x}_2 \\ \mathbf{x}_2 \cdot \mathbf{x}_1 & \mathbf{x}_2^2 \end{bmatrix}^{-1} = \frac{1}{\mathbf{x}_1^2 \mathbf{x}_2^2 - (\mathbf{x}_1 \cdot \mathbf{x}_2)^2} \begin{bmatrix} \mathbf{x}_2^2 & -\mathbf{x}_1 \cdot \mathbf{x}_2 \\ -\mathbf{x}_2 \cdot \mathbf{x}_1 & \mathbf{x}_1^2 \end{bmatrix}, \quad (2.26)$$

and multiplying by the $(1,0)$, and $(0,1)$ vectors picks out the respective columns, giving eq. (2.11).

Answer for Exercise 2.3

The bivector for the plane spanned by this basis is

$$\begin{aligned} \mathbf{x}_1 \wedge \mathbf{x}_2 &= (\mathbf{e}_1 + 2\mathbf{e}_2) \wedge (\mathbf{e}_2 - \mathbf{e}_3) \\ &= \mathbf{e}_{12} - \mathbf{e}_{13} - 2\mathbf{e}_{23} \\ &= \mathbf{e}_{12} + \mathbf{e}_{31} + 2\mathbf{e}_{32}. \end{aligned} \quad (2.28)$$

This has the square

$$\begin{aligned} (\mathbf{x}_1 \wedge \mathbf{x}_2)^2 &= (\mathbf{e}_{12} + \mathbf{e}_{31} + 2\mathbf{e}_{32}) \cdot (\mathbf{e}_{12} + \mathbf{e}_{31} + 2\mathbf{e}_{32}) \\ &= -1 - 1 - 4 \\ &= -6. \end{aligned} \quad (2.29)$$

Dotting $-\mathbf{x}_1$ with the bivector is

$$\begin{aligned} \mathbf{x}_1 \cdot (\mathbf{x}_2 \wedge \mathbf{x}_1) &= -(\mathbf{e}_1 + 2\mathbf{e}_2) \cdot (\mathbf{e}_{12} + \mathbf{e}_{31} + 2\mathbf{e}_{32}) \\ &= -(\mathbf{e}_2 - \mathbf{e}_3 - 2\mathbf{e}_1 - 4\mathbf{e}_3) \\ &= 2\mathbf{e}_1 - \mathbf{e}_2 + 5\mathbf{e}_3. \end{aligned} \quad (2.30)$$

For $\mathbf{x}_2$ the dot product with the bivector is

$$\begin{aligned}
\mathbf{x}_2 \cdot (\mathbf{x}_1 \wedge \mathbf{x}_2) &= (\mathbf{e}_2 - \mathbf{e}_3) \cdot (\mathbf{e}_{12} + \mathbf{e}_{31} + 2\mathbf{e}_{32}) \\
&= -\mathbf{e}_1 - 2\mathbf{e}_3 - \mathbf{e}_1 - 2\mathbf{e}_2 \\
&= -2\mathbf{e}_1 - 2\mathbf{e}_2 - 2\mathbf{e}_3,
\end{aligned} \quad (2.31)$$

so

$$\begin{aligned}
\mathbf{x}^1 &= \frac{1}{3}(\mathbf{e}_1 + \mathbf{e}_2 + \mathbf{e}_3) \\
\mathbf{x}^2 &= \frac{1}{6}(-2\mathbf{e}_1 + \mathbf{e}_2 - 5\mathbf{e}_3).
\end{aligned} \quad (2.32)$$

It is easy to verify that this has the desired semantics.

Answer for Exercise 2.4

Solution Part a. The curvilinear basis associated with this parameterization can be computed by inspection

$$\begin{aligned}
\mathbf{x}_1 &= \mathbf{e}_1 \cos u_2 + \beta \mathbf{e}_2 \sin u_2 \\
\mathbf{x}_2 &= u_1 (-\mathbf{e}_1 \sin u_2 + \beta \mathbf{e}_2 \cos u_2).
\end{aligned} \quad (2.43)$$

Solution Part b.
We need to compute the area element first

$$\begin{aligned}
\mathbf{x}_1 \wedge \mathbf{x}_2 &= (\mathbf{e}_1 \cos u_2 + \beta \mathbf{e}_2 \sin u_2) \wedge u_1 (-\mathbf{e}_1 \sin u_2 + \beta \mathbf{e}_2 \cos u_2) \\
&= u_1 \langle (\mathbf{e}_1 \cos u_2 + \beta \mathbf{e}_2 \sin u_2)(-\mathbf{e}_1 \sin u_2 + \beta \mathbf{e}_2 \cos u_2) \rangle_2 \\
&= u_1 \left(\beta \mathbf{e}_{12} \cos^2 u_2 - \beta \mathbf{e}_{21} \sin^2 u_2 \right) \\
&= u_1 \beta i,
\end{aligned} \quad (2.44)$$

where $i = \mathbf{e}_{12}$.
The reciprocal frame vectors are given by

$$\begin{aligned}
\mathbf{x}^1 &= \mathbf{x}_2 \cdot \frac{1}{\mathbf{x}_1 \wedge \mathbf{x}_2} \\
&= u_1 (-\mathbf{e}_1 \sin u_2 + \beta \mathbf{e}_2 \cos u_2) \frac{1}{u_1 \beta i} \\
&= \frac{1}{\beta} \mathbf{e}_2 \sin u_2 + \mathbf{e}_1 \cos u_2,
\end{aligned} \quad (2.45)$$

$$\begin{aligned}
\mathbf{x}^2 &= -\mathbf{x}_1 \cdot \frac{1}{\mathbf{x}_1 \wedge \mathbf{x}_2} \\
&= -(\mathbf{e}_1 \cos u_2 + \beta \mathbf{e}_2 \sin u_2) \frac{1}{u_1 \beta i} \\
&= \frac{1}{u_1}\left(\frac{1}{\beta}\mathbf{e}_2 \cos u_2 - \mathbf{e}_1 \sin u_2\right).
\end{aligned} \qquad (2.46)$$

Solution Part c. To verify that $\mathbf{x}_i \cdot \mathbf{x}^j = \delta_i{}^j$ we can compute each of the dot products

$$\begin{aligned}
\mathbf{x}^1 \cdot \mathbf{x}_1 &= \left\langle (\mathbf{e}_1 \cos u_2 + \beta \mathbf{e}_2 \sin u_2)\left(\frac{1}{\beta}\mathbf{e}_2 \sin u_2 + \mathbf{e}_1 \cos u_2\right)\right\rangle \\
&= \cos^2 u_2 + \sin^2 u_2 \\
&= 1,
\end{aligned} \qquad (2.47)$$

$$\begin{aligned}
\mathbf{x}^2 \cdot \mathbf{x}_2 &= \left\langle u_1(-\mathbf{e}_1 \sin u_2 + \beta \mathbf{e}_2 \cos u_2)\frac{1}{u_1}\left(\frac{1}{\beta}\mathbf{e}_2 \cos u_2 - \mathbf{e}_1 \sin u_2\right)\right\rangle \\
&= \sin^2 u_2 + \cos^2 u_2 \\
&= 1,
\end{aligned} \qquad (2.48)$$

$$\begin{aligned}
\mathbf{x}^1 \cdot \mathbf{x}_2 &= \left\langle \left(\frac{1}{\beta}\mathbf{e}_2 \sin u_2 + \mathbf{e}_1 \cos u_2\right)u_1(-\mathbf{e}_1 \sin u_2 + \beta \mathbf{e}_2 \cos u_2)\right\rangle \\
&= u_1 \sin u_2 \cos u_2 - u_1 \cos u_2 \sin u_2 \\
&= 0.
\end{aligned} \qquad (2.49)$$

$$\begin{aligned}
\mathbf{x}^2 \cdot \mathbf{x}_1 &= \left\langle \frac{1}{u_1}\left(\frac{1}{\beta}\mathbf{e}_2 \cos u_2 - \mathbf{e}_1 \sin u_2\right)(\mathbf{e}_1 \cos u_2 + \beta \mathbf{e}_2 \sin u_2)\right\rangle \\
&= \frac{1}{u_1}(\cos u_2 \sin u_2 - \sin u_2 \cos u_2) \\
&= 0.
\end{aligned} \qquad (2.50)$$

Answer for Exercise 2.5

$$\cosh(\mu - i\theta)\sinh(\mu + i\theta) = \frac{1}{4}\left(e^{\mu-i\theta} - e^{-\mu+i\theta}\right)\left(e^{\mu+i\theta} - e^{-\mu-i\theta}\right)$$
$$= \frac{1}{4}\left(e^{2\mu} - e^{-2\mu} + e^{2i\theta} - e^{-2i\theta}\right) \quad (2.54)$$
$$= \frac{1}{2}\left(\sinh(2\mu) + i\sin(2\theta)\right).$$

The second identity follows from the first, setting $\theta = 0$. Finally, for the third expanding the cosh in terms of exponentials, we find

$$\cosh(\mu + i\theta) = \frac{1}{2}\left(e^{\mu+i\theta} + e^{-\mu-i\theta}\right)$$
$$= \frac{e^{\mu}}{2}(\cos\theta + i\sin\theta) + \frac{e^{-\mu}}{2}(\cos\theta - i\sin\theta) \quad (2.55)$$
$$= \frac{e^{\mu} + e^{-\mu}}{2}\cos\theta + i\frac{e^{\mu} - e^{-\mu}}{2}\sin\theta$$
$$= \cosh\mu\cos\theta + i\sinh\mu\sin\theta.$$

Answer for Exercise 2.6

Solution Part a. Using the multiple angle cosh expansion, we find

$$\mathbf{e}_1\cosh(\mu + iu_2) = \mathbf{e}_1(\cosh\mu\cos u_2 + i\sinh\mu\sin u_2)$$
$$= \mathbf{e}_1\cosh\mu\cos u_2 + \mathbf{e}_2\sinh\mu\sin u_2, \quad (2.57)$$

so

$$\mathbf{x} = u_1\mathbf{e}_1\cosh(\mu + iu_2) = \mathbf{e}_1 a\cos u_2 + \mathbf{e}_2 b\sin u_2, \quad (2.58)$$

where

$$a = u_1\cosh\mu$$
$$b = u_1\sinh\mu, \quad (2.59)$$

are the semi-major and semi-minor axis values.

Solution Part b. The eccentricity (squared) is

$$\epsilon^2 = 1 - \tanh^2\mu$$
$$= \frac{\cosh^2\mu - \sinh^2\mu}{\cosh^2\mu} \quad (2.60)$$
$$= \frac{1}{\cosh^2\mu},$$

so the eccentricity is

$$\epsilon = \frac{1}{\cosh \mu}. \tag{2.61}$$

Solution Part c. Our curvilinear basis vectors are

$$\begin{aligned} \mathbf{x}_1 &= \mathbf{e}_1 \cosh (\mu + iu_2) \\ \mathbf{x}_2 &= \mathbf{e}_2 u_1 \sinh (\mu + iu_2) \end{aligned} \tag{2.62}$$

To compute the reciprocals we need the area element

$$\begin{aligned} \mathbf{x}_1 \wedge \mathbf{x}_2 &= \langle \mathbf{e}_1 \cosh (\mu + iu_2) \, \mathbf{e}_2 u_1 \sinh (\mu + iu_2) \rangle_2 \\ &= u_1 \langle i \cosh (\mu - iu_2) \sinh (\mu + iu_2) \rangle_2 \\ &= \frac{u_1}{2} \langle i \left(\sinh(2\mu) + i \sin(2u_2) \right) \rangle_2 \\ &= u_1 i \cosh \mu \sinh \mu. \end{aligned} \tag{2.63}$$

Our recipocal basis vectors are

$$\begin{aligned} \mathbf{x}^1 &= \mathbf{x}_2 \frac{1}{\mathbf{x}_1 \wedge \mathbf{x}_2} \\ &= \mathbf{e}_2 u_1 \sinh (\mu + iu_2) \, \frac{1}{u_1 i \cosh \mu \sinh \mu} \\ &= \mathbf{e}_1 \frac{\sinh (\mu + iu_2)}{\cosh \mu \sinh \mu}, \end{aligned} \tag{2.64}$$

and

$$\begin{aligned} \mathbf{x}^2 &= -\mathbf{x}_1 \frac{1}{\mathbf{x}_1 \wedge \mathbf{x}_2} \\ &= - \left(\mathbf{e}_1 \cosh (\mu + iu_2) \right) \frac{1}{u_1 i \cosh \mu \sinh \mu} \\ &= \frac{\mathbf{e}_2 \cosh (\mu + iu_2)}{u_1 \cosh \mu \sinh \mu}. \end{aligned} \tag{2.65}$$

Solution Part d.

$$\begin{aligned} \mathbf{x}_1 \cdot \mathbf{x}^1 &= \left\langle \mathbf{e}_1 \cosh (\mu + iu_2) \, \mathbf{e}_1 \frac{\sinh (\mu + iu_2)}{\cosh \mu \sinh \mu} \right\rangle \\ &= \frac{1}{\cosh \mu \sinh \mu} \left\langle \mathbf{e}_1^2 \cosh (\mu - iu_2) \left(\sinh (\mu + iu_2) \right) \right\rangle \\ &= \frac{1}{\cosh \mu \sinh \mu} \left\langle \frac{1}{2} \left(\sinh(2\mu) + i \sin(2u_2) \right) \right\rangle \\ &= \frac{\sinh(2\mu)}{2 \cosh \mu \sinh \mu} \\ &= 1. \end{aligned} \tag{2.66}$$

2.8 PROBLEM SOLUTIONS. 177

$$\begin{aligned}
\mathbf{x}_2 \cdot \mathbf{x}^2 &= \left\langle \mathbf{e}_2 u_1 \sinh(\mu + iu_2) \frac{\mathbf{e}_2 \cosh(\mu + iu_2)}{u_1 \cosh\mu \sinh\mu} \right\rangle \\
&= \frac{1}{\cosh\mu \sinh\mu} \left\langle \sinh(\mu + iu_2) \mathbf{e}_2^2 \cosh(\mu - iu_2) \right\rangle \\
&= 1.
\end{aligned} \qquad (2.67)$$

$$\begin{aligned}
\mathbf{x}_1 \cdot \mathbf{x}^2 &= \left\langle \mathbf{e}_1 \cosh(\mu + iu_2) \frac{\mathbf{e}_2 \cosh(\mu + iu_2)}{u_1 \cosh\mu \sinh\mu} \right\rangle \\
&= \frac{1}{u_1 \cosh\mu \sinh\mu} \left\langle \mathbf{e}_{12} \cosh(\mu - iu_2) \cosh(\mu + iu_2) \right\rangle \\
&= \frac{|\cosh(\mu + iu_2)|^2}{u_1 \cosh\mu \sinh\mu} \langle \mathbf{e}_{12} \rangle \\
&= 0.
\end{aligned} \qquad (2.68)$$

$$\begin{aligned}
\mathbf{x}^1 \cdot \mathbf{x}_2 &= \left\langle \mathbf{e}_1 \frac{\sinh(\mu + iu_2)}{\cosh\mu \sinh\mu} \mathbf{e}_2 u_1 \sinh(\mu + iu_2) \right\rangle \\
&= \frac{u_1 |\sinh(\mu + iu_2)|^2}{\cosh\mu \sinh\mu} \langle \mathbf{e}_{12} \rangle \\
&= 0.
\end{aligned} \qquad (2.69)$$

Answer for Exercise 2.7

Computing the various dot products is made easier by noting that $\mathbf{e}_3$ and $e^{i\phi}$ commute, whereas $e^{j\theta} \mathbf{e}_3 = \mathbf{e}_3 e^{-j\theta}$, $\mathbf{e}_1 e^{i\phi} = e^{-i\phi} \mathbf{e}_1$, $\mathbf{e}_2 e^{i\phi} = e^{-i\phi} \mathbf{e}_2$ (since $\mathbf{e}_3 j$, $\mathbf{e}_1 i$ and $\mathbf{e}_2 i$ all anticommute.) Also note that

$$\begin{aligned}
j\hat{\boldsymbol{\phi}} &= \mathbf{e}_{31} e^{i\phi} \mathbf{e}_2 e^{i\phi} \\
&= \mathbf{e}_{312} e^{-i\phi} e^{i\phi} \\
&= I.
\end{aligned} \qquad (2.108)$$

The dot products, working with the normalized vectors, are

$$\begin{aligned}
\hat{\mathbf{r}} \cdot \hat{\boldsymbol{\theta}} &= \langle \hat{\mathbf{r}} \hat{\mathbf{r}} j \rangle \\
&= \langle j \rangle \\
&= 0
\end{aligned} \qquad (2.109\text{a})$$

$$\hat{\mathbf{r}} \cdot \hat{\boldsymbol{\phi}} = \langle \mathbf{e}_3 e^{j\theta} \hat{\boldsymbol{\phi}} \rangle$$
$$= \langle \mathbf{e}_3 (\cos\theta + j\sin\theta) \hat{\boldsymbol{\phi}} \rangle$$
$$= \cos\theta \langle \mathbf{e}_3 \hat{\boldsymbol{\phi}} \rangle + \sin\theta \langle \mathbf{e}_3 j \hat{\boldsymbol{\phi}} \rangle \quad (2.109b)$$
$$= \cos\theta \langle \mathbf{e}_{32} \cos\phi + \mathbf{e}_{13} \sin\phi \rangle + \sin\theta \langle \mathbf{e}_{12} \rangle$$
$$= 0$$

$$\hat{\boldsymbol{\theta}} \cdot \hat{\boldsymbol{\phi}} = \langle \hat{\mathbf{r}} j \hat{\boldsymbol{\phi}} \rangle$$
$$= \langle \hat{\mathbf{r}} I \rangle \quad (2.109c)$$
$$= 0.$$

Answer for Exercise 2.8

$$\mathbf{e}_{31} e^{i\phi} \mathbf{e}_2 e i\phi = \mathbf{e}_{31} e^{i\phi} e^{-i\phi} \mathbf{e}_2$$
$$= \mathbf{e}_{312} \quad (2.110)$$
$$= I.$$

Answer for Exercise 2.9

A bit of shorthand is useful. We can write our Jacobian as

$$J = \begin{vmatrix} S_\theta C_\phi & S_\theta S_\phi & C_\theta \\ rC_\theta C_\phi & rC_\theta S_\phi & -rS_\theta \\ -rS_\theta S_\phi & rS_\theta C_\phi & 0 \end{vmatrix} = r^2 \begin{vmatrix} S_\theta C_\phi & S_\theta S_\phi & C_\theta \\ C_\theta C_\phi & C_\theta S_\phi & -S_\theta \\ -S_\theta S_\phi & S_\theta C_\phi & 0 \end{vmatrix}, \quad (2.111)$$

where the common factor of the two last rows has been factored out. Expanding the cofactors along the bottom row we have

$$J = -r^2 S_\theta S_\phi \begin{vmatrix} S_\theta S_\phi & C_\theta \\ C_\theta S_\phi & -S_\theta \end{vmatrix} - r^2 S_\theta C_\phi \begin{vmatrix} S_\theta C_\phi & C_\theta \\ C_\theta C_\phi & -S_\theta \end{vmatrix}$$
$$= -r^2 S_\theta S_\phi^2 \begin{vmatrix} S_\theta & C_\theta \\ C_\theta & -S_\theta \end{vmatrix} - r^2 S_\theta C_\phi^2 \begin{vmatrix} S_\theta C_\theta \\ C_\theta & -S_\theta \end{vmatrix} \quad (2.112)$$
$$= -r^2 S_\theta S_\phi^2 \left(-S_\theta^2 - C_\theta^2\right) - r^2 S_\theta C_\phi^2 \left(-S_\theta^2 - C_\theta^2\right)$$
$$= r^2 S_\theta \left(S_\phi^2 + C_\phi^2\right)$$
$$= r^2 S_\theta.$$

2.8 PROBLEM SOLUTIONS. 179

Had we done a traditional first row or column determinant expansion, things would have been considerably messier.

Answer for Exercise 2.10

$$\begin{aligned}\mathbf{x} &= r\left(\mathbf{e}_1 \sin\theta \cos\phi + \mathbf{e}_2 \sin\theta \sin\phi + \mathbf{e}_3 \cos\theta\right) \\ &= r\left(\sin\theta \mathbf{e}_1(\cos\phi + \mathbf{e}_{12}\sin\phi) + \mathbf{e}_3 \cos\theta\right) \\ &= r\left(\sin\theta \mathbf{e}_1 e^{\mathbf{e}_{12}\phi} + \mathbf{e}_3 \cos\theta\right) \\ &= r\mathbf{e}_3\left(\cos\theta + \sin\theta \mathbf{e}_3 \mathbf{e}_1 e^{\mathbf{e}_{12}\phi}\right).\end{aligned} \qquad (2.113)$$

Writing $j = \mathbf{e}_3 \mathbf{e}_1 e^{\mathbf{e}_{12}\phi}$, this is

$$\mathbf{x} = r\mathbf{e}_3 e^{j\theta}. \qquad (2.114)$$

Answer for Exercise 2.11

We'll only compute $\mathbf{x}_\theta$ here explicitly, as the other two vectors can be computed by inspection.

We start with a plain old chain rule expansion, with the cavaet that we must be careful not to commute j with anything but the $e^{\pm j\theta/2}$ terms.

$$\begin{aligned}\mathbf{x}_\theta &= \frac{\partial \mathbf{x}}{\partial \theta} \\ &= -\frac{j}{2} e^{-j\theta/2} \left(\rho \mathbf{e}_1 e^{i\phi} + R\mathbf{e}_3\right) e^{j\theta/2} + e^{-j\theta/2} \left(\rho \mathbf{e}_1 e^{i\phi} + R\mathbf{e}_3\right) e^{j\theta/2} \frac{j}{2}\end{aligned} \qquad (2.115)$$

Note that the bivector j commutes with $\mathbf{e}_1$, and then proceed to compute the ρ dependent part of $\mathbf{x}_\theta$

$$\begin{aligned}&\frac{\rho}{2} e^{-j\theta/2} \mathbf{e}_1 \left(-j e^{i\phi} + e^{i\phi} j\right) e^{j\theta/2} \\ &= \frac{\rho}{2} e^{-j\theta/2} \mathbf{e}_1 \left(-\mathbf{e}_{32}(\cos\phi + \mathbf{e}_{13}\sin\phi) + (\cos\phi + \mathbf{e}_{13}\sin\phi)\mathbf{e}_{32}\right) e^{j\theta/2} \\ &= \frac{\rho}{2} e^{-j\theta/2} \mathbf{e}_1 \left(-\mathbf{e}_{3213}\sin\phi + \mathbf{e}_{1332}\sin\phi\right) e^{j\theta/2} \\ &= \frac{\rho}{2} e^{-j\theta/2} \mathbf{e}_1 \left(-\mathbf{e}_{21}\sin\phi + \mathbf{e}_{12}\sin\phi\right) e^{j\theta/2} \\ &= \rho e^{-j\theta/2} \mathbf{e}_{112} \sin\phi e^{j\theta/2} \\ &= \rho e^{-j\theta/2} \mathbf{e}_2 \sin\phi e^{j\theta/2}.\end{aligned} \qquad (2.116)$$

Similarly, the R dependent contribution is

$$\begin{aligned}&\frac{R}{2}e^{-j\theta/2}\left(-j\mathbf{e}_3+\mathbf{e}_3 j\right)e^{j\theta/2}\\&\frac{R}{2}e^{-j\theta/2}\left(-\mathbf{e}_{323}+\mathbf{e}_{332}\right)e^{j\theta/2}\\&\frac{R}{2}e^{-j\theta/2}\left(\mathbf{e}_2+\mathbf{e}_2\right)e^{j\theta/2}\\&Re^{-j\theta/2}\mathbf{e}_2 e^{j\theta/2}.\end{aligned}\qquad(2.117)$$

Putting the pieces together, we have

$$\mathbf{x}_\theta = e^{-j\theta/2}\left(R+\rho\sin\phi\right)\mathbf{e}_2 e^{j\theta/2}, \qquad(2.118)$$

as claimed.

Answer for Exercise 2.12

Let $x = \rho\cos\phi$ and $y = \rho\sin\phi$, which provides the polar form of the field

$$\begin{aligned}\mathbf{v} &= -y\mathbf{e}_1 + x\mathbf{e}_2\\ &= -\rho\sin\phi\,\mathbf{e}_1 + \rho\cos\phi\,\mathbf{e}_2\\ &= \rho\mathbf{e}_2\left(\cos\phi - \mathbf{e}_2\mathbf{e}_1\sin\phi\right) \qquad(2.191)\\ &= \rho\mathbf{e}_2 e^{i\phi}\\ &= \rho\hat{\rho}i.\end{aligned}$$

The curl is

$$\begin{aligned}\nabla\wedge\mathbf{v} &= \left\langle\hat{\rho}\left(\frac{\partial}{\partial\rho}+\frac{i}{\rho}\frac{\partial}{\partial\phi}\right)\rho\hat{\rho}i\right\rangle_2\\ &= \left\langle\hat{\rho}\hat{\rho}i + \hat{\rho}\frac{i}{\rho}\rho\hat{\rho}i^2\right\rangle_2 \qquad(2.192)\\ &= i - \hat{\rho}i\hat{\rho}\\ &= 2i,\end{aligned}$$

as we also found from the Cartesian representation.

Answer for Exercise 2.13

Part a.

$$\begin{aligned}\nabla\wedge B &= (\mathbf{e}_1\partial_1 + \mathbf{e}_2\partial_2 + \mathbf{e}_3\partial_3)\wedge(x\mathbf{e}_{23} + y\mathbf{e}_{31} + z\mathbf{e}_{12})\\ &= \mathbf{e}_1\wedge\mathbf{e}_{23} + \mathbf{e}_2\wedge\mathbf{e}_{31} + \mathbf{e}_3\wedge\mathbf{e}_{12} \qquad(2.193)\\ &= 3\mathbf{e}_{123}.\end{aligned}$$

Part b.

$$\nabla \wedge B = (e_1\partial_1 + e_2\partial_2 + e_3\partial_3) \wedge (xe_{31})$$
$$= e_1 \wedge e_{31} \qquad (2.194)$$
$$= 0.$$

Part c.

$$\nabla \wedge B = (e_1\partial_1 + e_2\partial_2 + e_3\partial_3) \wedge (xyze_{23})$$
$$= yze_{123}. \qquad (2.195)$$

Answer for Exercise 2.14

With $\alpha = \sqrt{\omega_0^2 - k^2}$, we may factor the denominator

$$\omega^2 - 2j\omega k - \omega_0^2 = (\omega - (jk + \alpha))(\omega - (jk - \alpha)), \qquad (2.208)$$

showing that we have poles in the upper half plane at $jk \pm \alpha$.

It's important to understand the behaviour of the integral on the infinite semi-circular contours, which we can parameterize as $\omega = Re^{i\theta}$. The denominator is $O(1/R^2)$, but the exponential has the form

$$e^{j\omega\tau} = e^{j\tau R(\cos\theta + j\sin\theta)}$$
$$= e^{j\tau R\cos\theta} e^{-R\tau\sin\theta}. \qquad (2.209)$$

We see that the integral diverges on the upper half contour for $\tau < 0$, and diverges on the lower half contour for $\tau > 0$. There's a theorem (who's name I forget) that shows that the upper half contour integral evaluated for $\tau > 0$ will be zero on the infinite semicircle, as will the lower semicircular contour for $\tau < 0$, so if we compute the residues for the complete contours we find the value of the integral along the $[-\infty, \infty]$ horizontal.

We find

$$G(\tau < 0) = 0$$

$$G(\tau > 0) = \frac{-1}{2\pi} 2\pi j \left(\left.\frac{e^{j\omega\tau}}{\omega - (jk + \alpha)}\right|_{\omega = jk-\alpha} + \left.\frac{e^{j\omega\tau}}{\omega - (jk - \alpha)}\right|_{\omega = jk+\alpha} \right)$$

$$= \frac{1}{j}\left(\frac{e^{j\tau(jk-\alpha)}}{-2\alpha} + \frac{e^{j\tau(jk+\alpha)}}{2\alpha} \right)$$

$$= \frac{1}{\alpha} e^{-k\tau} \sin(\alpha\tau),$$

(2.210)

or

$$G(\tau) = \Theta(\tau)e^{-k\tau}\frac{\sin(\alpha\tau)}{\alpha}. \qquad (2.211)$$

Rather amusingly, when the system is supplied with an impulse function $f(t) = \delta(t)$, we see that the response to that infinite push on the swing is

$$\begin{aligned} x(t) &= \int_{-\infty}^{\infty} \Theta(t-t')e^{-k(t-t')}\frac{\sin(\alpha(t-t'))}{\alpha}\delta(t')dt' \\ &= \Theta(t)e^{-kt}\frac{\sin(\alpha t)}{\alpha}, \end{aligned} \qquad (2.212)$$

which describes an oscillation that starts at the point of the push, but decreases in amplitude steadily after that due to the damping term. Even with an infinite strength initial push, the child will eventually be exhorting the dad to supply another underdog, "Again, again, again!".

3

ELECTROMAGNETISM.

3.1 CONVENTIONAL FORMULATION.

Maxwell's equations provide an abstraction, the field, that aggregates the effects of an arbitrary electric charge and current distribution on a "test" charge distribution. The test charge is assumed to be small and isolated enough that it does not also appreciably change the fields themselves. Once the fields are determined, the Lorentz force equation can be used to determine the dynamics of the test particle. These dynamics can be determined without having to compute all the interactions of that charge with all the charges and currents in space, nor having to continually account for the interactions of those charge with each other.

We will use vector differential form of Maxwell's equations with antenna theory extensions (fictitious magnetic sources) as our starting point

$$\begin{aligned}\nabla \times \mathbf{E} &= -\mathbf{M} - \frac{\partial \mathbf{B}}{\partial t} \\ \nabla \times \mathbf{H} &= \mathbf{J} + \frac{\partial \mathbf{D}}{\partial t} \\ \nabla \cdot \mathbf{D} &= \rho \\ \nabla \cdot \mathbf{B} &= \rho_m. \end{aligned} \quad (3.1)$$

These equations relate the primary electric and magnetic fields

- $\mathbf{E}(\mathbf{x}, t)$: Electric field intensity [V/m] (Volts/meter)
- $\mathbf{H}(\mathbf{x}, t)$: Magnetic field intensity [A/m] (Amperes/meter),

and the induced electric and magnetic fields

- $\mathbf{D}(\mathbf{x}, t)$: Electric flux density (or displacement vector) [C/m] (Coulombs/meter)
- $\mathbf{B}(\mathbf{x}, t)$: Magnetic flux density [Wb/m^2] (Webers/square meter),

to the charge densities

- $\rho(\mathbf{x}, t)$: Electric charge density [C/m³] (Coulombs/cubic meter)
- $\rho_m(\mathbf{x}, t)$: Magnetic charge density [Wb/m³] (Webers/cubic meter),

and the current densities

- $\mathbf{J}(\mathbf{x}, t)$: Electric current density [A/m²] (Amperes/square meter),
- $\mathbf{M}(\mathbf{x}, t)$: Magnetic current density [V/m²] (Volts/square meter).

All of the fields and sources can vary in space and time, and are specified here in SI units. The sources $\mathbf{M}, \rho_m$ can be considered fictional, representing physical phenomena such as infinitesimal current loops.

In general, the relationship between the electric and magnetic fields (constitutivity relationships) may be complicated non-isotropic tensor operators, functions of all of $\mathbf{E}, \mathbf{D}, \mathbf{B}$ and $\mathbf{H}$. In this book, we will assume that the constitutive relationships between the electric and magnetic fields are independent

$$\mathbf{B} = \mu \mathbf{H}$$
$$\mathbf{D} = \epsilon \mathbf{E}, \tag{3.2}$$

where $\epsilon = \epsilon_r \epsilon_0$ is the permittivity of the medium [F/m] (Farads/meter), and $\mu = \mu_r \mu_0$ is the permeability of the medium [H/m] (Henries/meter). The permittivity and permeability may be functions of both time and position, and model the materials that the fields are propagating through. In free space $\mu_r = 1$ and $\epsilon_r = 1$ so these relationships are simply $\mathbf{B} = \mu_0 \mathbf{H}, \mathbf{D} = \epsilon_0 \mathbf{E}$, where

- $\epsilon_0 = 8.85 \times 10^{-12} C^2/N/m^2$: Permittivity of free space (Coulombs squared/Newton/square meter)
- $\mu_0 = 4\pi \times 10^{-7} N/A^2$: Permeability of free space (Newtons/Ampere-squared).

These constants are related to the speed of light, $c = 3.00 \times 10^8$ m/s by $\mu_0 \epsilon_0 = 1/c^2$.

Antenna theory extends Maxwell's equations with fictional magnetic charge and current densities that are useful to model real phenomena such as infinitesimal current loops. Antenna related problems are usually tackled in the frequency domain. We will use the engineering conventions for the frequency domain described in section 2.5.

Continuous models for charge and current distributions are used in Maxwell's equations, despite the fact that charges (i.e. electrons) are particles, and are not distributed in space. The discrete nature of electronic charge can be modeled using a delta function representation of the charge and current densities

$$\rho(\mathbf{x}, t) = \sum_a q_a \delta(\mathbf{x} - \mathbf{x}_a(t))$$
$$\mathbf{J}(\mathbf{x}, t) = \sum_a q_a \mathbf{v}_a(\mathbf{x}, \mathbf{t}).$$
(3.3)

This model is inherently non-quantum mechanical, as it assumes that it is possible to simultaneous measure the position and velocity of an electron.

The dynamics of particle interaction with the fields are provided by the Lorentz force and power equations

$$\frac{d\mathbf{p}}{dt} = q\left(\mathbf{E} + \mathbf{v} \times \mathbf{B}\right)$$
(3.4a)

$$\frac{d\mathcal{E}}{dt} = q\mathbf{E} \cdot \mathbf{v}.$$
(3.4b)

Both the energy and the momentum relations of eq. (3.4) are stated, since the simplest (relativistic) form of the Lorentz force equation directly encodes both. For readers unfamiliar with eq. (3.4b), exercise 3.1 provides a derivation method.

The quantities involved in the Lorentz equations are

- $\mathbf{p}(\mathbf{x}, t)$: Test particle momentum [kg m/s] (Kilogram meters/second)
- $\mathcal{E}(\mathbf{x}, t)$: Test particle kinetic energy [J] (Joules, kilogram meter^2/second^2)
- q : Test particle charge [C] (Coulombs)
- $\mathbf{v}$: Test particle velocity [m/s] (Meters/second)

The task of extracting all the physical meaning from the Maxwell and Lorentz equations is a difficult one. Our attempt to do so will use the formalism of geometric algebra.

3.1.1 Problems.

Exercise 3.1   Lorentz power and force relationship. (*§17 [17]*)

Using the relativistic definitions of momentum and energy

$$\mathbf{p}(\mathbf{x}, t) = \frac{m\mathbf{v}}{\sqrt{1 - \mathbf{v}^2/c^2}}$$

$$\mathcal{E}(\mathbf{x}, t) = \frac{mc^2}{\sqrt{1 - \mathbf{v}^2/c^2}},$$

show that $d\mathcal{E}/dt = \mathbf{v} \cdot d\mathbf{p}/dt$, and use this to derive eq. (3.4b) from eq. (3.4a).

3.2 MAXWELL'S EQUATION.

We will work with a multivector representation of the fields in isotropic media satisfying the constituency relationships from eq. (3.2), and define a multivector field that includes both electric and magnetic components

---
Definition 3.1: Electromagnetic field strength.

The *electromagnetic field strength* ([V/m] (Volts/meter)) is defined as

$$F = \mathbf{E} + I\eta\mathbf{H} \quad (= \mathbf{E} + Ic\mathbf{B}),$$

where

- $\eta = \sqrt{\mu/\epsilon}$ ([Ω] Ohms), is the impedance of the media.
- $c = 1/\sqrt{\epsilon\mu}$ ([m/s] meters/second), is the group velocity of a wave in the media. When $\epsilon = \epsilon_0, \mu = \mu_0$, c is the speed of light.

F is called the *F*araday by some authors.

The factors of η (or c) that multiply the magnetic fields are for dimensional consistency, since $[\sqrt{\epsilon}\mathbf{E}] = [\sqrt{\mu}\mathbf{H}] = [\mathbf{B}/\sqrt{\mu}]$. The justification for imposing a dual (or complex) structure on the electromagnetic field strength can be found in the historical development of Maxwell's equations, but we will also see such a structure arise naturally in short order.

No information is lost by imposing the complex structure of definition 3.1, since we can always obtain the electric field vector **E** and the magnetic field bivector $I\mathbf{H}$ by grade selection from the electromagnetic field strength when desired

$$\mathbf{E} = \langle F \rangle_1$$
$$I\mathbf{H} = \frac{1}{\eta}\langle F \rangle_2. \tag{3.5}$$

We will also define a multivector current containing all charge densities and current densities

---
Definition 3.2: Multivector current.

The *current* ($[V/m^2]$ (Volts/square meter)) is defined as

$$J = \eta\left(c\rho - \mathbf{J}\right) + I\left(c\rho_\mathrm{m} - \mathbf{M}\right).$$

When the fictitious magnetic source terms ($\rho_\mathrm{m}, \mathbf{M}$) are included, the current has one grade for each possible source (scalar, vector, bivector, trivector). With only conventional electric sources, the current is still a multivector, but contains only scalar and vector grades.

Given the multivector field and current, it is now possible to state Maxwell's equation (singular) in its geometric algebra form

---
Theorem 3.1: Maxwell's equation.

Maxwell's equation is a multivector equation relating the change in the electromagnetic field strength to charge and current densities and is written as

$$\left(\boldsymbol{\nabla} + \frac{1}{c}\frac{\partial}{\partial t}\right)F = J.$$

Maxwell's equation in this form will be the starting place for all the subsequent analysis in this book. As mentioned in section 2.6, the operator $\boldsymbol{\nabla} + (1/c)\partial_t$ will be called the *spacetime gradient*[1].

---
[1] This form of spacetime gradient is given a special symbol by a number of authors, but there is no general agreement on what to use. Instead of entering the fight, it will be written out in full in this book.

Proof. To prove theorem 3.1 we first insert the isotropic constituency relationships from eq. (3.2) into eq. (3.1), so that we are working with two field variables instead of four

$$\nabla \cdot \mathbf{E} = \frac{\rho}{\epsilon}$$
$$\nabla \times \mathbf{E} = -\mathbf{M} - \mu \frac{\partial \mathbf{H}}{\partial t}$$
$$\nabla \cdot \mathbf{H} = \frac{\rho_m}{\mu} \tag{3.6}$$
$$\nabla \times \mathbf{H} = \mathbf{J} + \epsilon \frac{\partial \mathbf{E}}{\partial t}$$

Inserting $\mathbf{a} = \nabla$ into eq. (1.85) the vector product of the gradient with another vector

$$\nabla \mathbf{b} = \nabla \cdot \mathbf{b} + I \nabla \times \mathbf{b}. \tag{3.7}$$

The dot and cross products for $\mathbf{E}$ and $\mathbf{H}$ in eq. (3.6) can be grouped using eq. (3.7) into multivector gradient equations

$$\nabla \mathbf{E} = \frac{\rho}{\epsilon} + I\left(-\mathbf{M} - \mu \frac{\partial \mathbf{H}}{\partial t}\right)$$
$$\nabla \mathbf{H} = \frac{\rho_m}{\mu} + I\left(\mathbf{J} + \epsilon \frac{\partial \mathbf{E}}{\partial t}\right). \tag{3.8}$$

Multiplying the gradient equation for the magnetic field by ηI so that both equations have the same dimensions, and so that the electric field appears in both equations as $\mathbf{E}$ and not $I\mathbf{E}$, we find

$$\nabla \mathbf{E} + \frac{1}{c}\frac{\partial}{\partial t}(I\eta \mathbf{H}) = \frac{1}{\epsilon}\rho - I\mathbf{M}$$
$$\nabla I\eta \mathbf{H} + \frac{1}{c}\frac{\partial \mathbf{E}}{\partial t} = Ic\rho_m - \eta \mathbf{J}, \tag{3.9}$$

where $\mu/\eta = \eta\epsilon = 1/c$ was used to simplify things slightly, and all the field contributions have been moved to the left hand side. The first multivector equation has only scalar and bivector grades, whereas the second has only vector and trivector grades. This means that if we add these equations, we can recover each by grade selection, and no information is lost. That sum is

$$\left(\nabla + \frac{1}{c}\frac{\partial}{\partial t}\right)(\mathbf{E} + I\eta \mathbf{H}) = \eta\left(c\rho - \mathbf{J}\right) + I\left(c\rho_m - \mathbf{M}\right). \tag{3.10}$$

3.3 WAVE EQUATION AND CONTINUITY.

Application of definition 3.1 and definition 3.2 to eq. (3.10) proves the theorem, verifying the assertion that Maxwell's equations can be consolidated into a single multivector equation. □

There is a lot of information packed into this single equation. Where possible, we want to work with the multivector form of Maxwell's equation, either in the compact form of theorem 3.1 or the explicit form of eq. (3.10), and not decompose Maxwell's equation into the conventional representation by grade selection operations.

3.2.0.1 *Problems.*

Exercise 3.2   Dot and cross product relation to vector product.

Using coordinate expansion, convince yourself of the validity of eq. (3.7).

Exercise 3.3   Extracting the conventional Maxwell's equations.

Apply grade 0,1,2, and 3 selection operations to eq. (3.10). Determine the multiplicative (scalar or trivector) constants required to obtain eq. (3.6) from the equations that result from such grade selection operations.

3.3 WAVE EQUATION AND CONTINUITY.

Some would argue that the conventional form eq. (3.1) of Maxwell's equations have built in redundancy since continuity equations on the charge and current densities couple some of these equations. We will take an opposing view, and show that such continuity equations are necessary consequences of Maxwell's equation in its wave equation form, and derive those conditions. This amounts to a statement that the multivector current J is not completely unconstrained.

Theorem 3.2: Wave equation and continuity conditions.

The electromagnetic field is a solution to the non-homogeneous wave equation ($\Box$: see definition 2.11)

$$\Box F = \left(\nabla - \frac{1}{c} \frac{\partial}{\partial t} \right) J.$$

In source free conditions, this reduces to a homogeneous wave equation, with group velocity c, the speed of the wave in the media. When expanded explicitly in terms of electric and magnetic fields, and charge and current densities, this single equation resolves to a non-homogeneous wave equation for each of the electric and magnetic fields

$$\Box \mathbf{E} = \frac{1}{\epsilon}\boldsymbol{\nabla}\rho + \mu\frac{\partial \mathbf{J}}{\partial t} + \boldsymbol{\nabla}\times\mathbf{M}$$

$$\Box \mathbf{H} = \frac{1}{\mu}\boldsymbol{\nabla}\rho_m + \epsilon\frac{\partial \mathbf{M}}{\partial t} - \boldsymbol{\nabla}\times\mathbf{J},$$

as well as a pair of continuity equations coupling the respective charge and current densities

$$\boldsymbol{\nabla}\cdot\mathbf{J} + \frac{\partial \rho}{\partial t} = 0$$

$$\boldsymbol{\nabla}\cdot\mathbf{M} + \frac{\partial \rho_m}{\partial t} = 0.$$

Proof. To prove, we operate on theorem 3.1 with $\boldsymbol{\nabla} - (1/c)\partial_t$, one of the factors, along with the spacetime gradient, of the d'Alembertian (wave equation) operator, which gives

$$\Box F = \left(\boldsymbol{\nabla} - \frac{1}{c}\frac{\partial}{\partial t}\right) J. \tag{3.20}$$

Since the left hand side has only grades 1 and 2, eq. (3.20) splits naturally into two equations, one for grades 1,2 and one for grades 0,3

$$\begin{aligned}\Box F &= \left\langle\left(\boldsymbol{\nabla} - \frac{1}{c}\frac{\partial}{\partial t}\right) J\right\rangle_{1,2} \\ 0 &= \left\langle\left(\boldsymbol{\nabla} - \frac{1}{c}\frac{\partial}{\partial t}\right) J\right\rangle_{0,3}.\end{aligned} \tag{3.21}$$

Unpacking these further, we find that there is information carried in the requirement that the grade 0,3 selection of eq. (3.21) is zero. In particular, grade 0 selection gives

$$\begin{aligned}0 &= \langle(\boldsymbol{\nabla} - (1/c)\partial_t)J\rangle \\ &= \left\langle\left(\boldsymbol{\nabla} - \frac{1}{c}\frac{\partial}{\partial t}\right)(\eta\,(c\rho - \mathbf{J}) + I\,(c\rho_m - \mathbf{M}))\right\rangle \\ &= -\eta\left(\boldsymbol{\nabla}\cdot\mathbf{J} + \frac{\partial \rho}{\partial t}\right),\end{aligned} \tag{3.22}$$

3.3 WAVE EQUATION AND CONTINUITY.

which demonstrates the continuity condition on the electric sources. Similarly, grade three selection gives

$$\begin{aligned}
0 &= \langle (\nabla - (1/c)\partial_t) J \rangle_3 \\
&= \left\langle \left(\nabla - \frac{1}{c}\frac{\partial}{\partial t} \right) (\eta (c\rho - \mathbf{J}) + I (c\rho_m - \mathbf{M})) \right\rangle_3 \\
&= -I \left(\nabla \cdot \mathbf{M} + \frac{\partial \rho_m}{\partial t} \right),
\end{aligned} \quad (3.23)$$

which demonstrates the continuity condition on the (fictitious) magnetic sources if included in the current.

For the non-homogeneous wave equation of theorem 3.2, the current derivatives may be expanded explicitly. For the wave equation for the electric field, this is

$$\begin{aligned}
\Box \mathbf{E} &= \left\langle \left(\nabla - \frac{1}{c}\frac{\partial}{\partial t} \right) J \right\rangle_1 \\
&= \left\langle \left(\nabla - \frac{1}{c}\frac{\partial}{\partial t} \right) \left(\frac{\rho}{\epsilon} - \eta \mathbf{J} + I(c\rho_m - \mathbf{M}) \right) \right\rangle_1 \\
&= \frac{1}{\epsilon}\nabla \rho - I(\nabla \wedge \mathbf{M}) + \frac{1}{c}\eta \frac{\partial \mathbf{J}}{\partial t} \\
&= \left\langle \left(\nabla - \frac{1}{c}\frac{\partial}{\partial t} \right) J \right\rangle_1 = \frac{1}{\epsilon}\nabla \rho + \mu \frac{\partial \mathbf{J}}{\partial t} + \nabla \times \mathbf{M},
\end{aligned} \quad (3.24)$$

as claimed. The forced magnetic field equation is

$$\begin{aligned}
\Box \mathbf{H} &= \frac{1}{\eta I}\left\langle \left(\nabla - \frac{1}{c}\frac{\partial}{\partial t} \right) J \right\rangle_2 \\
&= \frac{1}{\eta I}\left\langle \left(\nabla - \frac{1}{c}\frac{\partial}{\partial t} \right)\left(\frac{\rho}{\epsilon} - \eta \mathbf{J} + I(c\rho_m - \mathbf{M}) \right) \right\rangle_2 \\
&= \frac{1}{\eta I}\left(-\nabla \wedge \mathbf{J} + Ic\nabla \rho_m + \frac{I}{c}\frac{\partial \mathbf{M}}{\partial t} \right) \\
&= \frac{1}{I}\left(-I(\nabla \times \mathbf{J}) + I\frac{1}{\mu}\nabla \rho_m + I\epsilon \frac{\partial \mathbf{M}}{\partial t} \right) \\
&= \frac{1}{\mu}\nabla \rho_m + \epsilon \frac{\partial \mathbf{M}}{\partial t} - \nabla \times \mathbf{J}.
\end{aligned} \quad (3.25)$$

□

3.4 PLANE WAVES.

With all sources zero, the free space Maxwell's equation as given by theorem 3.1 for the electromagnetic field strength reduces to just

$$\left(\nabla + \frac{1}{c}\frac{\partial}{\partial t}\right) F(\mathbf{x}, t) = 0. \tag{3.26}$$

Utilizing a phasor representation of the form definition 2.10, we will define the phasor representation of the field as

> **Definition 3.3: Plane wave.**
>
> We represent the electromagnetic field strength *plane wave* solution of Maxwell's equation in phasor form as
>
> $$F(\mathbf{x}, t) = \mathrm{Re}\left(F(\mathbf{k})e^{j\omega t}\right),$$
>
> where the complex valued multivector $F(\mathbf{k})$ also has a presumed exponential dependence
>
> $$F(\mathbf{k}) = \tilde{F} e^{-j\mathbf{k}\cdot\mathbf{x}}.$$

We will now show that solutions of the electromagnetic field wave equation have the form

> **Theorem 3.3: Plane wave solutions to Maxwell's equation.**
>
> Single frequency *plane wave solutions of Maxwell's equation* have the form
>
> $$F(\mathbf{x}, t) = \mathrm{Re}\left(\left(1 + \hat{\mathbf{k}}\right)\hat{\mathbf{k}} \wedge \mathbf{E}\, e^{-j\mathbf{k}\cdot\mathbf{x} + j\omega t}\right),$$
>
> where $\|\mathbf{k}\| = \omega/c$, $\hat{\mathbf{k}} = \mathbf{k}/\|\mathbf{k}\|$ is the unit vector pointing along the propagation direction, and $\mathbf{E}$ is any complex-valued vector variable. When a $\mathbf{E} \cdot \mathbf{k} = 0$ constraint is imposed on the vector variable $\mathbf{E}$, that variable can be interpreted as the electric field, and the solution reduces to
>
> $$F(\mathbf{x}, t) = \mathrm{Re}\left(\left(1 + \hat{\mathbf{k}}\right) \mathbf{E}\, e^{-j\mathbf{k}\cdot\mathbf{x} + j\omega t}\right),$$

showing that the field phasor $F(\mathbf{k}) = \mathbf{E}(\mathbf{k}) + I\eta\mathbf{H}(\mathbf{k})$ splits naturally into electric and magnetic components

$$\mathbf{E}(\mathbf{k}) = \mathbf{E}\,e^{-j\mathbf{k}\cdot\mathbf{x}}$$
$$\eta\mathbf{H}(\mathbf{k}) = \hat{\mathbf{k}} \times \mathbf{E}\,e^{-j\mathbf{k}\cdot\mathbf{x}},$$

where the directions $\hat{\mathbf{k}}, \mathbf{E}, \mathbf{H}$ form a right handed triple.

Proof. We wish to act on $F(\mathbf{k})e^{-j\mathbf{k}\cdot\mathbf{x}+j\omega t}$ with the spacetime gradient $\nabla + (1/c)\partial_t$, but must take care of order when applying the gradient to a non-scalar valued function. In particular, if A is a multivector, then

$$\begin{aligned}\nabla A e^{-j\mathbf{k}\cdot\mathbf{x}} &= \sum_{m=1}^{3} \mathbf{e}_m \partial_m A e^{-j\mathbf{k}\cdot\mathbf{x}} \\ &= \sum_{m=1}^{3} \mathbf{e}_m A\,(-jk_m)\,e^{-j\mathbf{k}\cdot\mathbf{x}} \\ &= -j\mathbf{k}A.\end{aligned} \qquad (3.27)$$

Therefore, insertion of the presumed phasor solution of the field from definition 3.3 into eq. (3.26) gives

$$0 = -j\left(\mathbf{k} - \frac{\omega}{c}\right)F(\mathbf{k}). \qquad (3.28)$$

If $F(\mathbf{k})$ has a left multivector factor

$$F(\mathbf{k}) = \left(\mathbf{k} + \frac{\omega}{c}\right)\tilde{F}, \qquad (3.29)$$

where $\tilde{F}$ is a multivector to be determined, then

$$\begin{aligned}\left(\mathbf{k} - \frac{\omega}{c}\right)F(\mathbf{k}) &= \left(\mathbf{k} - \frac{\omega}{c}\right)\left(\mathbf{k} + \frac{\omega}{c}\right)\tilde{F} \\ &= \left(\mathbf{k}^2 - \left(\frac{\omega}{c}\right)^2\right)\tilde{F},\end{aligned} \qquad (3.30)$$

which is zero if $\|\mathbf{k}\| = \omega/c$. Let $\|\mathbf{k}\|\tilde{F} = F_0 + F_1 + F_2 + F_3$, where F_0, F_1, F_2, and F_3 respectively have grades 0,1,2,3, so that

$$\begin{aligned} F(\mathbf{k}) &= \left(1 + \hat{\mathbf{k}}\right)(F_0 + F_1 + F_2 + F_3) \\ &= F_0 + F_1 + F_2 + F_3 + \hat{\mathbf{k}}F_0 + \hat{\mathbf{k}}F_1 + \hat{\mathbf{k}}F_2 + \hat{\mathbf{k}}F_3 \\ &= F_0 + F_1 + F_2 + F_3 + \hat{\mathbf{k}}F_0 + \hat{\mathbf{k}} \cdot F_1 + \hat{\mathbf{k}} \cdot F_2 + \hat{\mathbf{k}} \cdot F_3 \\ &\quad + \hat{\mathbf{k}} \wedge F_1 + \hat{\mathbf{k}} \wedge F_2 \\ &= \left(F_0 + \hat{\mathbf{k}} \cdot F_1\right) + \left(F_1 + \hat{\mathbf{k}}F_0 + \hat{\mathbf{k}} \cdot F_2\right) \\ &\quad + \left(F_2 + \hat{\mathbf{k}} \cdot F_3 + \hat{\mathbf{k}} \wedge F_1\right) + \left(F_3 + \hat{\mathbf{k}} \wedge F_2\right). \end{aligned} \quad (3.31)$$

Since the field F has only vector and bivector grades, the grades zero and three components of the expansion above must be zero, or

$$\begin{aligned} F_0 &= -\hat{\mathbf{k}} \cdot F_1 \\ F_3 &= -\hat{\mathbf{k}} \wedge F_2, \end{aligned} \quad (3.32)$$

so

$$\begin{aligned} F(\mathbf{k}) &= \left(1 + \hat{\mathbf{k}}\right)\left(F_1 - \hat{\mathbf{k}} \cdot F_1 + F_2 - \hat{\mathbf{k}} \wedge F_2\right) \\ &= \left(1 + \hat{\mathbf{k}}\right)\left(F_1 - \hat{\mathbf{k}}F_1 + \hat{\mathbf{k}} \wedge F_1 + F_2 - \hat{\mathbf{k}}F_2 + \hat{\mathbf{k}} \cdot F_2\right). \end{aligned} \quad (3.33)$$

The multivector $1 + \hat{\mathbf{k}}$ has the projective property of gobbling any leading factors of $\hat{\mathbf{k}}$

$$\begin{aligned} (1 + \hat{\mathbf{k}})\hat{\mathbf{k}} &= \hat{\mathbf{k}} + 1 \\ &= 1 + \hat{\mathbf{k}}, \end{aligned} \quad (3.34)$$

so for $F_i \in F_1, F_2$

$$(1 + \hat{\mathbf{k}})(F_i - \hat{\mathbf{k}}F_i) = (1 + \hat{\mathbf{k}})(F_i - F_i) = 0, \quad (3.35)$$

leaving

$$F(\mathbf{k}) = \left(1 + \hat{\mathbf{k}}\right)\left(\hat{\mathbf{k}} \cdot F_2 + \hat{\mathbf{k}} \wedge F_1\right). \quad (3.36)$$

For $\hat{\mathbf{k}} \cdot F_2$ to be non-zero F_2 must be a bivector that lies in a plane containing $\hat{\mathbf{k}}$, and $\hat{\mathbf{k}} \cdot F_2$ is a vector in that plane that is perpendicular to $\hat{\mathbf{k}}$. On the other hand $\hat{\mathbf{k}} \wedge F_1$ is non-zero only if F_1 has a non-zero component that does not lie in along the $\hat{\mathbf{k}}$ direction, but $\hat{\mathbf{k}} \wedge F_1$, like F_2 describes a plane that containing $\hat{\mathbf{k}}$. This means that having both bivector and vector free variables F_2 and F_1 provide more degrees of freedom than required. For example, if $\mathbf{E}$ is any vector, and $F_2 = \hat{\mathbf{k}} \wedge \mathbf{E}$, then

$$\begin{aligned}\left(1 + \hat{\mathbf{k}}\right) \hat{\mathbf{k}} \cdot F_2 &= \left(1 + \hat{\mathbf{k}}\right) \hat{\mathbf{k}} \cdot \left(\hat{\mathbf{k}} \wedge \mathbf{E}\right) \\ &= \left(1 + \hat{\mathbf{k}}\right) \left(\mathbf{E} - \hat{\mathbf{k}} \left(\hat{\mathbf{k}} \cdot \mathbf{E}\right)\right) \\ &= \left(1 + \hat{\mathbf{k}}\right) \hat{\mathbf{k}} \left(\hat{\mathbf{k}} \wedge \mathbf{E}\right) \\ &= \left(1 + \hat{\mathbf{k}}\right) \hat{\mathbf{k}} \wedge \mathbf{E},\end{aligned} \quad (3.37)$$

which has the form $\left(1 + \hat{\mathbf{k}}\right)\left(\hat{\mathbf{k}} \wedge F_1\right)$, so the electromagnetic field strength phasor may be generally written

$$F(\mathbf{k}) = \left(1 + \hat{\mathbf{k}}\right) \hat{\mathbf{k}} \wedge \mathbf{E} \, e^{-j \mathbf{k} \cdot \mathbf{x}}, \quad (3.38)$$

Expanding the multivector factor $\left(1 + \hat{\mathbf{k}}\right) \hat{\mathbf{k}} \wedge \mathbf{E}$ we find

$$\begin{aligned}\left(1 + \hat{\mathbf{k}}\right) \hat{\mathbf{k}} \wedge \mathbf{E} &= \hat{\mathbf{k}} \cdot \left(\hat{\mathbf{k}} \wedge \mathbf{E}\right) + \cancel{\hat{\mathbf{k}} \wedge \left(\hat{\mathbf{k}} \wedge \mathbf{E}\right)} + \hat{\mathbf{k}} \wedge \mathbf{E} \\ &= \mathbf{E} - \hat{\mathbf{k}} \left(\hat{\mathbf{k}} \wedge \mathbf{E}\right) + \hat{\mathbf{k}} \wedge \mathbf{E}.\end{aligned} \quad (3.39)$$

The vector grade has the component of $\mathbf{E}$ along the propagation direction removed (i.e. it is the rejection), so there is no loss of generality should a $\mathbf{E} \cdot \mathbf{k} = 0$ constraint be imposed. Such as constraint let's us write the bivector as a vector product $\hat{\mathbf{k}} \wedge \mathbf{E} = \hat{\mathbf{k}} \mathbf{E}$, and then use the projective property eq. (3.34) to gobble the leading $\hat{\mathbf{k}}$ factor, leaving

$$F(\mathbf{k}) = \left(1 + \hat{\mathbf{k}}\right) \mathbf{E} \, e^{-j \mathbf{k} \cdot \mathbf{x}} = \left(\mathbf{E} + I \hat{\mathbf{k}} \times \mathbf{E}\right) e^{-j \mathbf{k} \cdot \mathbf{x}}. \quad (3.40)$$

It is also noteworthy that the directions $\hat{\mathbf{k}}, \hat{\mathbf{E}}, \hat{\mathbf{H}}$ form a right handed triple, which can be seen by computing their product

$$\begin{aligned}(\hat{\mathbf{k}} \hat{\mathbf{E}}) \hat{\mathbf{H}} &= (-\hat{\mathbf{E}} \hat{\mathbf{k}})(-I \hat{\mathbf{k}} \hat{\mathbf{E}}) \\ &= +I \hat{\mathbf{E}}^2 \hat{\mathbf{k}}^2 \\ &= I.\end{aligned} \quad (3.41)$$

These vectors must all be mutually orthonormal for their product to be a pseudoscalar multiple. Should there be doubt, explicit dot products may be computed with ease using grade selection operations

$$\begin{aligned}\hat{\mathbf{k}} \cdot \hat{\mathbf{H}} &= \left\langle \hat{\mathbf{k}}(-I \hat{\mathbf{k}} \hat{\mathbf{E}}) \right\rangle = -\left\langle I \hat{\mathbf{E}} \right\rangle = 0 \\ \hat{\mathbf{E}} \cdot \hat{\mathbf{H}} &= \left\langle \hat{\mathbf{E}}(-I \hat{\mathbf{k}} \hat{\mathbf{E}}) \right\rangle = -\left\langle I \hat{\mathbf{k}} \right\rangle = 0,\end{aligned} \quad (3.42)$$

where the zeros follow by noting that $I\hat{\mathbf{E}}, I\hat{\mathbf{k}}$ are both bivectors. The conventional representation of the right handed triple relationship between the propagation direction and fields is stated as a cross product, not as a pseudoscalar relationship as in eq. (3.41). These are easily seen to be equivalent

$$\begin{aligned}\hat{\mathbf{k}} &= I\hat{\mathbf{H}}\hat{\mathbf{E}} \\ &= I(\hat{\mathbf{H}} \wedge \hat{\mathbf{E}}) \\ &= I^2(\hat{\mathbf{H}} \times \hat{\mathbf{E}}) \\ &= \hat{\mathbf{E}} \times \hat{\mathbf{H}}.\end{aligned} \qquad (3.43)$$

□

3.5 STATICS.

3.5.1 Inverting the Maxwell statics equation.

Similar to electrostatics and magnetostatics, we can restrict attention to time invariant fields ($\partial_t F = 0$) and time invariant sources ($\partial_t J = 0$), but consider both electric and magnetic sources. In that case Maxwell's equation is reduced to an invertible first order gradient equation

$$\boldsymbol{\nabla} F(\mathbf{x}) = J(\mathbf{x}), \qquad (3.44)$$

Theorem 3.4: Maxwell's statics solution.

The solution to the Maxwell statics equation is given by

$$F(\mathbf{x}) = \frac{1}{4\pi} \int_V dV' \frac{\langle (\mathbf{x} - \mathbf{x}')J(\mathbf{x}')\rangle_{1,2}}{\|\mathbf{x} - \mathbf{x}'\|^3} + F_0,$$

where F_0 is any function for which $\boldsymbol{\nabla} F_0 = 0$. The explicit expansion in electric and magnetic fields and charge and current densities is given by

$$\mathbf{E}(\mathbf{x}) = \frac{1}{4\pi} \int_V dV' \frac{1}{\|\mathbf{x} - \mathbf{x}'\|^3} \left(\frac{1}{\epsilon}(\mathbf{x} - \mathbf{x}')\rho(\mathbf{x}') + (\mathbf{x} - \mathbf{x}') \times \mathbf{M}(\mathbf{x}') \right)$$

$$\mathbf{H}(\mathbf{x}) = \frac{1}{4\pi} \int_V dV' \frac{1}{\|\mathbf{x} - \mathbf{x}'\|^3} \left(\mathbf{J}(\mathbf{x}') \times (\mathbf{x} - \mathbf{x}') + \frac{1}{\mu}(\mathbf{x} - \mathbf{x}')\rho_m(\mathbf{x}') \right).$$

We see that the solution incorporates both a Coulomb's law contribution and a Biot-Savart law contribution, as well as their magnetic source analogues if applicable.

Proof. To prove theorem 3.4, we utilize the Green's function for the (first order) gradient eq. (2.232), finding immediately

$$F(\mathbf{x}) = \int_V dV' \, G(\mathbf{x}, \mathbf{x}') \boldsymbol{\nabla}' J(\mathbf{x}')$$
$$= \left\langle \int_V dV' \, G(\mathbf{x}, \mathbf{x}') \boldsymbol{\nabla}' J(\mathbf{x}') \right\rangle_{1,2} \quad (3.45)$$
$$= \frac{1}{4\pi} \int_V dV' \left\langle \frac{(\mathbf{x} - \mathbf{x}') J(\mathbf{x}')}{\|\mathbf{x} - \mathbf{x}'\|^3} \right\rangle_{1,2}.$$

Here a no-op grade selection has been inserted to simplify subsequent manipulation[2]. We are also free to add any grade 1,2 solution of the homogeneous gradient equation, which provides the multivector form of the solution.

To unpack the multivector result, let $\mathbf{s} = \mathbf{x} - \mathbf{x}'$, and expand the grade 1,2 selection

$$\langle \mathbf{s} J \rangle_{1,2} = \eta \langle \mathbf{s}(c\rho - \mathbf{J}) \rangle_{1,2} + \langle \mathbf{s} I(c\rho_m - \mathbf{M}) \rangle_{1,2}$$
$$= \eta c \mathbf{s} \rho - \eta (\mathbf{s} \wedge \mathbf{J}) + c I \mathbf{s} \rho_m - I(\mathbf{s} \wedge \mathbf{M}) \quad (3.46)$$
$$= \frac{1}{\epsilon} \mathbf{s} \rho + \eta I (\mathbf{J} \times \mathbf{s}) + \mathbf{s} c \rho_m I + \mathbf{s} \times \mathbf{M},$$

so the field is

$$F(\mathbf{x}) = \frac{1}{4\pi} \int_V dV' \frac{1}{\|\mathbf{x} - \mathbf{x}'\|^3} \left(\frac{1}{\epsilon} \mathbf{s} \rho + \mathbf{s} \times \mathbf{M} \right)$$
$$+ I \frac{1}{4\pi} \int_V dV' \frac{1}{\|\mathbf{x} - \mathbf{x}'\|^3} (\mathbf{s} c \rho_m + \eta \mathbf{J} \times \mathbf{s}). \quad (3.47)$$

Comparing this expansion to the field components $F = \mathbf{E} + \eta I \mathbf{H}$, our job is done. □

---

[2] If this grade selection filter is omitted, it is possible to show that the scalar and pseudoscalar contributions to the $(\mathbf{x} - \mathbf{x}') J$ product are zero on the boundary of the Green's integration volume. [16]

3.5.2 Enclosed charge.

In conventional electrostatics we obtain a relation between the normal electric field component and the enclosed charge by integrating the electric field divergence. The geometric algebra generalization of this relates the product of the normal and the electromagnetic field strength related to the enclosed multivector current

> **Theorem 3.5: Enclosed multivector current.**
>
> The *enclosed multivector current* in the volume is related to the surface integral of $\hat{\mathbf{n}}F$ over the boundary of the volume by
>
> $$\int_{\partial V} dA\, \hat{\mathbf{n}} F = \int_V dV\, J.$$
>
> This is a multivector equation, carrying information for each grade in the multivector current. That grade selection yeilds
>
> $$\int_{\partial V} dA\, \hat{\mathbf{n}} \cdot \mathbf{E} = \frac{1}{\epsilon} \int_V dV\, \rho$$
> $$\int_{\partial V} dA\, \hat{\mathbf{n}} \times \mathbf{H} = \int_V dV\, \mathbf{J}$$
> $$\int_{\partial V} dA\, \hat{\mathbf{n}} \times \mathbf{E} = -\int_V dV\, \mathbf{M}$$
> $$\int_{\partial V} dA\, \hat{\mathbf{n}} \cdot \mathbf{H} = \frac{1}{\mu} \int_V dV\, \rho_m.$$

Proof. To prove theorem 3.5 simply evaluate the volume integral of the gradient of the field using theorem 2.11

$$\int_V dV\, \boldsymbol{\nabla} F = \int_{\partial V} dA\, \hat{\mathbf{n}} F, \qquad (3.48)$$

and note that

$$\int_V dV\, \boldsymbol{\nabla} F = \int_V dV\, J. \qquad (3.49)$$

This is a multivector relationship, containing a substantial amount of information, which can be extracted by expanding $\hat{\mathbf{n}}F$

$$\begin{aligned}\hat{\mathbf{n}}F &= \hat{\mathbf{n}}\left(\mathbf{E} + I\eta\mathbf{H}\right) \\ &= \hat{\mathbf{n}} \cdot \mathbf{E} + I(\hat{\mathbf{n}} \times \mathbf{E}) + I\eta\left(\hat{\mathbf{n}} \cdot \mathbf{H} + I\hat{\mathbf{n}} \times \mathbf{H}\right) \\ &= \hat{\mathbf{n}} \cdot \mathbf{E} - \eta(\hat{\mathbf{n}} \times \mathbf{H}) + I(\hat{\mathbf{n}} \times \mathbf{E}) + I\eta(\hat{\mathbf{n}} \cdot \mathbf{H}).\end{aligned} \quad (3.50)$$

Inserting this into theorem 3.5, and equating grades, we find

$$\begin{aligned}\int_{\partial V} dA\, \hat{\mathbf{n}} \cdot \mathbf{E} &= \int_V dV \frac{\rho}{\epsilon} \\ -\int_{\partial V} dA\, \eta(\hat{\mathbf{n}} \times \mathbf{H}) &= -\int_V dV\, \eta\mathbf{J} \\ I\int_{\partial V} dA\, (\hat{\mathbf{n}} \times \mathbf{E}) &= -I\int_V dV\, \mathbf{M} \\ I\int_{\partial V} dA\, \eta(\hat{\mathbf{n}} \cdot \mathbf{H}) &= I\int_V dV\, c\rho_{\mathrm{m}},\end{aligned} \quad (3.51)$$

which completes the proof after cancelling common factors and some minor adjustments of the multiplicative constants. Of course eq. (3.51) could have obtained directly from Maxwell's equations in their conventional form eq. (3.1). However, had we integrated the conventional Maxwell's equations, it would not have been obvious that the crazy mix of fields, sources, dot and cross products in eq. (3.49) had a hidden structure as simple as $\int_{\partial V} dA\, \hat{\mathbf{n}}F = \int_V dV\, J$. □

3.5.3 *Enclosed current.*

In this section we will present the generalization of Ampere's law to line integrals of the total electromagnetic field strength.

> **Theorem 3.6: Line integral of the field.**
>
> The *line integral of the electromagnetic field strength* is
>
> $$\oint_{\partial A} d\mathbf{x}\, F = I \int_A dA \left(\hat{\mathbf{n}}J - \frac{\partial F}{\partial n}\right),$$

where $\partial F/\partial n = (\hat{\mathbf{n}} \cdot \boldsymbol{\nabla}) F$. Expressed in terms of the conventional consistent fields and sources, this multivector relationship expands to four equations, one for each grade

$$\oint_{\partial A} d\mathbf{x} \cdot \mathbf{E} = \int_A dA\, \hat{\mathbf{n}} \cdot \mathbf{M}$$

$$\oint_{\partial A} d\mathbf{x} \times \mathbf{H} = \int_A dA \left(-\hat{\mathbf{n}} \times \mathbf{J} + \frac{\hat{\mathbf{n}} \rho_m}{\mu} - \frac{\partial \mathbf{H}}{\partial n} \right)$$

$$\oint_{\partial A} d\mathbf{x} \times \mathbf{E} = \int_A dA \left(\hat{\mathbf{n}} \times \mathbf{M} + \frac{\hat{\mathbf{n}} \rho}{\epsilon} - \frac{\partial \mathbf{E}}{\partial n} \right)$$

$$\oint_{\partial A} d\mathbf{x} \cdot \mathbf{H} = - \int_A dA\, \hat{\mathbf{n}} \cdot \mathbf{J}.$$

The last of the scalar equations in theorem 3.6 is Ampere's law

$$\oint_{\partial A} d\mathbf{x} \cdot \mathbf{H} = \int_A \hat{\mathbf{n}} \cdot \mathbf{J} = I_{\text{enc}}, \tag{3.52}$$

and the first is the dual of Ampere's law for (fictitious) magnetic current density[3]. In eq. (3.52) the flux of the electric current density equals the enclosed current flowing through an open surface. This enclosed current equals the line integral of the magnetic field around the boundary of that surface.

Proof. To prove theorem 3.6 we compute the surface integral of the current $J = \boldsymbol{\nabla} F$

$$\int_A d^2\mathbf{x}\, J = \int_A d^2\mathbf{x}\, \boldsymbol{\nabla} F. \tag{3.53}$$

As we are working in $\mathbb{R}^3$ not $\mathbb{R}^2$, the gradient may not be replaced by the vector derivative in eq. (3.53). Instead we must split the gradient into its vector derivative component, the projection of the gradient onto the tangent plane of the integration surface, and its normal component

$$\boldsymbol{\nabla} = \partial + \hat{\mathbf{n}}(\hat{\mathbf{n}} \cdot \boldsymbol{\nabla}). \tag{3.54}$$

The surface integral form eq. (2.142) of the fundamental theorem of geometric calculus may be applied to the vector derivative portion of the field integral

$$\int_A d^2\mathbf{x}\, \boldsymbol{\nabla} F = \int_A d^2\mathbf{x}\, \partial F + \int_A d^2\mathbf{x}\, \hat{\mathbf{n}} (\hat{\mathbf{n}} \cdot \boldsymbol{\nabla}) F, \tag{3.55}$$

---

[3] Even without the fictitious magnetic sources, neither the name nor applications of the two cross product line integrals with the normal derivatives are familiar to the author.

3.5 STATICS.

so

$$\oint_{\partial A} d\mathbf{x}\, F = \int_A d^2\mathbf{x}\, (J - \hat{\mathbf{n}}\, (\hat{\mathbf{n}} \cdot \boldsymbol{\nabla})\, F)$$
$$= \int_A dA\, (I\hat{\mathbf{n}} J - (\hat{\mathbf{n}} \cdot \boldsymbol{\nabla})\, IF) \quad (3.56)$$
$$= \int_A dA \left(I\hat{\mathbf{n}} J - I\frac{\partial F}{\partial n} \right),$$

where the surface area bivector has been written in its dual form $d^2\mathbf{x} = I\hat{\mathbf{n}} dA$ in terms of a scalar area element, and the directional derivative has been written in scalar form with respect to a parameter n that represents the length along the normal direction. This proves the first part of theorem 3.6.

Observe that the $d\mathbf{x}\, F$ product has all possible grades

$$d\mathbf{x}\, F = d\mathbf{x}\, (\mathbf{E} + I\eta \mathbf{H})$$
$$= d\mathbf{x} \cdot \mathbf{E} + I\eta d\mathbf{x} \cdot \mathbf{H} + d\mathbf{x} \wedge \mathbf{E} + I\eta d\mathbf{x} \wedge \mathbf{H} \quad (3.57)$$
$$= d\mathbf{x} \cdot \mathbf{E} - \eta(d\mathbf{x} \times \mathbf{H}) + I(d\mathbf{x} \times \mathbf{E}) + I\eta(d\mathbf{x} \cdot \mathbf{H}),$$

as does the $I\hat{\mathbf{n}} J$ product (in general)

$$I\hat{\mathbf{n}} J = I\hat{\mathbf{n}} \left(\frac{\rho}{\epsilon} - \eta \mathbf{J} + I\, (c\rho_{\mathrm{m}} - \mathbf{M}) \right)$$
$$= \hat{\mathbf{n}} I \frac{\rho}{\epsilon} - \eta \hat{\mathbf{n}} I \mathbf{J} - \hat{\mathbf{n}} c\rho_{\mathrm{m}} + \hat{\mathbf{n}} \mathbf{M} \quad (3.58)$$
$$= \hat{\mathbf{n}} \cdot \mathbf{M} + \eta(\hat{\mathbf{n}} \times \mathbf{J}) - \hat{\mathbf{n}} c\rho_{\mathrm{m}} + I(\hat{\mathbf{n}} \times \mathbf{M}) + \hat{\mathbf{n}} I \frac{\rho}{\epsilon} - \eta I(\hat{\mathbf{n}} \cdot \mathbf{J}).$$

On the other hand $IF = I\mathbf{E} - \eta \mathbf{H}$ has only grades 1,2, like F itself. This allows the line integrals to be split by grade selection into components with and without a normal derivative

$$\oint_{\partial A} \langle d\mathbf{x}\, F \rangle_{0,3} = \int_A dA\, \langle I\hat{\mathbf{n}} J \rangle_{0,3}$$
$$\oint_{\partial A} \langle d\mathbf{x}\, F \rangle_{1,2} = \int_A dA\, (\langle I\hat{\mathbf{n}} J \rangle_{1,2} - (\hat{\mathbf{n}} \cdot \boldsymbol{\nabla})\, IF). \quad (3.59)$$

The first of eq. (3.59) contains Ampere's law and its dual as one multivector equation, which can be seen more readily by explicit expansion in the constituent fields and sources using eq. (3.57), eq. (3.58)

$$\oint_{\partial A} (d\mathbf{x} \cdot \mathbf{E} + I\eta(d\mathbf{x} \cdot \mathbf{H})) = \int_A dA \left(\hat{\mathbf{n}} \cdot \mathbf{M} - \eta I(\hat{\mathbf{n}} \cdot \mathbf{J}) \right)$$

$$\oint_{\partial A} (-\eta(d\mathbf{x} \times \mathbf{H}) + I(d\mathbf{x} \times \mathbf{E})) = \int_A dA \Big(\eta(\hat{\mathbf{n}} \times \mathbf{J}) - \hat{\mathbf{n}} c \rho_m$$
$$+ I(\hat{\mathbf{n}} \times \mathbf{M}) + \hat{\mathbf{n}} I \frac{\rho}{\epsilon} - \frac{\partial}{\partial n} (I\mathbf{E} - \eta \mathbf{H}) \Big).$$
(3.60)

Further grade selection operations, and minor adjustments of the leading constants completes the proof.

It is also worth pointing out that for pure magnetostatics problems where $J = \eta \mathbf{J}, F = I\eta \mathbf{H}$, that Ampere's law can be written in a trivector form

$$\oint_{\partial A} d\mathbf{x} \wedge F = I \int_A dA \, \hat{\mathbf{n}} \cdot \mathbf{J} = I\eta \int_A dA \, \hat{\mathbf{n}} \cdot \mathbf{J}. \tag{3.61}$$

This encodes the fact that the magnetic field component of the total electromagnetic field strength is most naturally expressed in geometric algebra as a bivector. □

3.5.4 Example field calculations.

Having seen a number of theoretical applications of the geometric algebra framework, let's now see how some of our new tools can be used to calculate the fields for specific static electromagnetism charge and current configurations.

3.5.4.1 Line segment.

In this example the (electric) field is calculated at a point on the z-axis, due to a finite line charge density of λ along a segment $[a, b]$ of the x-axis. The geometry of the problem is illustrated in fig. 3.1.

This is a fairly simple problem, and can be found in most introductory electromagnetic texts, usually set with the field observation point on the z-axis, and with a symmetric interval $[-l/2, l/2]$, which has the side effect of killing off all but the x-axis component of the field. For comparision

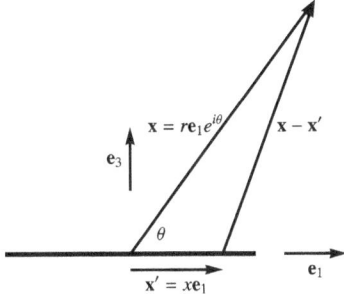

Figure 3.1: Line charge density.

purposes, this problem will be tackled first using conventional algebra, and then using geometric algebra.

Conventional approach. The integral we wish to evaluate is

$$\mathbf{E}(\mathbf{x}) = \frac{\lambda}{4\pi\epsilon} \int_a^b dx \frac{(r\cos\theta - x)\mathbf{e}_1 + r\sin\theta\mathbf{e}_3}{(r^2 + x^2 - 2rx\cos\theta)^{3/2}}. \tag{3.62}$$

This can be non-dimensionalized with a $u = x/r$ change of variables, and yields an integral for the x component and the z component of the field

$$E_x = \frac{\lambda}{4\pi\epsilon r} \int_{a/r}^{b/r} du \frac{\cos\theta - u}{(1 + u^2 - 2u\cos\theta)^{3/2}}$$
$$E_y = \frac{\lambda\sin\theta}{4\pi\epsilon r} \int_{a/r}^{b/r} du \left(1 + u^2 - 2u\cos\theta\right)^{-3/2}. \tag{3.63}$$

There is a common integral in the x and y components of the field. We can tidy this up a bit by writing

$$A = \int_{a/r}^{b/r} du \left(1 + u^2 - 2u\cos\theta\right)^{-3/2}$$
$$B = \int_{a/r}^{b/r} u du \left(1 + u^2 - 2u\cos\theta\right)^{-3/2}, \tag{3.64}$$

and then put the pieces back together again for the total field

$$\mathbf{E} = \frac{\lambda}{4\pi\epsilon r} \left((A\cos\theta - B)\mathbf{e}_1 + A\sin\theta\mathbf{e}_3\right). \tag{3.65}$$

Some additional structure can be imposed by introducing a rotation matrix to express the field observation point

$$\mathbf{x} = r\mathbf{R}_\theta \mathbf{e}_1, \tag{3.66}$$

where

$$\mathbf{R}_\theta = \begin{bmatrix} \cos\theta & 0 & -\sin\theta \\ 0 & 1 & 0 \\ \sin\theta & 0 & \cos\theta \end{bmatrix}. \tag{3.67}$$

Writing **1** for the $\mathbb{R}^3$ identity matrix, the field is

$$\mathbf{E} = \frac{\lambda}{4\pi\epsilon r} (A\mathbf{R}_\theta - B\mathbf{1}) \mathbf{e}_1. \tag{3.68}$$

In retrospect we could have started using eq. (3.66) and obtained this result more directly. The A integral above results in both scaling and rotation of the field, depending on the observation point and the limits of the integration. The B integral contributes only to the x-axis oriented component of the field.

Using geometric algebra. Introducing a unit imaginary $i = \mathbf{e}_{13}$ for the rotation from the x-axis to the z-axis, the field point observation point is

$$\mathbf{x} = r\mathbf{e}_1 e^{i\theta}. \tag{3.69}$$

The charge element point is $\mathbf{x}' = x\mathbf{e}_1$, so the difference can now be written with $\mathbf{e}_1$ factored to the left or to the right

$$\mathbf{x} - \mathbf{x}' = \mathbf{e}_1 \left(re^{i\theta} - x\right) = \left(re^{-i\theta} - x\right)\mathbf{e}_1. \tag{3.70}$$

These left and right factors can be used to convert the squared length of $\mathbf{x} - \mathbf{x}'$ into from a vector product into a product of conventional looking complex conjugates

$$\begin{aligned}(\mathbf{x} - \mathbf{x}')^2 &= \left(re^{-i\theta} - x\right)\mathbf{e}_1\mathbf{e}_1\left(re^{i\theta} - x\right) \\ &= \left(re^{-i\theta} - x\right)\left(re^{i\theta} - x\right),\end{aligned} \tag{3.71}$$

so the squared length of the difference is

$$\begin{aligned}(\mathbf{x} - \mathbf{x}')^2 &= r^2 + x^2 - rx\left(e^{i\theta} + e^{-i\theta}\right) \\ &= r^2 + x^2 - 2rx\cos\theta,\end{aligned} \tag{3.72}$$

and the total (electric) field is

$$\begin{aligned}F &= \frac{\lambda}{4\pi\epsilon} \int_a^b dx \frac{r\mathbf{e}_1 e^{i\theta} - x\mathbf{e}_1}{(r^2 + x^2 - 2xr\cos\theta)^{3/2}} \\ &= \frac{\lambda\mathbf{e}_1}{4\pi\epsilon r} \int_{a/r}^{b/r} du \frac{e^{i\theta} - u}{(1 + u^2 - 2u\cos\theta)^{3/2}}.\end{aligned} \tag{3.73}$$

We have replaced the matrix representation that had nine components, four zeros, and a lot of redundancy with a simple multivector result. Moreover, the integral factor has the appearance of a conventional complex integral, and we can toss it as is into any numerical or symbol integration systems capable of complex number integrals for evaluation. The end result is a single vector valued inverse radial factor $\lambda \mathbf{e}_1/(4\pi\epsilon r)$, multiplying by an integral that served to either scale or rotate-and-scale.

In particular, for $\theta = \pi/2$, plugging this integral into Mathematica, we find

$$\int du \frac{e^{i\theta} - u}{(1 + u^2 - 2u\cos\theta)^{3/2}} = \frac{1 + iu}{\sqrt{1 + u^2}}, \tag{3.74}$$

and for other angles $\theta \neq n\pi/2$

$$\int du \frac{e^{i\theta} - u}{(1 + u^2 - 2u\cos\theta)^{3/2}} = \frac{(1 - ue^{-i\theta})\sqrt{1 + u^2 - 2u\cos\theta}}{(1 + u^2)\sin(2\theta)}. \tag{3.75}$$

The numerator factors like $\mathbf{e}_1(1 + iu)$ and $\mathbf{e}_1(1 - ue^{-i\theta})$ compactly describe the direction of the vector field at the observation point. Either of these can be expanded explicitly in sines and cosines if desired

$$\begin{aligned}\mathbf{e}_1(1 + iu) &= \mathbf{e}_1 + u\mathbf{e}_3 \\ \mathbf{e}_1(1 - ue^{-i\theta}) &= \mathbf{e}_1(1 - u\cos\theta) + u\mathbf{e}_3\sin\theta.\end{aligned} \tag{3.76}$$

Perhaps more interesting than the precise form of the solution is the fact that geometric algebra allows for the introduction of a "complex plane" for many problems that have only two degrees of freedom. When such a complex plane is introduced, existing Computer Algebra Systems (CAS), like Mathematica, can be utilized for the grunt work of the evaluation.

3.5.4.2 Infinite line current.

Given a static line charge density and current density along the z-axis

$$\begin{aligned}\rho(\mathbf{x}) &= \lambda\delta(x)\delta(y) \\ \mathbf{J}(\mathbf{x}) &= \mathbf{v}\rho(\mathbf{x}) = v\lambda\mathbf{e}_3\delta(x)\delta(y),\end{aligned} \tag{3.77}$$

the total multivector current is

$$\begin{aligned}J &= \eta(c\rho - \mathbf{J}) \\ &= \eta(c - v\mathbf{e}_3)\lambda\delta(x)\delta(y) \\ &= \frac{\lambda}{\epsilon}\left(1 - \frac{v}{c}\mathbf{e}_3\right)\delta(x)\delta(y).\end{aligned} \tag{3.78}$$

We can find the field for this current using theorem 3.4. To do so, let the field observation point be $\mathbf{x} = \mathbf{x}_\perp + z\mathbf{e}_3$, so the total field is

$$F(\mathbf{x}) = \frac{\lambda}{4\pi\epsilon} \int_V dx' dy' dz' \frac{\langle (\mathbf{x} - \mathbf{x}')(1 - (v/c)\mathbf{e}_3) \rangle_{1,2}}{\|\mathbf{x} - \mathbf{x}'\|^3} \delta(x')\delta(y')$$

$$= \frac{\lambda}{4\pi\epsilon} \int_{-\infty}^{\infty} dz' \frac{\langle (\mathbf{x}_\perp + (z - z')\mathbf{e}_3)(1 - (v/c)\mathbf{e}_3) \rangle_{1,2}}{\left(\mathbf{x}_\perp^2 + (z - z')^2\right)^{3/2}}$$

$$= \frac{\lambda(\mathbf{x}_\perp - (v/c)\mathbf{x}_\perp\mathbf{e}_3)}{4\pi\epsilon} \int_{-\infty}^{\infty} \frac{dz'}{\left(\mathbf{x}_\perp^2 + (z - z')^2\right)^{3/2}} \qquad (3.79)$$

$$+ \frac{\lambda\mathbf{e}_3}{4\pi\epsilon} \int_{-\infty}^{\infty} \frac{(z - z')dz'}{\left(\mathbf{x}_\perp^2 + (z - z')^2\right)^{3/2}}.$$

The first integral is $2/\mathbf{x}_\perp^2$, whereas the second is zero (odd function, over even interval). The bivector term of the grade selection above had a $\mathbf{x}_\perp \wedge \mathbf{e}_3 = \mathbf{x}_\perp \mathbf{e}_3$ factor, which can be further reduced using cylindrical coordinates $\mathbf{x} = R\hat{\boldsymbol{\rho}} + z\mathbf{e}_3$, since $\mathbf{x}_\perp = R\hat{\boldsymbol{\rho}}$, which leaves

$$F(\mathbf{x}) = \frac{\lambda}{2\pi\epsilon R}\hat{\boldsymbol{\rho}}(1 - \mathbf{v}/c) = \mathbf{E}(1 - \mathbf{v}/c) = \mathbf{E} + I\left(\frac{\mathbf{v}}{c} \times \mathbf{E}\right), \qquad (3.80)$$

where $\mathbf{v} = v\mathbf{e}_3$. The vector component of this is the electric field, which is therefore directed radially, whereas the (dual) magnetic field $\eta I\mathbf{H}$ is a set of oriented planes spanning the radial and z-axis directions. We can also see that there is a constant proportionality factor that relates the electric and magnetic field components, namely

$$I\eta\mathbf{H} = -\mathbf{E}\mathbf{v}/c, \qquad (3.81)$$

or

$$\mathbf{H} = \mathbf{v} \times \mathbf{D}. \qquad (3.82)$$

Exercise 3.4 Linear magnetic density and currents.

Given magnetic charge density $\rho_m = \lambda_m \delta(x)\delta(y)$, and current density $\mathbf{M} = v\mathbf{e}_3\rho_m = \mathbf{v}\rho_m$, show that the field is given by

$$F(\mathbf{x}) = \frac{\lambda_m c}{2\pi R} I\hat{\boldsymbol{\rho}}\left(1 - \frac{\mathbf{v}}{c}\right),$$

or with $\mathbf{B} = \lambda_m \hat{\boldsymbol{\rho}}/(2\pi R)$,

$$F = \mathbf{B} \times \mathbf{v} + cI\mathbf{B}.$$

3.5.4.3 Infinite planar current.

A variation on the above example puts a uniform charge density $\rho(\mathbf{x}) = \sigma\delta(z)$ in a plane, along with an associated current density $\mathbf{J}(\mathbf{x}) = v\mathbf{e}_1 e^{i\theta}\rho(\mathbf{x})$, where $i = \mathbf{e}_{12}$. Letting $\mathbf{v} = v\mathbf{e}_1 e^{i\theta}$, the multivector current is

$$J(\mathbf{x}) = \sigma\eta\,(c - \mathbf{v})\,\delta(z), \tag{3.89}$$

so the field off the plane is

$$F(\mathbf{x}) = \frac{\sigma}{4\pi\epsilon} \iiint \frac{dz'\,dA'}{\|\mathbf{x} - \mathbf{x}'\|^3} \langle(\mathbf{x} - \mathbf{x}')(1 - \mathbf{v}/c)\rangle_{1,2}\delta(z'). \tag{3.90}$$

If $\mathbf{x}_\| = (\mathbf{x} \wedge \mathbf{e}_3)\mathbf{e}_3$, and $\mathbf{x}'_\| = (\mathbf{x}' \wedge \mathbf{e}_3)\mathbf{e}_3$, are the components of the vectors $\mathbf{x}, \mathbf{x}'$ in the x-y plane, then integration over z' and a change of variables $\mathbf{x}'_\| - \mathbf{x}_\| = u\mathbf{e}_1 e^{i\alpha}$ yields

$$F(\mathbf{x}) = \frac{\sigma}{4\pi\epsilon} \int_{u=0}^{\infty} \int_{\alpha=0}^{2\pi} \frac{u\,du\,d\alpha}{(z^2 + u^2)^{3/2}} \langle\left(z\mathbf{e}_3 + u\mathbf{e}_1 e^{i\alpha}\right)(1 - \mathbf{v}/c)\rangle_{1,2}. \tag{3.91}$$

The $e^{i\alpha}$ integrands are killed, so for $z \neq 0$, the field is

$$F(\mathbf{x}) = \frac{\sigma z}{2\epsilon|z|} \langle\mathbf{e}_3(1 - \mathbf{v}/c)\rangle_{1,2}. \tag{3.92}$$

Since $\mathbf{v} \in \text{span}\{\mathbf{e}_1, \mathbf{e}_2\}$ the product $\mathbf{e}_3\mathbf{v}$ is a bivector and the grade selection can be dropped, leaving

$$F(\mathbf{x}) = \frac{\sigma\,\text{sgn}(z)}{2\epsilon} \mathbf{e}_3\left(1 - \frac{\mathbf{v}}{c}\right). \tag{3.93}$$

This field toggles sign when crossing the plane, but is constant otherwise. The electric and magnetic field components are once again related by eq. (3.82).

It is worth pointing out that we can also compute this with Gauss's theorem, as stated in theorem 3.6, using the usual rectangular pillbox configuration, as illustrated in fig. 3.2, which gives us

$$A\mathbf{e}_3\,(F(z+) - F(z-)) = A\sigma\eta\,(c - \mathbf{v}) \int dz\delta(z). \tag{3.94}$$

Since $\eta c = 1/\epsilon$, after pre-multiplying with $\mathbf{e}_3/A$, we have

$$F(z+) - F(z-) = \frac{\sigma}{\epsilon}\mathbf{e}_3\left(1 - \frac{\mathbf{v}}{c}\right). \tag{3.95}$$

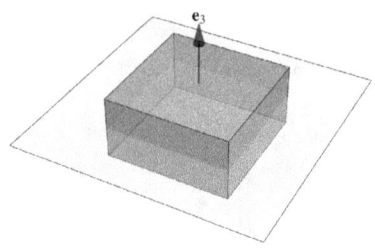

Figure 3.2: Rectangular Gaussian pillbox.

Assuming a symmetrical field, as is often required for Gauss's law applications, we set

$$F(z-) = -F(z+), \tag{3.96}$$

so

$$F(\mathbf{x}) = \frac{\sigma \operatorname{sgn}(z)}{2\epsilon} \mathbf{e}_3 \left(1 - \frac{\mathbf{v}}{c}\right), \tag{3.97}$$

as found by direct integration.

Exercise 3.5 Planar magnetic density and currents.

Given magnetic charge density $\rho_m = \sigma_m \delta(z)$, and current density $\mathbf{M} = \mathbf{v}\rho_m$, $\mathbf{v} = v\mathbf{e}_1 e^{i\theta}$, $i = \mathbf{e}_{12}$, show that the field is given by

$$F(\mathbf{x}) = \frac{\sigma_m c \operatorname{sgn}(z)}{2} i \left(1 - \frac{\mathbf{v}}{c}\right).$$

3.5.4.4 Arc line charge.

In this example we will examine the (electric) field due to a line charge density of λ along a circular arc segment $\phi' \in [a, b]$, of radius r in the x-y plane. The field will be evaluated at the spherical coordinate point (R, θ, ϕ), as illustrated in fig. 3.3.

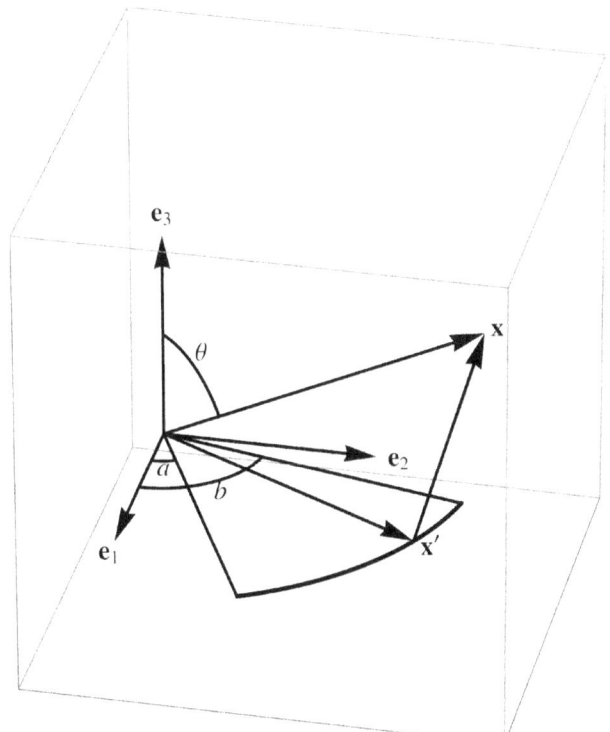

Figure 3.3: Circular line charge.

Using the GA spherical parameterization eq. (2.93), the observation point now has the simple representation

$$\mathbf{x} = R\mathbf{e}_3 e^{j\theta}, \tag{3.98}$$

and is the product of a polar directed vector with a complex exponential whose argument is the polar rotation angle. The bivector j is a function of the azimuthal angle ϕ, and encodes all the geometry of the rotation. To sum the contributions of the charge elements we need the distance between the charge element and the observation point. That vector difference is

$$\mathbf{x} - \mathbf{x}' = R\mathbf{e}_3 e^{j\theta} - r\mathbf{e}_1 e^{i\phi'}. \tag{3.99}$$

Compare this to the tuple representation

$$\mathbf{x} - \mathbf{x}' = (R\sin\theta\cos\phi - r\cos\phi', R\sin\theta\sin\phi - r\cos\phi', \cos\theta), \tag{3.100}$$

for which the prospect of working with is considerably less attractive. The squared length of eq. (3.99) is

$$(\mathbf{x} - \mathbf{x}')^2 = R^2 + r^2 - 2Rr\left(\mathbf{e}_3 e^{j\theta}\right)\cdot\left(\mathbf{e}_1 e^{i\phi'}\right). \tag{3.101}$$

The dot product of unit vectors in eq. (3.101) can be reduced using scalar grade selection

$$\begin{aligned}\left(\mathbf{e}_3 e^{j\theta}\right)\cdot\left(\mathbf{e}_1 e^{i\phi'}\right) &= \left\langle\left(\mathbf{e}_1 \sin\theta e^{i\phi}\right)\left(\mathbf{e}_1 e^{i\phi'}\right)\right\rangle \\ &= \sin\theta\left\langle e^{-i\phi} e^{i\phi'}\right\rangle \\ &= \sin\theta\cos(\phi' - \phi),\end{aligned} \tag{3.102}$$

so

$$\|\mathbf{x} - \mathbf{x}'\| = \sqrt{R^2 + r^2 - 2Rr\sin\theta\cos(\phi' - \phi)}. \tag{3.103}$$

The electric field is

$$F = \frac{1}{4\pi\epsilon_0}\int_a^b \lambda r d\phi' \frac{R\mathbf{e}_3 e^{j\theta} - r\mathbf{e}_1 e^{i\phi'}}{(R^2 + r^2 - 2Rr\sin\theta\cos(\phi' - \phi))^{3/2}}. \tag{3.104}$$

Non-dimensionalizing eq. (3.104) with $u = r/R$, a change of variables $\alpha = \phi' - \phi$, and noting that $i\hat{\boldsymbol{\phi}} = \mathbf{e}_1 e^{i\phi}$, the field is

$$\begin{aligned}F &= \frac{\lambda r}{4\pi\epsilon_0 R^2}\int_{a-\phi}^{b-\phi} d\alpha \frac{\mathbf{e}_3 e^{j\theta} - u\mathbf{e}_1 e^{i\phi} e^{i\alpha}}{(1 + u^2 - 2u\sin\theta\cos\alpha)^{3/2}} \\ &= \frac{\lambda r}{4\pi\epsilon_0 R^2}\int_{a-\phi}^{b-\phi} d\alpha \frac{\hat{\mathbf{r}} + \hat{\boldsymbol{\phi}} ui e^{i\alpha}}{(1 + u^2 - 2u\sin\theta\cos\alpha)^{3/2}},\end{aligned} \tag{3.105}$$

or

$$F = \hat{\mathbf{r}} \left(\frac{\lambda r}{4\pi\epsilon_0 R^2} \int_{a-\phi}^{b-\phi} \frac{d\alpha}{\left(1 + u^2 - 2u\sin\theta\cos\alpha\right)^{3/2}} \right)$$
$$+ \hat{\boldsymbol{\phi}} \left(\frac{\lambda r u i}{4\pi\epsilon_0 R^2} \int_{a-\phi}^{b-\phi} \frac{e^{i\alpha} d\alpha}{\left(1 + u^2 - 2u\sin\theta\cos\alpha\right)^{3/2}} \right). \quad (3.106)$$

Without CAS support for GA, this pair of integrals has to be evaluated separately. The first integral scales the radial component of the electric field. The second integral scales and rotates $\hat{\boldsymbol{\phi}}$ within the azimuthal plane, producing an electric field component in a $\hat{\boldsymbol{\phi}}' = \hat{\boldsymbol{\phi}} e^{i\Phi}$ direction.

3.5.4.5 Field of a ring current.

Let's now compute the field due to a static charge and current density on a ring of radius R as illustrated in fig. 3.4.

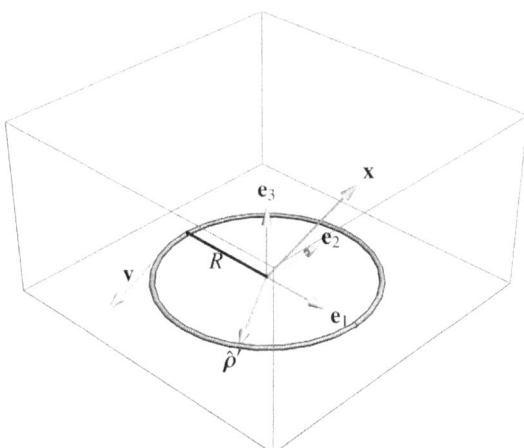

Figure 3.4: Field due to a circular distribution.

A static charge distribution on a ring at $z = 0$ has the form

$$\rho(\mathbf{x}) = \lambda \delta(z) \delta(r - R). \quad (3.107)$$

As always the current distribution is of the form $\mathbf{J} = \mathbf{v}\rho$, and in this case the velocity is azimuthal $\mathbf{v} = \mathbf{e}_2 e^{i\phi}$, $i = \mathbf{e}_{12}$. The total multivector current is

$$J = \frac{1}{\epsilon} \lambda \delta(z) \delta(r - R) \left(1 - \frac{\mathbf{v}}{c}\right). \quad (3.108)$$

Let the point that we observe the field, and the integration variables be

$$\begin{aligned}\mathbf{x} &= z\mathbf{e}_3 + r\hat{\boldsymbol{\rho}} \\ \mathbf{x}' &= z'\mathbf{e}_3 + r'\hat{\boldsymbol{\rho}}'.\end{aligned} \quad (3.109)$$

The field is

$$\begin{aligned}F(\mathbf{x}) &= \frac{\lambda}{4\pi\epsilon}\iiint dz'\, r'\, dr'\, d\phi'\, \delta(z')\delta(r' - R) \times \\ & \qquad \frac{\left\langle((z - z')\mathbf{e}_3 + r\hat{\boldsymbol{\rho}} - r'\hat{\boldsymbol{\rho}}')\left(1 - \frac{v}{c}\mathbf{e}_2 e^{i\phi'}\right)\right\rangle_{1,2}}{((z - z')^2 + (r\hat{\boldsymbol{\rho}} - r'\hat{\boldsymbol{\rho}}')^2)^{3/2}} \\ &= \frac{\lambda}{4\pi\epsilon}\int R d\phi' \frac{\left\langle(z\mathbf{e}_3 + r\hat{\boldsymbol{\rho}} - R\hat{\boldsymbol{\rho}}')\left(1 - \frac{v}{c}\mathbf{e}_2 e^{i\phi'}\right)\right\rangle_{1,2}}{(z^2 + (r\hat{\boldsymbol{\rho}} - R\hat{\boldsymbol{\rho}}')^2)^{3/2}}.\end{aligned} \quad (3.110)$$

Without loss of generality, we can align the axes so that $\hat{\boldsymbol{\rho}} = \mathbf{e}_1$, and introduce dimensionless variables

$$\begin{aligned}\tilde{z} &= z/R \\ \tilde{r} &= r/R,\end{aligned} \quad (3.111)$$

which gives

$$F = \frac{\lambda}{4\pi\epsilon R}\int_0^{2\pi} d\phi' \frac{\left\langle(\tilde{z}\mathbf{e}_3 + \tilde{r}\mathbf{e}_1 - \mathbf{e}_1 e^{i\phi'})\left(1 - \frac{v}{c}\mathbf{e}_2 e^{i\phi'}\right)\right\rangle_{1,2}}{(\tilde{z}^2 + (\tilde{r}\mathbf{e}_1 - \mathbf{e}_1 e^{i\phi'})^2)^{3/2}}. \quad (3.112)$$

In the denominator, the vector square expands as

$$\begin{aligned}(\tilde{r}\mathbf{e}_1 - \mathbf{e}_1 e^{i\phi'})^2 &= (\tilde{r} - e^{-i\phi'})\mathbf{e}_1^2(\tilde{r} - e^{i\phi'}) \\ &= \tilde{r}^2 + 1 - 2\tilde{r}\cos(\phi'),\end{aligned} \quad (3.113)$$

and the grade selection in the numerator is

$$\begin{aligned}\left\langle(\tilde{z}\mathbf{e}_3 + \tilde{r}\mathbf{e}_1 - \mathbf{e}_1 e^{i\phi'})\left(1 - \frac{v}{c}\mathbf{e}_2 e^{i\phi'}\right)\right\rangle_{1,2} &= \tilde{z}\mathbf{e}_3 + \tilde{r}\mathbf{e}_1 - \mathbf{e}_1 e^{i\phi'} \\ &\quad - \frac{v}{c}\left(\tilde{z}\mathbf{e}_{31}e^{i\phi'} + \tilde{r}i\cos(\phi') + i\right).\end{aligned} \quad (3.114)$$

Any of the exponential integrals terms are of the form

$$\int_0^{2\pi} d\phi'\, e^{i\phi'} f(\cos(\phi')) = \int_0^{2\pi} d\phi'\, \cos(\phi') f(\cos(\phi')), \quad (3.115)$$

since the $i \sin \phi' f(\cos(\phi'))$ contributions are odd functions around $\phi' = \pi$.

For general z, r the integrals above require numeric evaluation or special functions. Let

$$A = \int_0^{2\pi} d\phi' \frac{1}{(1 + \tilde{z}^2 + \tilde{r}^2 - 2\tilde{r}\cos(\phi'))^{3/2}}$$
$$= \frac{4E\left(-\frac{4\tilde{r}}{(\tilde{r}-1)^2+\tilde{z}^2}\right)}{\sqrt{\tilde{z}^2 + (\tilde{r}-1)^2}\left(\tilde{z}^2 + (\tilde{r}+1)^2\right)} \quad (3.116a)$$

$$B = \int_0^{2\pi} d\phi' \frac{\cos(\phi')}{(1 + \tilde{z}^2 + \tilde{r}^2 - 2\tilde{r}\cos(\phi'))^{3/2}}$$
$$= \frac{2\left((\tilde{z}^2 + \tilde{r}^2 + 1)E\left(-\frac{4\tilde{r}}{(\tilde{r}-1)^2+\tilde{z}^2}\right) - (\tilde{z}^2 + (\tilde{r}+1)^2)K\left(-\frac{4\tilde{r}}{(\tilde{r}-1)^2+\tilde{z}^2}\right)\right)}{\tilde{r}\sqrt{\tilde{z}^2 + (\tilde{r}-1)^2}\left(\tilde{z}^2 + (\tilde{r}+1)^2\right)},$$
$$(3.116b)$$

where $K(m)$, $E(m)$ are complete elliptic integrals of the first and second kind respectively. As seen in fig. 3.5, these functions are similar, both tailing off quickly with z, ρ, with largest values the ring.

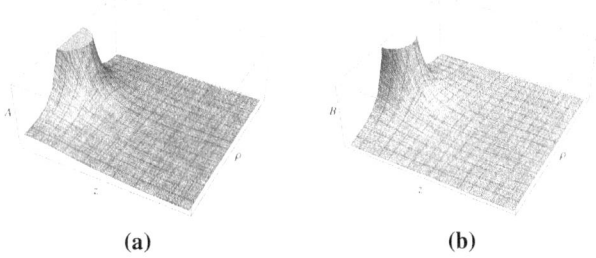

Figure 3.5: (a) $A(\tilde{z}, \tilde{\rho})$. (b) $B(\tilde{z}, \tilde{\rho})$.

Finally, restoring generality by making the transformation $\mathbf{e}_1 \to \mathbf{e}_1 e^{i\phi} = \hat{\rho}$, $\mathbf{e}_2 \to \mathbf{e}_2 e^{i\phi} = \hat{\phi}$, the field is now fully determined

$$F = \frac{\lambda}{4\pi\epsilon R}\left(\left(\tilde{z}\mathbf{e}_3 + \tilde{r}\hat{\rho} - \frac{vi}{c}\right)A - \left(\hat{\rho} + \frac{v}{c}(\tilde{z}\mathbf{e}_3\hat{\rho} + \tilde{r}i)\right)B\right). \quad (3.117)$$

The field directions are nicely parameterized as multivector expressions, with the relative weightings in different directions scaled by the position dependent integral coefficients of eq. (3.116). The multivector field can be separated into its respective electric and magnetic components by inspection

$$\mathbf{E} = \langle F \rangle_1 = \frac{\lambda}{4\pi R \epsilon} \left(\tilde{z} A \mathbf{e}_3 + \hat{\rho}(\tilde{r}A - B) \right)$$
$$\mathbf{H} = \frac{1}{\eta_0} \langle -IF \rangle_1 = \frac{\lambda v}{4\pi R} \left(-\mathbf{e}_3 (A + \tilde{r}B) - \hat{\phi} \tilde{z} A \right),$$
(3.118)

which, as expected, shows that the static charge distribution $\rho \propto \lambda^4$ only contributes to the electric field, and the static current distribution $\mathbf{J} \propto v\lambda$ only contributes to the magnetic field. See fig. 3.6, fig. 3.7 for plots of the electric and magnetic field directional variation near $\tilde{z} = 0$, and fig. 3.8 for larger z where the azimuthal component of the field dominates.

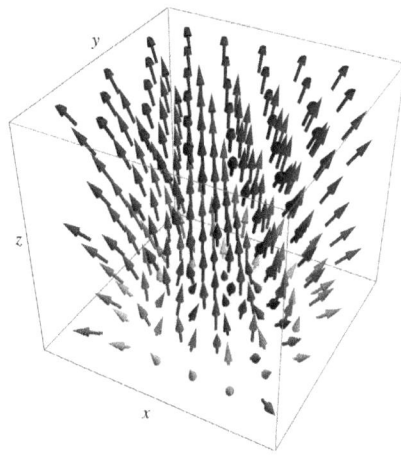

Figure 3.6: Electric field direction for circular charge density distribution near $z = 0$.

Exercise 3.6   Magnetic sources on a ring.

Given a multivector ring current

$$J = I\lambda_m \delta(z)\delta(r - R)(c - v\mathbf{e}_2 e^{i\phi}), i = \mathbf{e}_{12},$$

of constant magnitude, show that the field is

$$F = \frac{\lambda_m c}{4\pi R} \left(\left(\tilde{z}i + \tilde{r}\hat{\phi}\mathbf{e}_3 + \frac{v}{c}\mathbf{e}_3 \right) A + \left(\mathbf{e}_3 \hat{\phi} + \frac{v}{c}(\tilde{z}\hat{\rho} - \tilde{r}\mathbf{e}_3) \right) B \right).$$

---

4 $\propto$: proportional to.

3.5 STATICS.   215

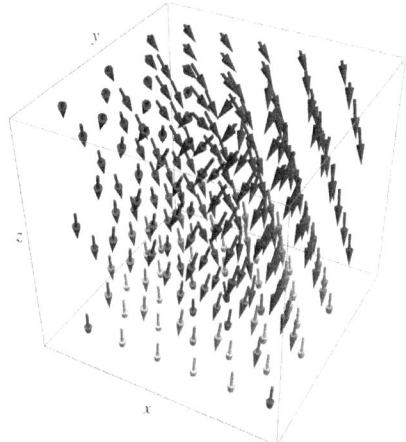

Figure 3.7: Magnetic field direction for circular current density distribution near $z = 0$.

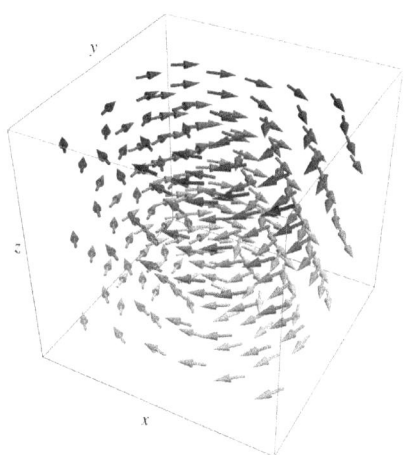

Figure 3.8: Magnetic field for larger z.

3.5.4.6 Ampere's law. Two current sources.

Let's try using Ampere's law as stated in theorem 3.6 two compute the field at a point in the blue region illustrated in fig. 3.9. This represents a pair of z-axis electric currents of magnitude I_1, I_2 flowing through the $z = 0$ points $\mathbf{p}_1, \mathbf{p}_2$ on the x-y plane.

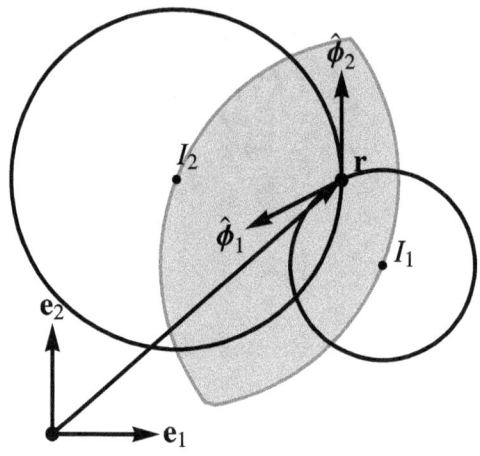

Figure 3.9: Magnetic field between two current sources.

Solving the system with superposition, let's consider first one source flowing through $\mathbf{p} = (p_x, p_y, 0)$ with current $\mathbf{J} = \mathbf{e}_3 I_e \delta(x - p_x)\delta(y - p_y)$, and evaluate the field due to this source at the point $\mathbf{r}$. With only magnetic sources in the multivector current, Ampere's law takes the form

$$\oint_{\partial A} d\mathbf{x}\, F = -I \int_A dA\, \mathbf{e}_3(-\eta \mathbf{J}) = I\eta I_e. \tag{3.119}$$

The field F must be a bivector satisfying $d\mathbf{x} \cdot F = 0$. The circle is parameterized by

$$\mathbf{r} = \mathbf{p} + R\mathbf{e}_1 e^{i\phi}, \tag{3.120}$$

so

$$d\mathbf{x} = R\mathbf{e}_2 e^{i\phi} d\phi = R\hat{\boldsymbol{\phi}} d\phi. \tag{3.121}$$

With the line element having only a $\hat{\boldsymbol{\phi}}$ component, F must be a bivector proportional to $\mathbf{e}_3\hat{\mathbf{r}}$. Let $F = F_0 \mathbf{e}_{31} e^{i\phi}$, where F_0 is a scalar, so that $d\mathbf{r}F$ is a constant multiple of the unit pseudoscalar

$$\begin{aligned}\int_0^{2\pi} d\mathbf{r}F &= RF_0 \int_0^{2\pi} \mathbf{e}_2 e^{i\phi} d\phi \, \mathbf{e}_{31} e^{i\phi} \\ &= RF_0 \int_0^{2\pi} \mathbf{e}_{231} e^{-i\phi} e^{i\phi} d\phi \\ &= 2\pi I R F_0,\end{aligned} \tag{3.122}$$

so

$$\begin{aligned}F_0 &= \frac{1}{I 2\pi R} I I_e \\ &= \frac{I_e}{2\pi R}.\end{aligned} \tag{3.123}$$

The field strength relative to the point $\mathbf{p}$ is

$$\begin{aligned}F &= \frac{\eta I_e}{2\pi R} \mathbf{e}_3 \hat{\mathbf{r}} \\ &= \frac{\eta I_e}{2\pi R} \mathbf{e}_3 \hat{\mathbf{r}}.\end{aligned} \tag{3.124}$$

Switching to an origin relative coordinate system, removing the $z = 0$ restriction for $\mathbf{r}$ and $\mathbf{p}_k$, and summing over both currents, the total field at any point $\mathbf{r}$ strictly between the currents is

$$\begin{aligned}F &= \sum_{k=1,2} \frac{\eta I_k}{2\pi} \mathbf{e}_3 \frac{1}{\mathbf{e}_3 \left(\mathbf{e}_3 \wedge (\mathbf{r} - \mathbf{p}_k)\right)} \\ &= \sum_{k=1,2} \frac{\eta I_k}{2\pi} \frac{1}{\mathbf{e}_3 \wedge (\mathbf{r} - \mathbf{p}_k)}.\end{aligned} \tag{3.125}$$

The bivector nature of a field with only electric current density sources is naturally represented by the wedge product $\mathbf{e}_3 \wedge (\mathbf{r} - \mathbf{p}_k)$ which is a vector product of $\mathbf{e}_3$ and the projection of $\mathbf{r} - \mathbf{p}_k$ onto the x-y plane.

3.6 DYNAMICS.

3.6.1 Inverting Maxwell's equation.

Maxwell's equation (theorem 3.1) is invertable using the Green's function for the spacetime gradient theorem 2.19. That solution is

Theorem 3.7: Jefimenkos solution.

The solution of Maxwell's equation is given by

$$F(\mathbf{x}, t) = F_0(\mathbf{x}, t) + \frac{1}{4\pi} \int dV' \left(\frac{\hat{\mathbf{r}}}{r^2} J(\mathbf{x}', t_r) + \frac{1}{cr} (1 + \hat{\mathbf{r}}) \dot{J}(\mathbf{x}', t_r) \right),$$

where $F_0(\mathbf{x}, t)$ is any specific solution of the homogeneous equation $(\nabla + (1/c)\partial_t) F_0 = 0$, time derivatives are denoted by overdots, and all times are evaluated at the retarded time $t_r = t - r/c$. When expanded in terms of the electric and magnetic fields (ignoring magnetic sources), the non-homogeneous portion of this solution is known as Jefimenkos' equations [10].

$$\mathbf{E} = \frac{1}{4\pi} \int dV' \left(\frac{\hat{\mathbf{r}}}{\epsilon r} \left(\frac{\rho(\mathbf{x}', t_r)}{r} + \frac{\dot{\rho}(\mathbf{x}', t_r)}{c} \right) - \frac{\eta}{cr} \dot{J}(\mathbf{x}', t_r) \right)$$
$$\mathbf{H} = \frac{1}{4\pi} \int dV' \left(\frac{1}{cr} \dot{J}(\mathbf{x}', t_r) + \frac{1}{r^2} J(\mathbf{x}', t_r) \right) \times \hat{\mathbf{r}},$$

(3.126)

Proof. The full solution is

$$F(\mathbf{x}, t) = F_0(\mathbf{x}, t) + \int dV' dt' \, G(\mathbf{x} - \mathbf{x}', t - t') J(\mathbf{x}', t')$$
$$= F_0(\mathbf{x}, t)$$
$$+ \frac{1}{4\pi} \int dV' dt' \left(\left(-\frac{\hat{\mathbf{r}}}{r^2} \frac{\partial}{\partial r} + \frac{\hat{\mathbf{r}}}{r^2} + \frac{1}{cr} \frac{\partial}{\partial t} \right) \delta(-r/c + t - t') \right) J(\mathbf{x}', t')$$
(3.127)

where $\mathbf{r} = \mathbf{x} - \mathbf{x}'$, $r = \|\mathbf{r}\|$ and $\hat{\mathbf{r}} = \mathbf{r}/r$. With the help of eq. (C.18), the derivatives in the Green's function eq. (3.127) can be evaluated, and the convolution reduces to

$$\int dt' G(\mathbf{x} - \mathbf{x}', t - t') J(\mathbf{x}', t')$$
$$= \frac{1}{4\pi} \left(\frac{\hat{\mathbf{r}}}{r^2} J(\mathbf{x}', t_r) - \frac{\hat{\mathbf{r}}}{r} \left(-\frac{1}{c} \right) \frac{d}{dt_r} J(\mathbf{x}', t_r) + \frac{1}{cr} \frac{d}{dt_r} J(\mathbf{x}', t_r) \right) \Big|_{t_r = t - r/c}.$$
(3.128)

There have been lots of opportunities to mess up with signs and factors of c, so let's expand this out explicitly for a non-magnetic current source

$J = \rho/\epsilon - \eta \mathbf{J}$. Neglect the contribution of the homogeneous solution F_0, and utilizing our freedom to insert a no-op grade 1,2 selection operation, that removes any scalar and pseudoscalar terms that are necessarily killed over the full integration range, we find

$$\begin{aligned} F &= \frac{1}{4\pi} \int dV' \left\langle \frac{\hat{\mathbf{r}}}{r^2} \left(\frac{\rho}{\epsilon} - \eta \mathbf{J}\right) + \frac{1}{cr} (1+\hat{\mathbf{r}}) \left(\frac{\dot{\rho}}{\epsilon} - \eta \dot{\mathbf{J}}\right) \right\rangle_{1,2} \\ &= \frac{1}{4\pi} \int dV' \left(\frac{\hat{\mathbf{r}}}{\epsilon r^2}\rho - \eta \frac{\hat{\mathbf{r}}}{r^2} \wedge \mathbf{J} - \frac{\eta}{cr}\dot{\mathbf{J}} + \frac{1}{cr}\hat{\mathbf{r}}\frac{\dot{\rho}}{\epsilon} - \frac{\eta}{cr}\hat{\mathbf{r}} \wedge \dot{\mathbf{J}}\right) \quad (3.129) \\ &= \frac{1}{4\pi} \int dV' \left(\frac{\hat{\mathbf{r}}}{\epsilon r^2}\rho + \frac{\hat{\mathbf{r}}\dot{\rho}}{\epsilon cr} - \frac{\eta \dot{\mathbf{J}}}{cr} - I\frac{\eta}{cr}\hat{\mathbf{r}} \times \dot{\mathbf{J}} - I\frac{\eta}{r^2}\hat{\mathbf{r}} \times \mathbf{J}\right). \end{aligned}$$

As $F = \mathbf{E} + I\eta \mathbf{H}$, the respective electric and magnetic fields by inspection. After re-inserting the space and time parameters that we suppressed temporarily, our work is done. □

The disadvantages of separating the field and current components into their constituent components is also made painfully obvious by the complexity of the conventional statement of the solution compared to the equivalent multivector form.

3.7 ENERGY AND MOMENTUM.

3.7.1 Field energy and momentum density and the energy momentum tensor.

It is assumed here that the conventional definitions of the field energy and momentum density are known to the reader, as well as the conservation equations relating their space and time derivatives. For reference, the conventional definitions of those densities follow.

Definition 3.4: Energy and momentum density, Poynting vector.

The quantities $\mathcal{E}$ and $\mathcal{P}$ defined as

$$\mathcal{E} = \frac{1}{2}\left(\epsilon \mathbf{E}^2 + \mu \mathbf{H}^2\right)$$

$$\mathcal{P}c = \frac{1}{c}\mathbf{E} \times \mathbf{H},$$

are known respectively as the *field energy density* and the *momentum density*. $\mathbf{S} = c^2 \mathscr{P} = \mathbf{E} \times \mathbf{H}$ is called the Poynting vector.

We will derive the conservation relationships that justify calling $\mathcal{E}, \mathscr{P}$ the energy and momentum densities, and will also show that the Poynting vector represents the energy flux through a surface per unit time.

In geometric algebra, it is arguably more natural to write the Poynting vector as a bivector-vector dot product such as

$$\mathbf{S} = \frac{1}{\eta}(I\eta\mathbf{H}) \cdot \mathbf{E}, \tag{3.130}$$

since this involves only components of the total electromagnetic field strength $F = \mathbf{E} + I\eta\mathbf{H}$. However, we can do better, representing both $\mathcal{E}$ and $\mathbf{S}$ in terms of F directly. The key to doing so is making use of the fact that the energy and momentum densities are themselves components of a larger symmetric rank-2 energy momentum tensor, which can in turn be represented compactly in geometric algebra.

Definition 3.5: Energy-momentum and Maxwell stress tensors.

The rank-2 symmetric tensor $\Theta^{\mu\nu}$, with components

$$\Theta^{00} = \frac{\epsilon}{2}\left(\mathbf{E}^2 + \eta^2 \mathbf{H}^2\right)$$

$$\Theta^{0i} = \frac{1}{c}(\mathbf{E} \times \mathbf{H}) \cdot \mathbf{e}_i$$

$$\Theta^{ij} = -\epsilon\left(E_i E_j + \eta^2 H_i H_j - \frac{1}{2}\delta_{ij}\left(\mathbf{E}^2 + \eta^2 \mathbf{H}^2\right)\right),$$

is called the energy momentum tensor. The spatial index subset of this tensor is known as the Maxwell stress tensor, and is often represented in dyadic notation

$$\left(\mathbf{a} \cdot \overleftrightarrow{\mathbf{T}}\right) \cdot \mathbf{b} = \sum_{i,j} a_i T_{ij} b_j,$$

or

$$\mathbf{a} \cdot \overleftrightarrow{\mathbf{T}} \equiv \sum_{i,j} a_i T_{ij} \mathbf{e}_j$$

where $T_{ij} = -\Theta^{ij}$.

Here we use the usual convention of Greek indices such as μ, ν for ranging over both time (0) and spatial $\{1, 2, 3\}$ indexes, and Latin letters such as i, j for the "spatial" indexes $\{1, 2, 3\}$. The names and notation for the tensors vary considerably[5].

In geometric algebra the energy momentum tensor, and the Maxwell stress tensor may be represented as linear grade $(0, 1)$-multivector valued functions of a grade $(0, 1)$-multivector.

> **Definition 3.6: Energy momentum and Maxwell stress tensors.**
>
> We define the *energy momentum tensor* as
>
> $$T(a) = \frac{1}{2}\epsilon F a F^{\dagger},$$
>
> where a is a grade $(0, 1)$-multivector parameter. We introduce a shorthand notation for grade one selection with vector valued parameters
>
> $$\mathbf{T}(\mathbf{a}) = \langle T(\mathbf{a})\rangle_1,$$
>
> and call this the *Maxwell stress tensor*.

> **Theorem 3.8: Expansion of the energy momentum tensor.**
>
> Given a scalar parameter α, and a vector parameter $\mathbf{a} = \sum_k a_k \mathbf{e}_k$, the *energy momentum tensor* of definition 3.6 is a grade $(0, 1)$-multivector, and may be expanded in terms of $\mathcal{E}, \mathbf{S}$ and $\mathbf{T}(\mathbf{a})$ as
>
> $$T(\alpha + \mathbf{a}) = \alpha\left(\mathcal{E} + \frac{\mathbf{S}}{c}\right) - \mathbf{a} \cdot \frac{\mathbf{S}}{c} + \mathbf{T}(\mathbf{a}),$$
>
> where $\mathbf{T}(\mathbf{e}_i) \cdot \mathbf{e}_j = -\Theta^{ij}$, or $\mathbf{T}(\mathbf{a}) = \mathbf{a} \cdot \overleftrightarrow{\mathbf{T}}$.

---

[5] $\Theta^{\mu\nu}$ in definition 3.5 is called the symmetric stress tensor by some authors [15], and the energy momentum tensor by others, and is sometimes written $T^{\mu\nu}$ ([17], [6]). The sign conventions and notation for the spatial components Θ^{ij} vary as well, but all authors appear to call this subset the Maxwell stress tensor. The Maxwell stress tensor may be written as $\sigma_{ij}(= -\Theta^{ij})$ [17], or as $T_{ij}(= -\Theta^{ij})$ ([10], [15].)

Theorem 3.8 relates the geometric algebra definition of the energy momentum tensor to the quantities found in the conventional electromagnetism literature.

Proof. Because T is a linear function of its parameter, we may prove this in parts, starting with $\alpha = 1, \mathbf{a} = 0$, which gives

$$\begin{aligned} T(1) &= \frac{1}{2}\epsilon F F^\dagger \\ &= \frac{1}{2}\epsilon \left(\mathbf{E} + I\eta\mathbf{H} \right) \left(\mathbf{E} - I\eta\mathbf{H} \right) \\ &= \frac{1}{2}\epsilon \left(\mathbf{E}^2 + \eta^2\mathbf{H}^2 \right) + \frac{1}{2}I\epsilon\eta \left(\mathbf{HE} - \mathbf{EH} \right) \\ &= \frac{1}{2}\left(\epsilon\mathbf{E}^2 + \mu\mathbf{H}^2 \right) + \frac{I}{c}\mathbf{H} \wedge \mathbf{E} \\ &= \frac{1}{2}\left(\epsilon\mathbf{E}^2 + \mu\mathbf{H}^2 \right) + \frac{1}{c}\mathbf{E} \times \mathbf{H} \\ &= \mathcal{E} + \frac{\mathbf{S}}{c}. \end{aligned} \qquad (3.131)$$

An immediate take away from this expansion is that we may dispense with any requirement to refer to electric or magnetic field components in isolation and can express the energy and momentum densities (and Poynting) vector in terms of only the total electromagnetic field strength

$$\begin{aligned} \mathcal{E} &= \frac{1}{2}\epsilon\langle FF^\dagger \rangle \\ \mathcal{P}c &= \frac{1}{2}\epsilon\langle FF^\dagger \rangle_1 \\ \mathbf{S} &= \frac{1}{2\eta}\langle FF^\dagger \rangle_1. \end{aligned} \qquad (3.132)$$

The power of this simple construction will be illustrated later when we compute the field energy and momentum densities for a number of Maxwell equation solutions in their geometric algebra form.

An expansion of $T(\mathbf{e}_k)$ is harder to do algebraically than eq. (3.131), but doing so will demonstrate that $T(a)$ is a 0,1 grade multivector parameter for any grade 0,1 parameter[6]. Cheating a bit, here are the results of a

---

6 Such an expansion is a worthwhile problem to develop GA manipulation skills. The reader is encouraged to try this independently first, and to refer to chapter D for hints if required.

brute force expansion of $T(a)$ using a Mathematica GA computer algebra package

$$T(1) = \frac{\epsilon}{2}\left(E_1^2 + E_2^2 + E_3^2\right) + \frac{\epsilon\eta^2}{2}\left(H_1^2 + H_2^2 + H_3^2\right)$$
$$+ \mathbf{e}_1\eta\epsilon\left(E_2 H_3 - E_3 H_2\right)$$
$$+ \mathbf{e}_2\eta\epsilon\left(E_3 H_1 - E_1 H_3\right) \qquad (3.133\text{a})$$
$$+ \mathbf{e}_3\eta\epsilon\left(E_1 H_2 - E_2 H_1\right).$$

$$T(\mathbf{e}_1) = \eta\epsilon\left(E_3 H_2 - E_2 H_3\right)$$
$$+ \frac{1}{2}\mathbf{e}_1\epsilon\left(E_1^2 - E_2^2 - E_3^2\right) + \frac{\epsilon\eta^2}{2}\left(H_1^2 - H_2^2 - H_3^2\right)$$
$$+ \mathbf{e}_2\epsilon\left(E_1 E_2 + \eta^2 H_1 H_2\right) \qquad (3.133\text{b})$$
$$+ \mathbf{e}_3\epsilon\left(E_1 E_3 + \eta^2 H_1 H_3\right)$$

$$T(\mathbf{e}_2) = \eta\epsilon\left(E_1 H_3 - E_3 H_1\right)$$
$$+ \mathbf{e}_1\epsilon\left(E_1 E_2 + \eta^2 H_1 H_2\right)$$
$$+ \frac{1}{2}\mathbf{e}_2\epsilon\left(-E_1^2 + E_2^2 - E_3^2\right) + \frac{\epsilon\eta^2}{2}\left(-H_1^2 + H_2^2 - H_3^2\right) \qquad (3.133\text{c})$$
$$+ \mathbf{e}_3\epsilon\left(E_2 E_3 + \eta^2 H_2 H_3\right)$$

$$T(\mathbf{e}_3) = \eta\epsilon\left(E_2 H_1 - E_1 H_2\right)$$
$$+ \mathbf{e}_1\epsilon\left(E_1 E_3 + \eta^2 H_1 H_3\right)$$
$$+ \mathbf{e}_2\epsilon\left(E_2 E_3 + \eta^2 H_2 H_3\right) \qquad (3.133\text{d})$$
$$+ \frac{1}{2}\mathbf{e}_3\epsilon\left(-E_1^2 - E_2^2 + E_3^2\right) + \frac{\epsilon\eta^2}{2}\left(-H_1^2 - H_2^2 + H_3^2\right)$$

Comparison to definition 3.5 shows that multivector energy momentum tensor is related to the conventional tensor representation by

$$\begin{aligned} \langle T(1) \rangle &= \Theta_0{}^0 = \Theta^{00} \\ \langle T(1) \rangle_1 \cdot \mathbf{e}_i &= \Theta_0{}^i = \Theta^{0i} \\ \langle T(\mathbf{e}_i) \rangle &= \Theta_i{}^0 = -\Theta^{i0} \\ \mathbf{T}(\mathbf{e}_i) \cdot \mathbf{e}_j &= \Theta_i{}^j = -\Theta^{ij} = T_{ij}. \end{aligned} \qquad (3.134)$$

The only thing left to show is that how **T(a)** is equivalent to the dyadic notation found in ([10], [15]).

$$\begin{aligned} \mathbf{T}(\mathbf{a}) &= \sum_i a_i \mathbf{T}(\mathbf{e}_i) \\ &= \sum_{i,j} a_i \left(\mathbf{T}(\mathbf{e}_i) \cdot \mathbf{e}_j \right) \mathbf{e}_j \\ &= \sum_{i,j} a_i T_{ij} \mathbf{e}_j \\ &= \mathbf{a} \cdot \overleftrightarrow{\mathbf{T}}. \end{aligned} \qquad (3.135)$$

The dyadic notation is really just a clumsy way of expressing the fact that **T(a)** is a linear vector valued function of a vector, which naturally has a matrix representation. □

3.7.2 Poynting's theorem (prerequisites.)

Poynting's theorem is a set of conservation relationships between relating space and time change of energy density and momentum density, or more generally between related components of the energy momentum tensor. The most powerful way of stating Poynting's theorem using geometric algebra requires a few new concepts, differential operator valued linear functions, and the adjoint.

Definition 3.7: Operator valued multivector functions.

Given a multivector valued linear functions of the form $f(x) = AxB$, where A, B, x are multivectors, and a linear operator D such as ∇, ∂_t, or $\nabla + (1/c)\partial_t$, the operator valued linear function $f(D)$ is defined as

$$f(D) = A \overleftrightarrow{D} B = (A \overleftarrow{D})B + A(\overrightarrow{D} B),$$

where $\overleftrightarrow{D}$ indicates that D is acting bidirectionally to the left and to the right.

Perhaps counter intuitively, using operator valued parameters for the energy momentum tensor T or the Maxwell stress tensor **T** will be particularly effective to express the derivatives of the tensor. There are a few cases of interest, all related to evaluation of the tensor with a parameter value of the spacetime gradient.

3.7 ENERGY AND MOMENTUM. 225

> **Theorem 3.9: Energy momentum tensor operator parameters.**
>
> $$T((1/c)\partial_t) = \frac{1}{c}\frac{\partial T(1)}{\partial t} = \frac{1}{c}\frac{\partial}{\partial t}\left(\mathcal{E} + \frac{\mathbf{S}}{c}\right)$$
>
> $$\langle T(\boldsymbol{\nabla}) \rangle = -\boldsymbol{\nabla} \cdot \frac{\mathbf{S}}{c}$$
>
> $$\langle T(\boldsymbol{\nabla}) \rangle_1 = \mathbf{T}(\boldsymbol{\nabla}) = \sum_{k=1}^{3} (\boldsymbol{\nabla} \cdot \mathbf{T}(\mathbf{e}_k))\, \mathbf{e}_k.$$

Proof. We will proceed to prove each of the results of theorem 3.9 in sequence, starting with the time partial, which is a scalar operator

$$\begin{aligned}
T(\partial_t) &= \frac{\epsilon}{2} F \stackrel{\leftrightarrow}{\partial_t} F^\dagger \\
&= \frac{\epsilon}{2}\left((\partial_t F)F^\dagger + F(\partial_t F^\dagger)\right) \\
&= \frac{\epsilon}{2}\partial_t F F^\dagger \\
&= \partial_t T(1).
\end{aligned} \qquad (3.136)$$

To evaluate the tensor at the gradient we have to take care of order. This is easiest in a scalar selection where we may cyclically permute any multivector factors

$$\begin{aligned}
\langle T(\boldsymbol{\nabla}) \rangle &= \frac{\epsilon}{2}\left\langle F \stackrel{\leftrightarrow}{\boldsymbol{\nabla}} F^\dagger \right\rangle \\
&= \frac{\epsilon}{2}\left\langle \boldsymbol{\nabla} F^\dagger F \right\rangle \\
&= \frac{\epsilon}{2}\boldsymbol{\nabla}\left\langle F^\dagger F \right\rangle_1,
\end{aligned} \qquad (3.137)$$

but

$$\begin{aligned}
F^\dagger F &= (\mathbf{E} - I\eta\mathbf{H})(\mathbf{E} + I\eta\mathbf{H}) \\
&= \mathbf{E}^2 + \eta^2 \mathbf{H}^2 + I\eta(\mathbf{E}\mathbf{H} - \mathbf{H}\mathbf{E}) \\
&= \mathbf{E}^2 + \eta^2 \mathbf{H}^2 - 2\eta \mathbf{E} \times \mathbf{H}.
\end{aligned} \qquad (3.138)$$

Plugging eq. (3.138) into eq. (3.137) proves the result.

Finally, we want to evaluate the Maxwell stress tensor of the gradient

$$\begin{aligned}
\mathbf{T}(\boldsymbol{\nabla}) &= \sum_{k=1}^{3} \mathbf{e}_k \left(\mathbf{T}(\boldsymbol{\nabla}) \right) \cdot \mathbf{e}_k \\
&= \sum_{k,m=1}^{3} \mathbf{e}_k \partial_m \left(\mathbf{T}(\mathbf{e}_m) \cdot \mathbf{e}_k \right) \\
&= \sum_{k,m=1}^{3} \mathbf{e}_k \partial_m \left(\mathbf{T}(\mathbf{e}_k) \cdot \mathbf{e}_m \right) \\
&= \sum_{k=1}^{3} \mathbf{e}_k \left(\boldsymbol{\nabla} \cdot \mathbf{T}(\mathbf{e}_k) \right),
\end{aligned} \quad (3.139)$$

as claimed. $\square$

Will want to integrate $\mathbf{T}(\boldsymbol{\nabla})$ over a volume, which is essentially a divergence operation.

Theorem 3.10: Divergence integral for the Maxwell stress tensor.

$$\int_V dV\, \mathbf{T}(\boldsymbol{\nabla}) = \int_{\partial V} dA\, \mathbf{T}(\hat{\mathbf{n}}).$$

Proof. To prove theorem 3.10, we make use of the symmetric property of the Maxwell stress tensor

$$\begin{aligned}
\int_V dV\, \mathbf{T}(\boldsymbol{\nabla}) &= \sum_k \int_V dV\, \mathbf{e}_k \boldsymbol{\nabla} \cdot \mathbf{T}(\mathbf{e}_k) \\
&= \sum_k \int_{\partial V} dA\, \mathbf{e}_k \hat{\mathbf{n}} \cdot \mathbf{T}(\mathbf{e}_k) \\
&= \sum_{k,m} \int_{\partial V} dA\, \mathbf{e}_k n_m \mathbf{T}(\mathbf{e}_k) \cdot \mathbf{e}_m \\
&= \sum_{k,m} \int_{\partial V} dA\, \mathbf{e}_k n_m \mathbf{T}(\mathbf{e}_m) \cdot \mathbf{e}_k \\
&= \sum_k \int_{\partial V} dA\, \mathbf{e}_k \mathbf{T}(\hat{\mathbf{n}}) \cdot \mathbf{e}_k \\
&= \int_{\partial V} dA\, \mathbf{T}(\hat{\mathbf{n}}),
\end{aligned} \quad (3.140)$$

as claimed. □

Finally, before stating Poynting's theorem, we want to introduce the concept of an adjoint.

Definition 3.8: Adjoint.

The *adjoint* $\overline{A}(x)$ of a linear operator $A(x)$ is defined implicitly by the scalar selection

$$\left\langle y\overline{A}(x) \right\rangle = \langle xA(y) \rangle.$$

The adjoint of the energy momentum tensor is particularly easy to calculate.

Theorem 3.11: Adjoint of the energy momentum tensor.

The *adjoint of the energy momentum tensor* is

$$\overline{T}(x) = \frac{\epsilon}{2} F^{\dagger} x F.$$

The adjoint $\overline{T}$ and T satisfy the following relationships

$$\left\langle \overline{T}(1) \right\rangle = \langle T(1) \rangle = \mathcal{E}$$

$$\left\langle \overline{T}(1) \right\rangle_1 = -\langle T(1) \rangle_1 = -\frac{\mathbf{S}}{c}$$

$$\left\langle \overline{T}(\mathbf{a}) \right\rangle = -\langle T(\mathbf{a}) \rangle = \mathbf{a} \cdot \frac{\mathbf{S}}{c}$$

$$\left\langle \overline{T}(\mathbf{a}) \right\rangle_1 = \langle T(\mathbf{a}) \rangle_1 = \mathbf{T}(\mathbf{a}).$$

Proof. Using the cyclic scalar selection permutation property $\langle ABC \rangle = \langle CAB \rangle$ we form

$$\begin{aligned}\langle xT(y) \rangle &= \frac{\epsilon}{2} \left\langle xFyF^{\dagger} \right\rangle \\ &= \frac{\epsilon}{2} \left\langle yF^{\dagger}xF \right\rangle.\end{aligned} \quad (3.141)$$

Referring back to definition 3.8 we see that the adjoint must have the stated form. Proving the grade selection relationships of eq. (3.141) has been left as an exercise for the reader. A brute force symbolic algebra proof using Mathematica is also available in stressEnergyTensorValues.nb.

□

As in theorem 3.9, the adjoint may also be evaluated with differential operator parameters.

Theorem 3.12: Adjoint energy-momentum tensor.

$$\langle \bar{T}((1/c)\partial_t) \rangle = \frac{1}{c}\frac{\partial T(1)}{\partial t} = \frac{1}{c}\frac{\partial \mathcal{E}}{\partial t}$$

$$\langle \bar{T}((1/c)\partial_t) \rangle_1 = -\frac{1}{c^2}\frac{\partial \mathbf{S}}{\partial t}$$

$$\langle \bar{T}(\boldsymbol{\nabla}) \rangle = \boldsymbol{\nabla} \cdot \frac{\mathbf{S}}{c}$$

$$\langle \bar{T}(\boldsymbol{\nabla}) \rangle_1 = \mathbf{T}(\boldsymbol{\nabla}).$$

Proof. The proofs of each of the statements in theorem 3.12 are all fairly simple

$$\begin{aligned}\bar{T}((1/c)\partial_t) &= \frac{1}{c}\frac{\epsilon}{2}\frac{\partial}{\partial t}\left(F^\dagger F\right) \\ &= \frac{1}{c}\frac{\partial}{\partial t}\left(\mathcal{E} - \frac{\mathbf{S}}{c}\right).\end{aligned} \qquad (3.142)$$

$$\begin{aligned}\langle \bar{T}(\boldsymbol{\nabla}) \rangle &= \langle 1\bar{T}(\boldsymbol{\nabla}) \rangle \\ &= \langle \boldsymbol{\nabla} T(1) \rangle \\ &= \boldsymbol{\nabla} \cdot \langle T(1) \rangle_1 \\ &= \boldsymbol{\nabla} \cdot \frac{\mathbf{S}}{c}.\end{aligned} \qquad (3.143)$$

$$\begin{aligned}\left\langle \overline{T}(\boldsymbol{\nabla})\right\rangle_1 &= \sum_k \mathbf{e}_k \left(\mathbf{e}_k \cdot \left\langle \overline{T}(\boldsymbol{\nabla})\right\rangle_1 \right) \\ &= \sum_k \mathbf{e}_k \left\langle \mathbf{e}_k \overline{T}(\boldsymbol{\nabla})\right\rangle \\ &= \sum_k \mathbf{e}_k \left\langle \boldsymbol{\nabla} T(\mathbf{e}_k)\right\rangle \\ &= \sum_k \mathbf{e}_k \boldsymbol{\nabla} \cdot \left\langle T(\mathbf{e}_k)\right\rangle_1 \\ &= \sum_k \mathbf{e}_k \boldsymbol{\nabla} \cdot \mathbf{T}(\mathbf{e}_k) \\ &= \mathbf{T}(\boldsymbol{\nabla}). \end{aligned} \qquad (3.144)$$

□

3.7.3 Poynting theorem.

All the prerequisites for stating Poynting's theorem are now finally complete.

> **Theorem 3.13: Poynting's theorem (differential form.)**
>
> The adjoint energy momentum tensor of the spacetime gradient satisfies the following multivector equation
>
> $$\overline{T}(\boldsymbol{\nabla} + (1/c)\partial_t) = \frac{\epsilon}{2}\left(F^\dagger J + J^\dagger F\right).$$
>
> The multivector $F^\dagger J + J^\dagger F$ can only have scalar and vector grades, since it equals its reverse. This equation can be put into a form that is more obviously a conservation law by stating it as a set of scalar grade identities
>
> $$\boldsymbol{\nabla} \cdot \langle T(a)\rangle_1 + \frac{1}{c}\frac{\partial}{\partial t}\langle T(a)\rangle = \frac{\epsilon}{2}\langle a(F^\dagger J + J \dagger F)\rangle,$$

or as a pair of scalar and vector grade conservation relationships

$$\frac{1}{c}\frac{\partial \mathcal{E}}{\partial t} + \nabla \cdot \frac{\mathbf{S}}{c} = -\frac{1}{c}(\mathbf{E} \cdot \mathbf{J} + \mathbf{H} \cdot \mathbf{M})$$

$$-\frac{1}{c^2}\frac{\partial \mathbf{S}}{\partial t} + \mathbf{T}(\nabla) = \rho \mathbf{E} + \epsilon \mathbf{E} \times \mathbf{M} + \rho_m \mathbf{H} + \mu \mathbf{J} \times \mathbf{H}.$$

Conventionally, only the scalar grade relating the time rate of change of the energy density to the flux of the Poynting vector, is called Poynting's theorem. Here the more general multivector (adjoint) relationship is called *Poynting's theorem*, which includes conservation laws relating for the field energy and momentum densities and conservation laws relating the Poynting vector components and the Maxwell stress tensor.

Proof. The conservation relationship of theorem 3.13 follows from

$$F^\dagger \left(\overset{\leftrightarrow}{\nabla} + \frac{1}{c}\overset{\leftrightarrow}{\partial_t} \right) F = \left(\left(\nabla + \frac{1}{c}\partial_t \right) F \right)^\dagger F + F^\dagger \left(\left(\nabla + \frac{1}{c}\partial_t \right) F \right) \quad (3.145)$$
$$= J^\dagger F + F^\dagger J.$$

The scalar form of theorem 3.13 follows from

$$\langle a \overline{T}(\nabla + (1/c)\partial_t) \rangle = \langle (\nabla + (1/c)\partial_t) T(a) \rangle$$
$$= \nabla \cdot \langle T(a) \rangle_1 + \frac{1}{c}\frac{\partial}{\partial t}\langle T(a) \rangle. \quad (3.146)$$

We may use the scalar form of the theorem to extract the scalar grade, by setting $a = 1$, for which the right hand side can be reduced to a single term since scalars are reversion invariant

$$\langle F^\dagger J \rangle = \langle F^\dagger J \rangle^\dagger = \langle J^\dagger F \rangle, \quad (3.147)$$

so

$$\nabla \cdot \langle T(1) \rangle_1 + \frac{1}{c}\frac{\partial}{\partial t}\langle T(1) \rangle = \nabla \cdot \frac{\mathbf{S}}{c} + \frac{1}{c}\frac{\partial \mathcal{E}}{\partial t}$$
$$= \frac{\epsilon}{2}\langle F^\dagger J + J^\dagger F \rangle$$
$$= \epsilon \langle F^\dagger J \rangle$$
$$= \epsilon \langle (\mathbf{E} - I\eta \mathbf{H})(\eta(c\rho - \mathbf{J}) + I(c\rho_m - \mathbf{M})) \rangle$$
$$= \epsilon (-\eta \mathbf{E} \cdot \mathbf{J} - \eta \mathbf{H} \cdot \mathbf{M})$$
$$= -\frac{1}{c}\mathbf{E} \cdot \mathbf{J} - \frac{1}{c}\mathbf{H} \cdot \mathbf{M},$$

$$\tag{3.148}$$

which proves the claimed explicit expansion of the scalar grade selection of Poynting's theorem.

The left hand side of the vector grade selection follows by linearity using theorem 3.12

$$\left\langle \overline{T}(\nabla + (1/c)\partial_t) \right\rangle_1 = \left\langle \overline{T}(\nabla) + \overline{T}((1/c)\partial_t) \right\rangle_1$$
$$= \mathbf{T}(\nabla) - \frac{1}{c^2}\frac{\partial \mathbf{S}}{\partial t}. \tag{3.149}$$

The right hand side is a bit messier to simplify. Let's do this in pieces by superposition, first considering just electric sources

$$\begin{aligned}\frac{\epsilon}{2}\left\langle \mathbf{e}_k \left(F^\dagger J + J^\dagger F \right) \right\rangle &= \frac{\epsilon\eta}{2}\left\langle \mathbf{e}_k \left((\mathbf{E} - I\eta\mathbf{H})(c\rho - \mathbf{J}) + (c\rho - \mathbf{J})(\mathbf{E} + I\eta\mathbf{H}) \right) \right\rangle \\ &= \frac{1}{2c}\mathbf{e}_k \cdot \left\langle (\mathbf{E} - I\eta\mathbf{H})(c\rho - \mathbf{J}) + (c\rho - \mathbf{J})(\mathbf{E} + I\eta\mathbf{H}) \right\rangle_1 \\ &= \frac{1}{c}\mathbf{e}_k \cdot (c\rho\mathbf{E} + I\eta\mathbf{H} \wedge \mathbf{J}) \\ &= \frac{1}{c}\mathbf{e}_k \cdot (c\rho\mathbf{E} - \eta\mathbf{H} \times \mathbf{J}) \\ &= \mathbf{e}_k \cdot (\rho\mathbf{E} + \mu\mathbf{J} \times \mathbf{H}),\end{aligned}$$
$$\tag{3.150}$$

and then magnetic sources

$$\begin{aligned}\frac{\epsilon}{2}\left\langle \mathbf{e}_k \left(F^\dagger J + J^\dagger F \right) \right\rangle &\\ &= \frac{\epsilon}{2}\left\langle \mathbf{e}_k \left((\mathbf{E} - I\eta\mathbf{H})I(c\rho_m - \mathbf{M}) - I(c\rho_m - \mathbf{M})(\mathbf{E} + I\eta\mathbf{H}) \right) \right\rangle \\ &= \frac{\epsilon}{2}\mathbf{e}_k \cdot \left\langle (I\mathbf{E} + \eta\mathbf{H})(c\rho_m - \mathbf{M}) + (c\rho_m - \mathbf{M})(-I\mathbf{E} + \eta\mathbf{H}) \right\rangle_1 \\ &= \epsilon\mathbf{e}_k \cdot (\eta c\rho_m\mathbf{H} - I\mathbf{E} \wedge \mathbf{M}) \\ &= \mathbf{e}_k \cdot (\rho_m\mathbf{H} + \epsilon\mathbf{E} \times \mathbf{M}).\end{aligned} \tag{3.151}$$

Jointly, eq. (3.149), eq. (3.150), eq. (3.151) complete the proof. □

The integral form of theorem 3.13 submits nicely to physical interpretation.

Theorem 3.14: Poynting's theorem (integral form.)

$$\frac{\partial}{\partial t} \int_V dV\, \mathcal{E} = -\int_{\partial V} dA\, \hat{n} \cdot \mathbf{S} - \int_V dV\, (\mathbf{J} \cdot \mathbf{E} + \mathbf{M} \cdot \mathbf{H})$$

$$\int_V dV\, (\rho \mathbf{E} + \mathbf{J} \times \mathbf{B})$$

$$+ \int_V dV\, (\rho_m \mathbf{H} - \epsilon \mathbf{M} \times \mathbf{E}) = -\frac{\partial}{\partial t} \int_V dV\, \mathcal{P} + \int_{\partial V} dA\, \mathbf{T}(\hat{n}).$$

Proof of theorem 3.14 is left to the reader, but requires only the divergence theorem, theorem 3.10, and definition 3.4.

The scalar integral in theorem 3.14 relates the rate of change of total energy in a volume to the flux of the Poynting through the surface bounding the volume. If the energy in the volume increases(decreases), then in a current free region, there must be Poynting flux into(out-of) the volume. The direction of the Poynting vector is the direction that the energy is leaving the volume, but only the projection of the Poynting vector along the normal direction contributes to this energy loss.

The right hand side of the vector integral in theorem 3.14 is a continuous representation of the Lorentz force (or dual Lorentz force for magnetic charges), the mechanical force on the charges in the volume. This can be seen by setting $\mathbf{J} = \rho \mathbf{v}$ (or $\mathbf{M} = \rho_m \mathbf{M}$)

$$\int_V dV\, (\rho \mathbf{E} + \mathbf{J} \times \mathbf{B}) = \int_V dV\, \rho\, (\mathbf{E} + \mathbf{v} \times \mathbf{B})$$
$$= \int_V dq\, (\mathbf{E} + \mathbf{v} \times \mathbf{B}). \tag{3.152}$$

As the field in the volume is carrying the (electromagnetic) momentum $\mathbf{p}_{em} = \int_V dV\, \mathcal{P}$, we can identify the sum of the Maxwell stress tensor normal component over the bounding integral as time rate of change of the mechanical and electromagnetic momentum

$$\boxed{\frac{d\mathbf{p}_{mech}}{dt} + \frac{d\mathbf{p}_{em}}{dt} = \int_{\partial V} dA\, \mathbf{T}(\hat{n}).} \tag{3.153}$$

3.7.4 Examples: Some static fields.

We've found solutions for a number of static charge and current distributions.

(a) For constant electric sources along the z-axis (eq. (3.80)), with current $\mathbf{J}$ moving with velocity $\mathbf{v} = v\mathbf{e}_3$, the field had the form $F = E\hat{\rho}(1 - \mathbf{v}/c)$.

(b) For constant magnetic sources along the z-axis (exercise 3.4), with current $\mathbf{M}$ moving with velocity $\mathbf{v} = v\mathbf{e}_3$, the field had the form $F = \eta H I\hat{\rho}(1 - \mathbf{v}/c)$.

(c) For constant electric sources in the x-y plane (eq. (3.93)), with current $\mathbf{J}$ moving with velocity $\mathbf{v} = v\mathbf{e}_1 e^{i\theta}$, $i = \mathbf{e}_{12}$, the field had the form $F = E\mathbf{e}_3 (1 - \mathbf{v}/c)$.

(d) For constant magnetic sources in the x-y plane (exercise 3.5), with current $\mathbf{M}$ moving with velocity $\mathbf{v} = v\mathbf{e}_1 e^{i\theta}$, $i = \mathbf{e}_{12}$, the field had the form $F = \eta H i (1 - \mathbf{v}/c)$.

In all cases the field has the form $F = A(1 - \mathbf{v}/c)$, where A is either a vector or a bivector that anticommutes with the current velocity $\mathbf{v}$, so the energy momentum tensor $T(1)$ has the form

$$\begin{aligned} T(1) &= \frac{\epsilon}{2} A (1 - \mathbf{v}/c)^2 A^\dagger \\ &= \frac{\epsilon}{2} A A^\dagger (1 + \mathbf{v}/c)^2 \\ &= \frac{\epsilon}{2} A A^\dagger \left(1 + \left(\frac{\mathbf{v}}{c}\right)^2 + 2\frac{\mathbf{v}}{c} \right). \end{aligned} \quad (3.154)$$

For the electric sources this is

$$\mathcal{E} + \frac{\mathbf{S}}{c} = \frac{\epsilon}{2} E^2 \left(1 + \left(\frac{\mathbf{v}}{c}\right)^2 + 2\frac{\mathbf{v}}{c} \right), \quad (3.155)$$

or

$$\mathcal{E} = \frac{\epsilon}{2} E^2 \left(1 + \left(\frac{\mathbf{v}}{c}\right)^2 \right) \quad (3.156)$$

$$\mathbf{S} = \epsilon E^2 \mathbf{v}.$$

For the magnetic sources this is

$$\mathcal{E} + \frac{\mathbf{S}}{c} = \frac{\mu}{2} H^2 \left(1 + \left(\frac{\mathbf{v}}{c}\right)^2 + 2\frac{\mathbf{v}}{c} \right), \quad (3.157)$$

or

$$\mathcal{E} = \frac{\mu}{2} H^2 \left(1 + \left(\frac{\mathbf{v}}{c}\right)^2 \right) \quad (3.158)$$

$$\mathbf{S} = \mu H^2 \mathbf{v}.$$

There are three terms in the multivector $(1 - \mathbf{v}/c)^2 = 1 + (\mathbf{v}/c)^2 + 2\mathbf{v}/c$. For electric sources, the first scalar term is due to the charge distribution, and provides the electric field contribution to the energy density. The second scalar term is due to the current distribution, and provides the magnetic field contribution to the energy density. The final vector term, proportional to the current velocity contributes to the Poynting vector, showing that the field momentum travels along the direction of the current in these static configurations.

Calculation of the $T(\mathbf{e}_k)$ tensor components is generally more involved. Let's do this calculation for each of the fields above in turn to illustrate.

(a): To calculate $T(\mathbf{e}_3)$ we can reduce the following products

$$\begin{aligned} F\mathbf{e}_3 F^\dagger &= E^2 \hat{\boldsymbol{\rho}} (1 - \mathbf{v}/c) \mathbf{e}_3 (1 - \mathbf{v}/c) \hat{\boldsymbol{\rho}} \\ &= -E^2 \mathbf{e}_3 \hat{\boldsymbol{\rho}} (1 - \mathbf{v}/c)^2 \hat{\boldsymbol{\rho}} \\ &= -E^2 \mathbf{e}_3 \hat{\boldsymbol{\rho}} \left(1 + \mathbf{v}^2/c^2 - 2\mathbf{v}/c\right) \hat{\boldsymbol{\rho}} \\ &= -E^2 \mathbf{e}_3 \hat{\boldsymbol{\rho}}^2 \left(1 + \mathbf{v}^2/c^2 + 2\mathbf{v}/c\right) \\ &= -E^2 \mathbf{e}_3 \left(1 + \mathbf{v}^2/c^2 + 2\mathbf{v}/c\right). \end{aligned} \quad (3.159)$$

Since

$$T(\mathbf{e}_k) = -\frac{\mathbf{S}}{c} \cdot \mathbf{e}_k + \mathbf{T}(\mathbf{e}_k). \quad (3.160)$$

This means that $\mathbf{S} \cdot \mathbf{e}_3 = \epsilon E^2 \mathbf{v}$, as already found. The vector component of this tensor element is

$$\mathbf{T}(\mathbf{e}_3) = -\frac{\epsilon}{2} E^2 \mathbf{e}_3 \left(1 + \mathbf{v}^2/c^2\right). \quad (3.161)$$

This component of the stress tensor is aligned along the same axis as the velocity. Calculation of the other stress tensor components is easiest in cylindrical coordinates. Along the radial direction

$$\begin{aligned} \hat{\boldsymbol{\rho}} (1 - \mathbf{v}/c) \hat{\boldsymbol{\rho}} (1 - \mathbf{v}/c) \hat{\boldsymbol{\rho}} &= \hat{\boldsymbol{\rho}}^2 (1 + \mathbf{v}/c)(1 - \mathbf{v}/c) \hat{\boldsymbol{\rho}} \\ &= \left(1 - \mathbf{v}^2/c^2\right) \hat{\boldsymbol{\rho}}, \end{aligned} \quad (3.162)$$

and along the azimuthal direction

$$\begin{aligned} \hat{\boldsymbol{\rho}} (1 - \mathbf{v}/c) \hat{\boldsymbol{\theta}} (1 - \mathbf{v}/c) \hat{\boldsymbol{\rho}} &= \hat{\boldsymbol{\rho}} \hat{\boldsymbol{\theta}} (1 + \mathbf{v}/c)(1 - \mathbf{v}/c) \hat{\boldsymbol{\rho}} \\ &= -\hat{\boldsymbol{\theta}} \hat{\boldsymbol{\rho}}^2 \left(1 - \mathbf{v}^2/c^2\right) \\ &= -\hat{\boldsymbol{\theta}} \left(1 - \mathbf{v}^2/c^2\right). \end{aligned} \quad (3.163)$$

Since $T(\mathbf{a})$ is a linear operator for any vector parameters $\mathbf{a}$, it cannot have any grade zero component along any directions $\mathbf{e} \cdot \mathbf{e}_3 = 0$. No grade zero component of $T(\mathbf{e}_1), T(\mathbf{e}_2)$ implies that the Poynting vector is zero along the $\mathbf{e}_1$ and $\mathbf{e}_2$ directions respectively, as we saw above in eq. (3.156).

In summary

$$\mathbf{T}(\hat{\boldsymbol{\rho}}) = \frac{\epsilon}{2} E^2 \left(1 - \mathbf{v}^2/c^2\right) \hat{\boldsymbol{\rho}}$$
$$\mathbf{T}(\hat{\boldsymbol{\theta}}) = -\frac{\epsilon}{2} E^2 \left(1 - \mathbf{v}^2/c^2\right) \hat{\boldsymbol{\theta}} \qquad (3.164)$$
$$\mathbf{T}(\mathbf{e}_3) = -\frac{\epsilon}{2} E^2 \left(1 + \mathbf{v}^2/c^2\right) \mathbf{e}_3.$$

For this field that $\mathbf{T}(\hat{\boldsymbol{\rho}})$ is entirely radial, whereas $\mathbf{T}(\hat{\boldsymbol{\theta}})$ is entirely azimuthal.

In terms of an arbitrary vector in cylindrical coordinates

$$\mathbf{a} = a_\rho \hat{\boldsymbol{\rho}} + a_\theta \hat{\boldsymbol{\theta}} + a_z \mathbf{e}_3, \qquad (3.165)$$

the grade one component of the tensor is

$$\mathbf{T}(\mathbf{a}) = \frac{\epsilon}{2} E^2 \left(1 - \mathbf{v}^2/c^2\right) \left(a_\rho \hat{\boldsymbol{\rho}} - a_\theta \hat{\boldsymbol{\theta}}\right)$$
$$- \frac{\epsilon}{2} E^2 \left(1 + \mathbf{v}^2/c^2\right) a_z \mathbf{e}_3. \qquad (3.166)$$

(b): For $F = \eta H I \hat{\boldsymbol{\rho}} \left(1 - \mathbf{v}/c\right)$, and $\mathbf{v} = v \mathbf{e}_3$ we have

$$F \mathbf{a} F^\dagger = \eta^2 H^2 I \hat{\boldsymbol{\rho}} \left(1 - \mathbf{v}/c\right) \mathbf{a} \left(1 - \mathbf{v}/c\right) \hat{\boldsymbol{\rho}} (-I)$$
$$= \eta^2 H^2 \hat{\boldsymbol{\rho}} \left(1 - \mathbf{v}/c\right) \mathbf{a} \left(1 - \mathbf{v}/c\right) \hat{\boldsymbol{\rho}}. \qquad (3.167)$$

We can write the tensor components immediately, since eq. (3.167) has exactly the same structure as the tensor components computed in part (a) above. That is

$$\mathbf{T}(\mathbf{a}) = \frac{\mu}{2} H^2 \left(1 - \mathbf{v}^2/c^2\right) \left(a_\rho \hat{\boldsymbol{\rho}} - a_\theta \hat{\boldsymbol{\theta}}\right)$$
$$- \frac{\mu}{2} H^2 \left(1 + \mathbf{v}^2/c^2\right) a_z \mathbf{e}_3. \qquad (3.168)$$

(c): For $F = E \mathbf{e}_3 \left(1 - \mathbf{v}/c\right)$, and $\mathbf{v} = v \hat{\boldsymbol{\rho}}$, we have

$$F \mathbf{a} F^\dagger = E^2 \mathbf{e}_3 \left(1 - (v/c)\hat{\boldsymbol{\rho}}\right) \mathbf{a} \left(1 - (v/c)\hat{\boldsymbol{\rho}}\right) \mathbf{e}_3, \qquad (3.169)$$

so we need the following grade selections

$$\begin{aligned}
\langle \mathbf{e}_3 (1-(v/c)\hat{\rho}) \hat{\rho} (1-(v/c)\hat{\rho}) \mathbf{e}_3 \rangle_1 &= \langle \mathbf{e}_3 \hat{\rho} (1-(v/c)\hat{\rho})^2 \mathbf{e}_3 \rangle_1 \\
&= \langle \mathbf{e}_3 \hat{\rho} \left(1 + (v^2/c^2) - 2(v/c)\hat{\rho}\right) \mathbf{e}_3 \rangle_1 \\
&= \left(1 + (v^2/c^2)\right) \mathbf{e}_3 \hat{\rho} \mathbf{e}_3 \\
&= -\left(1 + (v^2/c^2)\right) \hat{\rho} \\
\langle \mathbf{e}_3 (1-(v/c)\hat{\rho}) \hat{\theta} (1-(v/c)\hat{\rho}) \mathbf{e}_3 \rangle_1 &= \langle \mathbf{e}_3 \hat{\theta} (1+(v/c)\hat{\rho}) (1-(v/c)\hat{\rho}) \mathbf{e}_3 \rangle_1 \\
&= \langle \mathbf{e}_3 \hat{\theta} \left(1 - (v^2/c^2)\right) \mathbf{e}_3 \rangle_1 \\
&= -\left(1 - (v^2/c^2)\right) \hat{\theta} \\
\langle \mathbf{e}_3 (1-(v/c)\hat{\rho}) \mathbf{e}_3 (1-(v/c)\hat{\rho}) \mathbf{e}_3 \rangle_1 &= \langle (1+(v/c)\hat{\rho}) (1-(v/c)\hat{\rho}) \mathbf{e}_3 \rangle_1 \\
&= \left(1 - (v^2/c^2)\right) \mathbf{e}_3.
\end{aligned}$$
(3.170)

So the Maxwell stress tensor components of interest are

$$\begin{aligned}
\mathbf{T}(\mathbf{a}) = &-\frac{\epsilon}{2} E^2 \left(1 + (\mathbf{v}^2/c^2)\right) a_\rho \hat{\rho} \\
&+ \frac{\epsilon}{2} E^2 \left(1 - (\mathbf{v}^2/c^2)\right) \left(a_z \mathbf{e}_3 - a_\theta \hat{\theta}\right).
\end{aligned}$$
(3.171)

(d): For $F = \eta H i (1 - \mathbf{v}/c)$, $i = \mathbf{e}_{12}$, and $\mathbf{v} = v\hat{\rho}$, we can use a duality transformation for the unit bivector i

$$F = \eta H I \mathbf{e}_3 (1 - \mathbf{v}/c),$$
(3.172)

so

$$F\mathbf{a}F^\dagger = \eta^2 H^2 \mathbf{e}_3 (1 - \mathbf{v}/c) \, \mathbf{a} \, (1 - \mathbf{v}/c) \, \mathbf{e}_3.$$
(3.173)

Equation (3.173) has the structure found in part (c) above, so

$$\begin{aligned}
\mathbf{T}(\mathbf{a}) = &-\frac{\mu}{2} H^2 \left(1 + (\mathbf{v}^2/c^2)\right) a_\rho \hat{\rho} \\
&+ \frac{\mu}{2} H^2 \left(1 - (\mathbf{v}^2/c^2)\right) \left(a_z \mathbf{e}_3 - a_\theta \hat{\theta}\right).
\end{aligned}$$
(3.174)

3.7.5 Complex energy and power.

3.7 ENERGY AND MOMENTUM.

> **Theorem 3.15: Complex power representation.**
>
> Given a time domain representation of a phasor based field $F = F(\omega)$
>
> $$F(t) = \operatorname{Re}\left(Fe^{j\omega t}\right),$$
>
> the energy momentum tensor multivector $T(1)$ has the representation
>
> $$T(1) = \mathcal{E} + \frac{\mathbf{S}}{c} = \frac{\epsilon}{4}\operatorname{Re}\left(F^*F^\dagger + FF^\dagger e^{2j\omega t}\right).$$
>
> With the usual definition of the complex Poynting vector
>
> $$\mathcal{S} = \frac{1}{2}\mathbf{E}\times\mathbf{H}^* = \frac{1}{2}(I\mathbf{H}^*)\cdot\mathbf{E},$$
>
> the energy and momentum components of $T(1)$, for real μ, ϵ are
>
> $$\mathcal{E} = \frac{1}{4}\left(\epsilon|\mathbf{E}|^2 + \mu|\mathbf{H}|^2\right) + \frac{1}{4}\operatorname{Re}\left(\left(\epsilon\mathbf{E}^2 + \mu\mathbf{H}^2\right)e^{2j\omega t}\right)$$
>
> $$\mathbf{S} = \operatorname{Re}\mathcal{S} + \frac{1}{2}\operatorname{Re}\left((\mathbf{E}\times\mathbf{H})\,e^{2j\omega t}\right).$$

Proof. To prove theorem 3.15 we start by expanding the real part operation explicitly

$$\begin{aligned}F(t) &= \operatorname{Re}\left(Fe^{j\omega t}\right)\\ &= \frac{1}{2}\left(Fe^{j\omega t} + F^*e^{-j\omega t}\right).\end{aligned} \tag{3.175}$$

The energy momentum multivector for the field is therefore

$$\begin{aligned}\frac{1}{2}\epsilon F(t)F(t)^\dagger &= \frac{\epsilon}{8}\left(Fe^{j\omega t} + F^*e^{-j\omega t}\right)\left(F^\dagger e^{j\omega t} + (F^*)^\dagger e^{-j\omega t}\right)\\ &= \frac{\epsilon}{8}\left(FF^\dagger e^{2j\omega t} + \left(FF^\dagger e^{2j\omega t}\right)^* + F^*F^\dagger + \left(F^*F^\dagger\right)^*\right),\end{aligned} \tag{3.176}$$

so we have

$$\begin{aligned}\mathcal{E} + \frac{\mathbf{S}}{c} &= \frac{1}{2}\epsilon F(t)F(t)^\dagger\\ &= \frac{\epsilon}{4}\operatorname{Re}\left(F^*F^\dagger + FF^\dagger e^{2j\omega t}\right),\end{aligned} \tag{3.177}$$

which proves the first part of the theorem.

Next, we'd like to expand $T(1)$

$$\begin{aligned}\frac{1}{4}\epsilon F^* F^\dagger &= \frac{1}{4}\epsilon \left(\mathbf{E}^* + I\eta \mathbf{H}^*\right)\left(\mathbf{E} - I\eta \mathbf{H}\right) \\ &= \frac{1}{4}\left(\mathbf{E}^* \epsilon \mathbf{E} + \epsilon\eta^2 \mathbf{H}^* \mathbf{H} + I\epsilon\eta \left(\mathbf{H}^*\mathbf{E} - \mathbf{E}^*\mathbf{H}\right)\right) \\ &= \frac{1}{4}\left(\epsilon|\mathbf{E}|^2 + \mu|\mathbf{H}|^2 + \frac{I}{c}\left(\mathbf{H}^*\mathbf{E} - \mathbf{E}^*\mathbf{H}\right)\right).\end{aligned} \quad (3.178)$$

The scalar terms are already real, but the real part of the vector term is

$$\begin{aligned}\frac{I}{4c}\operatorname{Re}\left(\mathbf{H}^*\mathbf{E} - \mathbf{E}^*\mathbf{H}\right) &= \frac{I}{8c}\left(\mathbf{H}^*\mathbf{E} - \mathbf{E}^*\mathbf{H} + \mathbf{H}\mathbf{E}^* - \mathbf{E}\mathbf{H}^*\right) \\ &= \frac{I}{8c}\left(2\mathbf{H}^* \wedge \mathbf{E} + 2\mathbf{H} \wedge \mathbf{E}^*\right) \\ &= \frac{1}{4c}\left(\mathbf{E} \times \mathbf{H}^* + \mathbf{E}^* \times \mathbf{H}\right) \\ &= \frac{1}{2c}\operatorname{Re}\left(\mathbf{E} \times \mathbf{H}^*\right).\end{aligned} \quad (3.179)$$

The $\epsilon F F^\dagger$ factor of $e^{2j\omega t}$ above was expanded in eq. (3.131), so the energy momentum multivector is

$$\begin{aligned}\mathcal{E} + \frac{\mathbf{S}}{c} &= \frac{1}{4}\left(\epsilon|\mathbf{E}|^2 + \mu|\mathbf{H}|^2\right) + \frac{1}{2c}\operatorname{Re}\left(\mathbf{E} \times \mathbf{H}^*\right) \\ &\quad + \operatorname{Re}\left(\left(\frac{1}{4}\left(\epsilon\mathbf{E}^2 + \mu\mathbf{H}^2\right) + \frac{1}{2c}\mathbf{E} \times \mathbf{H}\right)e^{2j\omega t}\right).\end{aligned} \quad (3.180)$$

Expressing eq. (3.180) in terms of the complex Poynting vector $\mathcal{S}$, completes the proof. □

Observe that averaging over one period T kills any sinusoidal contributions, so the steady state energy and Poynting vectors are just

$$\begin{aligned}\frac{1}{T}\int_\tau^{\tau+T} \mathcal{E}(t)dt &= \frac{1}{4}\left(\epsilon|\mathbf{E}|^2 + \mu|\mathbf{H}|^2\right) \\ \frac{1}{T}\int_\tau^{\tau+T} \mathbf{S}(t)dt &= \operatorname{Re}\mathcal{S}.\end{aligned} \quad (3.181)$$

3.8 LORENTZ FORCE.

3.8.1 Statement.

We now wish to express the Lorentz force equation eq. (3.4a) in its geometric algebra form. A few definitions are helpful.

3.8 LORENTZ FORCE. 239

> **Definition 3.9: Energy momentum multivector.**
>
> For a particle with energy $\mathcal{E}$ and momentum $\mathbf{p}$, we define the *energy momentum multivector* as
>
> $$T = \mathcal{E} + c\mathbf{p}.$$

> **Definition 3.10: Multivector charge.**
>
> We may define a *multivector charge* that includes both the magnitude and velocity (relative to the speed of light) of the charged particle.
>
> $$Q = \int_V J dV,$$
>
> where $\mathbf{J} = \rho_e \mathbf{v}_e, \mathbf{M} = \rho_m \mathbf{v}_m$. For electric charges this is
>
> $$Q = q_e \left(1 + \mathbf{v}_e/c\right),$$
>
> and for magnetic charges
>
> $$Q = I q_m \left(1 + \mathbf{v}_m/c\right),$$
>
> where $q_e = \int_V \rho_e dV, q_m = \int_V \rho_m dV$.

With a multivector charge defined, the Lorentz force equation can be stated in terms of the total electromagnetic field strength

> **Theorem 3.16: Lorentz force and power.**
>
> The respective power and force experienced by particles with electric (and/or magnetic) charges is described by definition 3.10 is
>
> $$\frac{1}{c}\frac{dT}{dt} = \left\langle FQ^\dagger \right\rangle_{0,1} = \frac{1}{2}\left(F^\dagger Q + FQ^\dagger\right).$$
>
> where $\langle dT/dt \rangle = d\mathcal{E}/dt$ is the power and $\langle dT/dt \rangle_1 = cd\mathbf{p}/dt$ is the force on the particle, and $Q^\dagger$ is the electric or magnetic charge/veloc-

ity multivector of definition 3.10. The conventional representation of the Lorentz force/power equations

$$\langle FQ^\dagger \rangle_1 = \frac{d\mathbf{p}}{dt} = q\left(\mathbf{E} + \mathbf{v} \times \mathbf{B}\right)$$

$$c\langle FQ^\dagger \rangle = \frac{d\mathcal{E}}{dt} = q\mathbf{E} \cdot \mathbf{v}.$$

may be recovered by grade selection operations. For magnetic particles, such a grade selection gives

$$\langle FQ^\dagger \rangle_1 = \frac{d\mathbf{p}}{dt} = q_m\left(c\mathbf{B} - \frac{1}{c}\mathbf{v}_m \times \mathbf{E}\right)$$

$$c\langle FQ^\dagger \rangle = \frac{d\mathcal{E}}{dt} = \frac{1}{\eta}q_m\mathbf{B} \cdot \frac{\mathbf{v}_m}{c}.$$

Proof. To prove theorem 3.16, we can expand the multivector product $Fq(1 + \mathbf{v}/c)$ into its constituent grades

$$qF\left(1 + \frac{\mathbf{v}}{c}\right) = q\left(\mathbf{E} + Ic\mathbf{B}\right)\left(1 + \frac{\mathbf{v}}{c}\right)$$

$$= q\mathbf{E} + qI\mathbf{B}\mathbf{v} + \frac{q}{c}\mathbf{E}\mathbf{v} + qcI\mathbf{B}$$

$$= \frac{q}{c}\mathbf{E} \cdot \mathbf{v} + q\left(\mathbf{E} + \mathbf{v} \times \mathbf{B}\right) + q\left(cI\mathbf{B} + \frac{1}{c}\mathbf{E} \wedge \mathbf{v}\right) + q(I\mathbf{B}) \wedge \mathbf{v}.$$

(3.182)

We see the (c-scaled) particle power relationship eq. (3.4b) in the grade zero component and the Lorentz force eq. (3.4b) in the grade 1 component. A substitution $q \to -Iq_m$, $\mathbf{v} \to \mathbf{v}_m$, and subsequent grade 0,1 selection gives

$$\left\langle -Iq_m F\left(1 + \frac{\mathbf{v}_m}{c}\right)\right\rangle_{0,1} = -Iq_m\left(cI\mathbf{B} + \frac{1}{c}\mathbf{E} \wedge \mathbf{v}_m\right) - Iq_m I\mathbf{B} \cdot \mathbf{v}_m$$

$$= q_m\left(c\mathbf{B} - \frac{1}{c}\mathbf{v}_m \times \mathbf{E}\right) + q_m\mathbf{B} \cdot \mathbf{v}_m.$$

(3.183)

The grade one component of this multivector has the required form for the dual Lorentz force equation from theorem 3.14. Scaling the grade zero component by c completes the proof. □

3.8.2 Constant magnetic field.

The Lorentz force equation that determines the dynamics of a charged particle in an external field F has been restated as a multivector differential equation, but how to solve such an equation is probably not obvious. Given a constant external magnetic field bivector $F = Ic\mathbf{B}$, the Lorentz force equation, for small velocities[7], is

$$m\frac{d\mathbf{v}}{dt} = qF \cdot \frac{\mathbf{v}}{c}, \tag{3.184}$$

or

$$\begin{aligned} \Omega &= -\frac{qF}{mc} \\ \frac{d\mathbf{v}}{dt} &= \mathbf{v} \cdot \Omega, \end{aligned} \tag{3.185}$$

where Ω is a bivector containing all the constant factors.

This can be solved by introducing a multivector integration factor R and its reverse $R^\dagger$ on the left and right respectively

$$\begin{aligned} R\frac{d\mathbf{v}}{dt}R^\dagger &= R\mathbf{v} \cdot \Omega R^\dagger \\ &= \frac{1}{2}R\left(\mathbf{v}\Omega - \Omega\mathbf{v}\right)R^\dagger \\ &= \frac{1}{2}R\mathbf{v}\Omega R^\dagger - \frac{1}{2}R\Omega\mathbf{v}R^\dagger, \end{aligned} \tag{3.186}$$

or

$$0 = R\frac{d\mathbf{v}}{dt}R^\dagger + \frac{1}{2}R\Omega\mathbf{v}R^\dagger - \frac{1}{2}R\mathbf{v}\Omega R^\dagger. \tag{3.187}$$

Let

$$\dot{R} = R\Omega/2. \tag{3.188}$$

Since Ω is a bivector $\dot{R}^\dagger = -\Omega R^\dagger/2$, so by chain rule

$$0 = \frac{d}{dt}\left(R\mathbf{v}R^\dagger\right). \tag{3.189}$$

The integrating factor has solution

$$R = e^{\Omega t/2}, \tag{3.190}$$

---

[7] The (relativistically) correct Lorentz force equation for zero electric field is $d(m\gamma\mathbf{v})/dt = qF \cdot \mathbf{v}/c$ where $\gamma^{-1} = \sqrt{1 - \mathbf{v}^2/c^2}$.

a "complex exponential", so the solution of eq. (3.184) is

$$\mathbf{v}(t) = e^{-\Omega t/2}\mathbf{v}(0)e^{\Omega t/2}. \tag{3.191}$$

The velocity of the charged particle traces out a helical path. Any component of the initial velocity $\mathbf{v}(0)_\perp$ perpendicular to the Ω plane is untouched by this rotation operation, whereas components of the initial velocity $\mathbf{v}(0)_\parallel$ that lie in the Ω plane will trace out a circular path. If $\hat{\Omega}$ is the unit bivector for this plane, that velocity is

$$\begin{aligned}\mathbf{v}(0)_\parallel &= \left(\mathbf{v}(0) \cdot \hat{\Omega}\right)\hat{\Omega}^{-1} = (\mathbf{v}(0) \wedge \hat{\mathbf{B}}) \cdot \hat{\mathbf{B}} \\ \mathbf{v}(0)_\perp &= \left(\mathbf{v}(0) \wedge \hat{\Omega}\right)\hat{\Omega}^{-1} = (\mathbf{v}(0) \cdot \hat{\mathbf{B}})\hat{\mathbf{B}} \\ \mathbf{v}(t) &= \mathbf{v}(0)_\parallel e^{\Omega t} + \mathbf{v}(0)_\perp \\ &= \mathbf{v}(0)_\parallel \cos(qBt/m) + \mathbf{v}(0)_\parallel \times \hat{\mathbf{B}}\sin(qBt/m) + \mathbf{v}(0)_\perp,\end{aligned} \tag{3.192}$$

where $\mathbf{B} = B\hat{\mathbf{B}}$.

A multivector integration factor method for solving the Lorentz force equation in constant external electric and magnetic fields can be found in [11]. Other examples, solved using a relativistic formulation of GA, can be found in [6], [12], and [13].

3.8.2.1 Problems.

Exercise 3.7 Constant magnetic field.

In eq. (3.192), each of $\left(\mathbf{v}(0) \cdot \hat{\Omega}\right)\hat{\Omega}^{-1}$, $\left(\mathbf{v}(0) \wedge \hat{\Omega}\right)\hat{\Omega}^{-1}$, and $\mathbf{v}(0)_\parallel e^{\Omega t} + \mathbf{v}(0)_\perp$, was expanded by setting $\hat{\Omega} = I\hat{\mathbf{B}}$. Perform those calculations.

3.9 POLARIZATION.

3.9.1 Phasor representation.

In a discussion of polarization, it is convenient to align the propagation direction along a fixed direction, usually the z-axis. Setting $\hat{\mathbf{k}} = \mathbf{e}_3, \beta z = \mathbf{k} \cdot \mathbf{x}$ in a plane wave representation from theorem 3.3 the field is

$$\begin{aligned}F(\mathbf{x}, \omega) &= (1 + \mathbf{e}_3)\mathbf{E}e^{-j\beta z} \\ F(\mathbf{x}, t) &= \mathrm{Re}\left(F(\mathbf{x}, \omega)e^{j\omega t}\right),\end{aligned} \tag{3.198}$$

where $\mathbf{E} \cdot \mathbf{e}_3 = 0$, (i.e. $\mathbf{E}$ is an electric field, and not just a free parameter).

Here the imaginary j has no intrinsic geometrical interpretation, $\mathbf{E} = \mathbf{E}_r + j\mathbf{E}_i$ is allowed to have complex values, and all components of $\mathbf{E}$ is perpendicular to the propagation direction ($\mathbf{E}_r \cdot \mathbf{e}_3 = \mathbf{E}_i \cdot \mathbf{e}_3 = 0$). Stated explicitly, this means that the electric field phasor may have real or complex components in either of the transverse plane basis directions, as in

$$\mathbf{E} = (\alpha_1 + j\beta_1)\mathbf{e}_1 + (\alpha_2 + j\beta_2)\mathbf{e}_2. \tag{3.199}$$

The total time domain field for this general phasor field is easily found to be

$$F(\mathbf{x}, t) = (1 + \mathbf{e}_3)\left((\alpha_1\mathbf{e}_1 + \alpha_2\mathbf{e}_2)\cos(\omega t - \beta z) - (\beta_1\mathbf{e}_1 + \beta_2\mathbf{e}_2)\sin(\omega t - \beta z)\right). \tag{3.200}$$

Different combinations of $\alpha_1, \alpha_2, \beta_1, \beta_2$ lead to linear, circular, or elliptically polarized plane wave states to be discussed shortly. Before doing so, we want to find natural multivector representations of eq. (3.200). Such representations are possible using either the pseudoscalar for the transverse plane $\mathbf{e}_{12}$, or the $\mathbb{R}^3$ pseudoscalar I.

3.9.2 Transverse plane pseudoscalar.

3.9.2.1 Statement.

In this section the pseudoscalar of the transverse plane, written $i = \mathbf{e}_{12}$, is used as an imaginary.

Definition 3.11: Phase angle.

Define the total phase as

$$\phi(z, t) = \omega t - \beta z.$$

We seek a representation of the field utilizing complex exponentials of the phase, instead of signs and cosines. It will be helpful to define the coordinates of the Jones vector to state that representation.

> **Definition 3.12: Jones vectors.**
>
> The coordinates of the Jones vector, conventionally defined as a tuple of complex values (c_1, c_2), are
>
> $$c_1 = \alpha_1 + i\beta_1$$
> $$c_2 = \alpha_2 + i\beta_2.$$
>
> In this definition we have used $i = \mathbf{e}_{12}$, the pseudoscalar of the transverse plane, as the imaginary.

We will not use the Jones vector as a tuple, but will use c_1, c_2 as stated above.

> **Theorem 3.17: Circular polarization coefficients.**
>
> The time domain representation of the field in eq. (3.200) can be stated in terms of the total phase as
>
> $$F = (1 + \mathbf{e}_3)\, \mathbf{e}_1 \left(\alpha_R e^{i\phi} + \alpha_L e^{-i\phi} \right),$$
>
> where
>
> $$\alpha_R = \frac{1}{2}(c_1 + ic_2)$$
> $$\alpha_L = \frac{1}{2}(c_1 - ic_2)^\dagger,$$
>
> where c_1, c_2 are the 0,2 grade multivector representation of the Jones vector coordinates from definition 3.12.

Proof. To prove theorem 3.17, we have only to factor $\mathbf{e}_1$ out of eq. (3.200) and then substitute complex exponentials for the sine and cosine

$$(\alpha_1 \mathbf{e}_1 + \alpha_2 \mathbf{e}_2) \cos(\phi) - (\beta_1 \mathbf{e}_1 + \beta_2 \mathbf{e}_2) \sin(\phi)$$
$$= \mathbf{e}_1 \left((\alpha_1 + \alpha_2 i) \cos(\phi) - (\beta_1 + \beta_2 i) \sin(\phi) \right)$$
$$= \frac{\mathbf{e}_1}{2} \Big((\alpha_1 + \alpha_2 i) \left(e^{i\phi} + e^{-i\phi} \right)$$
$$\quad + (\beta_1 + \beta_2 i) i \left(e^{i\phi} - e^{-i\phi} \right) \Big)$$
$$= \frac{\mathbf{e}_1}{2} \Big((\alpha_1 + i\beta_1 + i(\alpha_2 + i\beta_2)) e^{i\phi}$$
$$\quad + \left((\alpha_1 + i\beta_1)^\dagger + i(\alpha_2 + i\beta_2)^\dagger \right) e^{-i\phi} \Big)$$
$$= \frac{\mathbf{e}_1}{2} \left((c_1 + ic_2) e^{i\phi} + (c_1 - ic_2)^\dagger e^{-i\phi} \right). \qquad \square$$

3.9.2.2 Linear polarization.

Linear polarization is described by

$$\begin{aligned} \alpha_R &= \frac{1}{2} \|\mathbf{E}\| \, e^{i(\psi + \theta)} \\ \alpha_L &= \frac{1}{2} \|\mathbf{E}\| \, e^{i(\psi - \theta)}, \end{aligned} \quad (3.201)$$

so the field is

$$F = (1 + \mathbf{e}_3) \|\mathbf{E}\| \, \mathbf{e}_1 e^{i\psi} \cos(\omega t - \beta z + \theta). \quad (3.202)$$

Here θ is an arbitrary initial phase. The electric field $\mathbf{E}$ traces out all the points along the line spanning the points between $\pm \mathbf{e}_1 e^{i\psi} \|\mathbf{E}\|$, whereas the magnetic field $\mathbf{H}$ traces out all the points along $\pm \mathbf{e}_2 e^{i\psi} \|\mathbf{E}\|/\eta$ as illustrated (with $\eta = 1$) in fig. 3.10.

3.9.2.3 Circular polarization.

A field for which the change in phase results in the electric field tracing out a (clockwise,counterclockwise) circle

$$\begin{aligned} \mathbf{E}_R &= \|\mathbf{E}\| \left(\mathbf{e}_1 \cos\phi + \mathbf{e}_2 \sin\phi \right) = \|\mathbf{E}\| \, \mathbf{e}_1 \exp(\mathbf{e}_{12}\phi) \\ \mathbf{E}_L &= \|\mathbf{E}\| \left(\mathbf{e}_1 \cos\phi - \mathbf{e}_2 \sin\phi \right) = \|\mathbf{E}\| \, \mathbf{e}_1 \exp(-\mathbf{e}_{12}\phi), \end{aligned} \quad (3.203)$$

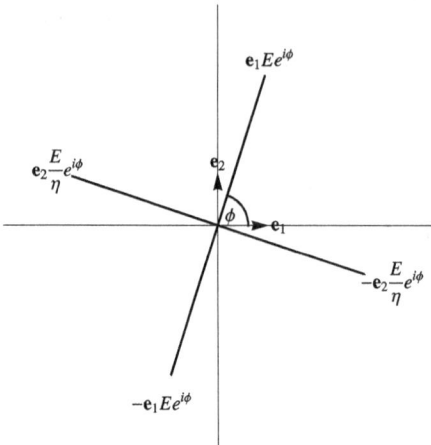

Figure 3.10: Linear polarization.

is referred to as having (right,left) circular polarization, so the choice $\alpha_R = \|\mathbf{E}\|, \alpha_L = 0$ results in a right polarized wave

$$F = (1 + \mathbf{e}_3)\|\mathbf{E}\|\,\mathbf{e}_1 e^{i(\omega t - kz)}, \tag{3.204}$$

and $\alpha_L = \|\mathbf{E}\|, \alpha_R = 0$ results in a left polarized wave

$$F = (1 + \mathbf{e}_3)\|\mathbf{E}\|\,\mathbf{e}_1 e^{-i(\omega t - kz)}, \tag{3.205}$$

There are different conventions for the polarization orientation, and here the IEEE antenna convention discussed in [3] are used.

3.9.2.4 Elliptical parameterization.

An elliptical polarized electric field can be parameterized as

$$\mathbf{E} = E_a \mathbf{e}_1 \cos\phi + E_b \mathbf{e}_2 \sin\phi, \tag{3.206}$$

which corresponds to circular polarization coefficients with values

$$\begin{aligned} \alpha_R &= \frac{1}{2}(E_a - E_b) \\ \alpha_L &= \frac{1}{2}(E_a + E_b). \end{aligned} \tag{3.207}$$

Therefore an elliptically polarized field can be represented as

$$F = \frac{1}{2}(1 + \mathbf{e}_3)\mathbf{e}_1\left((E_a + E_b)e^{i\phi} + (E_a - E_b)e^{-i\phi}\right). \tag{3.208}$$

An interesting variation of the elliptical polarization uses a hyperbolic parameterization. If a, b are the semi-major/minor axes of the ellipse (i.e. $a > b$), and $\mathbf{a} = a\mathbf{e}_1 e^{i\psi}$ is the vectoral representation of the semi-major axis (not necessarily placed along $\mathbf{e}_1$), and $e = \sqrt{1-(b/a)^2}$ is the eccentricity of the ellipse, then it can be shown ([11]) that an elliptic parameterization can be written in the compact form

$$\mathbf{r}(\phi) = e\mathbf{a}\cosh(\tanh^{-1}(b/a) + i\phi). \tag{3.209}$$

When the bivector imaginary $i = \mathbf{e}_{12}$ is used then this parameterization is real and has only vector grades, so the electromagnetic field for a general elliptic wave has the form

$$\begin{aligned} F &= eE_a\left(1+\mathbf{e}_3\right)\mathbf{e}_1 e^{i\psi}\cosh\left(m+i\phi\right) \\ m &= \tanh^{-1}\left(E_b/E_a\right) \\ e &= \sqrt{1-(E_b/E_a)^2}, \end{aligned} \tag{3.210}$$

where $E_a(E_b)$ are the magnitudes of the electric field components lying along the semi-major(minor) axes, and the propagation direction $\mathbf{e}_3$ is orthogonal to both the major and minor axis directions. An elliptic electric field polarization is illustrated in fig. 3.11, where the vectors representing the major and minor axes are $\mathbf{E}_a = E_a\mathbf{e}_1 e^{i\psi}$, $\mathbf{E}_b = E_b\mathbf{e}_1 e^{i\psi}$. Observe that setting $E_b = 0$ results in the linearly polarized field of eq. (3.202).

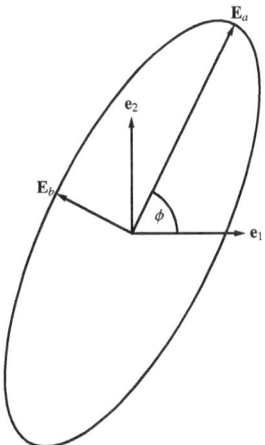

Figure 3.11: Electric field with elliptical polarization.

3.9.2.5 Energy and momentum.

Each polarization considered above (linear, circular, elliptical) have the same general form

$$F = (1 + \mathbf{e}_3) \mathbf{e}_1 e^{i\psi} f(\phi), \qquad (3.211)$$

where $f(\phi)$ is a complex valued function (i.e. grade 0,2). The structure of eq. (3.211) could be more general than considered so far. For example, a Gaussian modulation could be added into the mix with $f(\phi) = e^{i\phi - (\phi/\sigma)^2/2}$. Independent of the form of f, we may compute the energy, momentum and Maxwell stress tensor for the plane wave given by eq. (3.211).

Theorem 3.18: Plane wave energy momentum tensor components.

The *energy momentum tensor components* for the plane wave given by eq. (3.211) are

$$T(1) = -T(\mathbf{e}_3) = \epsilon (1 + \mathbf{e}_3) f f^\dagger \quad \left(= \mathcal{E} + \frac{\mathbf{S}}{c} \right)$$

$$T(\mathbf{e}_1) = T(\mathbf{e}_2) = 0.$$

Only the propagation direction of a plane wave, regardless of its polarization (or even whether or not there are Gaussian or other damping factors), carries any energy or momentum, and only the propagation direction component of the Maxwell stress tensor $\mathbf{T}(\mathbf{a})$ is non-zero.

Proof. To prove theorem 3.18, we may compute $T(a)$ separately for each of $a = 1, \mathbf{e}_1, \mathbf{e}_2, \mathbf{e}_3$. Key to all of these computations is the fact that $\mathbf{e}_3$ commutes with scalars and i, and $\mathbf{e}_1, \mathbf{e}_2$ both anticommute with i, and more generally $\begin{bmatrix} \mathbf{e}_1 \\ \mathbf{e}_2 \end{bmatrix} (a + ib) = (a - ib) \begin{bmatrix} \mathbf{e}_1 \\ \mathbf{e}_2 \end{bmatrix}$. For $T(1)$ we need the product of the field and its reverse

$$FF^\dagger = (1 + \mathbf{e}_3) \underbrace{\mathbf{e}_1 e^{i\psi} f f^\dagger e^{-i\psi} \mathbf{e}_1}_{\text{scalar}} (1 + \mathbf{e}_3) \qquad (3.212)$$

$$= (1 + \mathbf{e}_3)^2 f f^\dagger$$

$$= 2 (1 + \mathbf{e}_3) f f^\dagger,$$

so $T(1) = \epsilon (1 + \mathbf{e}_3) f f^\dagger$. For $T(\mathbf{e}_3)$ we have

$$\begin{aligned}
F\mathbf{e}_3 F^\dagger &= (1 + \mathbf{e}_3) \mathbf{e}_1 e^{i\psi} f \mathbf{e}_3 f^\dagger e^{-i\psi} \mathbf{e}_1 (1 + \mathbf{e}_3) \\
&= -(1 + \mathbf{e}_3) \mathbf{e}_3 \mathbf{e}_1 e^{i\psi} f f^\dagger e^{-i\psi} \mathbf{e}_1 (1 + \mathbf{e}_3) \\
&= -(1 + \mathbf{e}_3) \mathbf{e}_1 e^{i\psi} f f^\dagger e^{-i\psi} \mathbf{e}_1 (1 + \mathbf{e}_3) \\
&= -2 (1 + \mathbf{e}_3) f f^\dagger,
\end{aligned} \quad (3.213)$$

so $T(\mathbf{e}_3) = -T(1)$. For $T(\mathbf{e}_1)$, we have

$$\begin{aligned}
F\mathbf{e}_1 F^\dagger &= (1 + \mathbf{e}_3) \mathbf{e}_1 e^{i\psi} f \mathbf{e}_1 f^\dagger e^{-i\psi} \mathbf{e}_1 (1 + \mathbf{e}_3) \\
&= (1 + \mathbf{e}_3) \mathbf{e}_1 e^{i\psi} f^2 e^{i\psi} \mathbf{e}_1^2 (1 + \mathbf{e}_3) \\
&= (1 + \mathbf{e}_3) \mathbf{e}_1 f^2 e^{2i\psi} (1 + \mathbf{e}_3) \\
&= (1 + \mathbf{e}_3) \mathbf{e}_1 (1 + \mathbf{e}_3) f^2 e^{2i\psi} \\
&= (1 + \mathbf{e}_3) (1 - \mathbf{e}_3) \mathbf{e}_1 f^2 e^{2i\psi} \\
&= \left(1 - \mathbf{e}_3^2\right) \mathbf{e}_1 f^2 e^{2i\psi} \\
&= 0.
\end{aligned} \quad (3.214)$$

Clearly $F\mathbf{e}_2 F^\dagger = 0$ as well, so $T(\mathbf{e}_1) = T(\mathbf{e}_2) = 0$. □

Using theorem 3.18 the energy momentum vector for the linearly polarized wave of eq. (3.202) is

$$T(1) = \frac{\epsilon}{2} (1 + \mathbf{e}_3) \|\mathbf{E}\|^2 \cos^2(\phi + \theta), \quad (3.215)$$

and for the circularly polarized wave of eq. (3.204), or eq. (3.205) is

$$T(1) = \frac{\epsilon}{2} (1 + \mathbf{e}_3) \|\mathbf{E}\|^2. \quad (3.216)$$

A circularly polarized wave carries maximum energy and momentum, whereas the energy and momentum of a linearly polarized wave oscillates with the phase angle.

For the elliptically polarized wave of eq. (3.210) we have

$$f(\phi) = eE_a \cosh(m + i\phi). \quad (3.217)$$

The absolute value of f is

$$\begin{aligned}
f f^\dagger &= e^2 E_a^2 \cosh(m + i\phi) (\cosh(m + i\phi))^\dagger \\
&= e^2 E_a^2 (\cosh(2m) + \cos(2\phi)) \\
&= e^2 E_a^2 \left(\frac{E_b^2}{E_a^2} + 2 \left(1 - \frac{E_b^2}{E_a^2}\right) \cos^2 \phi \right)
\end{aligned} \quad (3.218)$$

The simplification above made use of the identity

$$(1 - (b/a)^2)\cosh(2\operatorname{atanh}(b/a)) = 1 + (b/a)^2. \tag{3.219}$$

The energy momentum for an elliptically polarized wave is therefore

$$T(1) = \frac{\epsilon}{2}(1+\mathbf{e}_3)e^2 E_a^2 \left(\frac{E_b^2}{E_a^2} + 2\left(1 - \frac{E_b^2}{E_a^2}\right)\cos^2\phi\right). \tag{3.220}$$

As expected, the phase dependent portion of the energy momentum tensor vanishes as the wave function approaches circular polarization.

3.9.3 Pseudoscalar imaginary.

In this section we use the $\mathbb{R}^3$ pseudoscalar as an imaginary. As before, we seek a representation of the field utilizing complex exponentials of the phase, instead of signs and cosines, and as before the we wish to define Jones vector coordinates as a go-between.

Definition 3.13: Jones vectors.

The coordinates of the Jones vector, conventionally defined as a tuple of complex values (c_1, c_2), are

$$c_1 = \alpha_1 + I\beta_1$$
$$c_2 = \alpha_2 + I\beta_2.$$

In this definition we have used the $\mathbb{R}^3$ pseudoscalar I as the imaginary.

We will not use the Jones vector as a tuple, but will use c_1, c_2 as stated above.

Theorem 3.19: Circular polarization coefficients.

The time domain representation of the field in eq. (3.200) can be stated in terms of the total phase as

$$F = (1+\mathbf{e}_3)\,\mathbf{e}_1\left(\alpha_R e^{-I\phi} + \alpha_L e^{I\phi}\right),$$

where

$$\alpha_R = \frac{1}{2}(c_1 + Ic_2)^{\dagger}$$

$$\alpha_L = \frac{1}{2}(c_1 - Ic_2),$$

where c_1, c_2 are the 0,2 grade multivector representation of the Jones vector coordinates from definition 3.13.

Notice that the signs of the exponentials have flipped for the left and right handed circular polarizations. It may not obvious that the electric and magnetic fields in this representation have the desired transverse properties. To see why that is still the case, and to understand the conjugation in the complex exponentials, consider the right circular polarization case with $\alpha_R = \|\mathbf{E}\|, \alpha_L = 0$

$$\begin{aligned} F &= (1+\mathbf{e}_3)\,\mathbf{e}_1\,\|\mathbf{E}\|\,e^{-I\phi} \\ &= (1+\mathbf{e}_3)\,\|\mathbf{E}\|\,(\mathbf{e}_1\cos\phi - \mathbf{e}_{23}\sin\phi) \\ &= (1+\mathbf{e}_3)\,\|\mathbf{E}\|\,(\mathbf{e}_1\cos\phi + \mathbf{e}_{32}\sin\phi), \end{aligned} \quad (3.221)$$

but since $(1+\mathbf{e}_3)\,\mathbf{e}_3 = 1+\mathbf{e}_3$, we have

$$F = (1+\mathbf{e}_3)\,\|\mathbf{E}\|\,(\mathbf{e}_1\cos\phi + \mathbf{e}_2\sin\phi), \quad (3.222)$$

which has the claimed right circular polarization.

Proof. To prove theorem 3.19 itself, the sine and cosine in eq. (3.200) can be expanded in complex exponentials

$$\begin{aligned} &2(\alpha_1\mathbf{e}_1 + \alpha_2\mathbf{e}_2)\cos\phi - 2(\beta_1\mathbf{e}_1 + \beta_2\mathbf{e}_2)\sin\phi \\ &= (\alpha_1\mathbf{e}_1 + \alpha_2\mathbf{e}_2)\left(e^{I\phi} + e^{-I\phi}\right) + (\beta_1\mathbf{e}_1 + \beta_2\mathbf{e}_2)I\left(e^{I\phi} - e^{-I\phi}\right) \\ &= (\alpha_1\mathbf{e}_1 - I\alpha_2(I\mathbf{e}_2))\left(e^{I\phi} + e^{-I\phi}\right) + (\beta_1\mathbf{e}_1 - I\beta_2(I\mathbf{e}_2))I\left(e^{I\phi} - e^{-I\phi}\right). \end{aligned} \quad (3.223)$$

Since the leading $1+\mathbf{e}_3$ gobbles any $\mathbf{e}_3$ factors, its action on the dual of $\mathbf{e}_2$ is

$$\begin{aligned} (1+\mathbf{e}_3)\,I\mathbf{e}_2 &= (1+\mathbf{e}_3)\,\mathbf{e}_{31} \\ &= (1+\mathbf{e}_3)\,\mathbf{e}_1. \end{aligned} \quad (3.224)$$

This allows for unconditionally factoring of $\mathbf{e}_1$ from eq. (3.223), so the field is

$$\begin{aligned}F &= \frac{1}{2}\left(1+\mathbf{e}_3\right)\mathbf{e}_1\left((\alpha_1 - I\alpha_2)\left(e^{I\phi}+e^{-I\phi}\right)+(\beta_1 - I\beta_2)I\left(e^{I\phi}-e^{-I\phi}\right)\right)\\ &= \frac{1}{2}\left(1+\mathbf{e}_3\right)\mathbf{e}_1 \times\\ &\quad \left((\alpha_1 + I\beta_1 - I(\alpha_2 + I\beta_2))e^{I\phi}+(\alpha_1 - I\beta_1 - I(\alpha_2 - I\beta_2))e^{-I\phi}\right)\\ &= \frac{1}{2}\left(1+\mathbf{e}_3\right)\mathbf{e}_1\left((c_1 - Ic_2)e^{I\phi}+\left(c_1^\dagger - Ic_2^\dagger\right)e^{-I\phi}\right)\\ &= \left(1+\mathbf{e}_3\right)\mathbf{e}_1\left(\alpha_R e^{-I\phi}+\alpha_L e^{-I\phi}\right).\end{aligned} \quad (3.225)$$

$\square$

Observe that there are some advantages to the pseudoscalar plane wave form, especially for computing energy momentum tensor components since I commutes with all grades. For example, we can see practically by inspection that

$$T(1) = \mathcal{E} + \frac{\mathbf{S}}{v} = \epsilon\left(1+\mathbf{e}_3\right)\left(|\alpha_R|^2 + |\alpha_L|^2\right), \quad (3.226)$$

where the absolute value is computed using the reverse as the conjugation operation $|z|^2 = zz^\dagger$.

3.10 TRANSVERSE FIELDS IN A WAVEGUIDE.

We now wish to consider more general solutions to the source free Maxwell's equation than the plane wave solutions derived in section 3.4. One way of tackling this problem is to assume the solution exists, but ask how the field components that lie strictly along the propagation direction are related to the transverse components of the field. Without loss of generality, it can be assumed that the propagation direction is along the z-axis.

Theorem 3.20: Transverse and propagation field components.

If $\mathbf{e}_3$ is the propagation direction, the components of a field F in the propagation direction and in the transverse plane are respectively

$$F_z = \frac{1}{2}(F + \mathbf{e}_3 F \mathbf{e}_3)$$
$$F_t = \frac{1}{2}(F - \mathbf{e}_3 F \mathbf{e}_3),$$

where $F = F_z + F_t$.

Proof. To determine the components of the field that lie in the propagation direction and transverse planes, we state the field in the propagation direction, building it from the electric and magnetic field projections along the z-axis

$$\begin{aligned} F_z &= (\mathbf{E} \cdot \mathbf{e}_3)\mathbf{e}_3 + I\eta(\mathbf{H} \cdot \mathbf{e}_3)\mathbf{e}_3 \\ &= \frac{1}{2}(\mathbf{E}\mathbf{e}_3 + \mathbf{e}_3\mathbf{E})\mathbf{e}_3 + \frac{1}{2}I\eta(\mathbf{H}\mathbf{e}_3 + \mathbf{e}_3\mathbf{H})\mathbf{e}_3 \\ &= \frac{1}{2}(\mathbf{E} + \mathbf{e}_3\mathbf{E}\mathbf{e}_3) + \frac{1}{2}I\eta(\mathbf{H} + \mathbf{e}_3\mathbf{H}\mathbf{e}_3) \\ &= \frac{1}{2}(F + \mathbf{e}_3 F \mathbf{e}_3). \end{aligned} \quad (3.227)$$

The difference $F - F_z$ is the transverse component

$$\begin{aligned} F_t &= F - F_z \\ &= F - \frac{1}{2}(F + \mathbf{e}_3 F \mathbf{e}_3) \\ &= \frac{1}{2}(F - \mathbf{e}_3 F \mathbf{e}_3). \end{aligned} \quad (3.228)$$

$\square$

We wish to split the gradient into transverse and propagation direction components.

Definition 3.14: Transverse and propagation direction gradients.

Define the *propagation direction gradient* as $\mathbf{e}_3 \partial_z$, and *transverse gradient* by

$$\boldsymbol{\nabla}_t = \boldsymbol{\nabla} - \mathbf{e}_3 \partial_z.$$

Given this definition, we seek to show that

> **Theorem 3.21: Transverse and propagation field solutions.**
>
> Given a field propagating along the z-axis (either forward or backwards), with angular frequency ω, represented by the real part of
>
> $$F(x,y,z,t) = F(x,y)e^{j\omega t - jkz},$$
>
> the field components that solve the source free Maxwell's equation are related by
>
> $$F_t = j\frac{1}{\frac{\omega}{c} - k\mathbf{e}_3}\nabla_t F_z$$
> $$F_z = j\frac{1}{\frac{\omega}{c} - k\mathbf{e}_3}\nabla_t F_t.$$
>
> Written out explicitly, the transverse field component expands as
>
> $$\mathbf{E}_t = \frac{j}{\omega^2\mu\epsilon - k^2}\left(-k\nabla_t E_z + \mu\omega \mathbf{e}_3 \times \nabla_t H_z\right)$$
> $$\mathbf{H}_t = -\frac{j}{\omega^2\mu\epsilon - k^2}\left(\epsilon\omega \mathbf{e}_3 \times \nabla_t E_z + k\eta\nabla_t H_z\right),$$
>
> and the propagation fields are
>
> $$\mathbf{E}_z = -\frac{j}{\omega^2\mu\epsilon - k^2}\left(\omega\mu\left(\nabla_t \times \mathbf{H}_t\right) + k\left(\mathbf{e}_3 \times \nabla_t\right) \times \mathbf{E}_t\right)$$
> $$\mathbf{H}_z = \frac{j}{\omega^2\mu\epsilon - k^2}\left(\omega\epsilon\left(\nabla_t \times \mathbf{E}_t\right) - k\left(\mathbf{e}_3 \times \nabla_t\right) \times \mathbf{H}_t\right).$$

Proof. To prove we first insert the assumed phasor representation into Maxwell's equation, which gives

$$\left(\nabla_t + j\left(\frac{\omega}{c} - k\mathbf{e}_3\right)\right)F(x,y) = 0. \tag{3.229}$$

Dropping the x, y dependence for now (i.e. $F(x,y) \to F$), we find a relation between the transverse gradient of F and the propagation direction gradient of F

$$\nabla_t F = -j\left(\frac{\omega}{c} - k\mathbf{e}_3\right)F. \tag{3.230}$$

From this we now seek to determine the relationships between F_t and F_z.

Since ∇_t has only $\mathbf{e}_1, \mathbf{e}_2$ components, $\mathbf{e}_3$ anticommutes with the transverse gradient

$$\mathbf{e}_3 \nabla_t = -\nabla_t \mathbf{e}_3, \tag{3.231}$$

This means that

$$\frac{1}{2} \left(\nabla_t F \pm \mathbf{e}_3 \left(\nabla_t F \right) \mathbf{e}_3 \right) = \frac{1}{2} \left(\nabla_t F \mp \nabla_t \mathbf{e}_3 F \mathbf{e}_3 \right) \\ = \nabla_t \frac{1}{2} \left(F \mp \mathbf{e}_3 F \mathbf{e}_3 \right), \tag{3.232}$$

or

$$\frac{1}{2} \left(\nabla_t F + \mathbf{e}_3 \left(\nabla_t F \right) \mathbf{e}_3 \right) = \nabla_t F_t \\ \frac{1}{2} \left(\nabla_t F - \mathbf{e}_3 \left(\nabla_t F \right) \mathbf{e}_3 \right) = \nabla_t F_z, \tag{3.233}$$

so Maxwell's equation eq. (3.230) becomes

$$\nabla_t F_t = -j \left(\frac{\omega}{c} - k\mathbf{e}_3 \right) F_z \\ \nabla_t F_z = -j \left(\frac{\omega}{c} - k\mathbf{e}_3 \right) F_t. \tag{3.234}$$

Provided $\omega^2 \neq (kc)^2$, these can be inverted. Such an inversion allows an application of the transverse gradient to whichever one of F_z, F_t is known, to compute the other, as stated in theorem 3.21.

The grunt work of that inversion is left to exercise 3.8 and exercise 3.9. □

3.10.1 Problems.

Exercise 3.8   Transverse electric and magnetic field components.

Find $\mathbf{E}_t, \mathbf{H}_t$, in terms of E_z, H_z, as stated in theorem 3.21.

Exercise 3.9   Propagation direction components.

Find $\mathbf{E}_z, \mathbf{H}_z$, in terms of $\mathbf{E}_t, \mathbf{H}_t$, as stated in theorem 3.21.

3.11 MULTIVECTOR POTENTIAL.

3.11.1 *Definitions.*

We know from conventional electromagnetism (given no fictitious magnetic sources) that we can represent the six components of the electric and magnetic fields in terms of four scalar fields

$$\mathbf{E} = -\nabla\phi - \frac{\partial \mathbf{A}}{\partial t}$$
$$\mathbf{H} = \frac{1}{\mu}\nabla \times \mathbf{A}. \tag{3.251}$$

where

1. ϕ is the scalar potential V (Volts), and

2. $\mathbf{A}$ is the vector potential Wb/m (Webers/meter).

The conventional way of constructing these potentials makes use of the identities,

$$\nabla \cdot (\nabla \times \mathbf{A}) = 0$$
$$\nabla \times (\nabla\phi) = 0, \tag{3.252}$$

which are special cases of $\nabla \wedge \nabla \wedge \chi = 0$ (for blades χ.) Applying those to the source free Maxwell's equations to find representations of $\mathbf{E}, \mathbf{H}$ that automatically satisfy those equations. For that conventional analysis, see §18-6 [8], §10.1 [10], or §6.4 [15]. We can also find such a potential representation using geometric algebra methods that are cross product free (exercise 3.10.)

For Maxwell's equations with fictitious magnetic sources, it can be shown that a potential representation of the field

$$\mathbf{H} = -\nabla\phi_m - \frac{\partial \mathbf{F}}{\partial t}$$
$$\mathbf{E} = -\frac{1}{\epsilon}\nabla \times \mathbf{F}. \tag{3.253}$$

satisfies the source-free grades of Maxwell's equation. Here

1. ϕ_m is the scalar potential for (fictitious) magnetic sources A (Amperes), and

2. **F** is the vector potential for (fictitious) magnetic sources C (Coulombs).

See [4], and [19] for such derivations. As with the conventional source potentials, we can also apply our geometric algebra toolbox to easily find these results (exercise 3.11.)

In eq. (3.251), and eq. (3.253) we have a mix of time partials and curls that is reminiscent of Maxwell's equation itself. It's obvious to wonder whether there is a more coherent integrated form for the potential. This is in fact the case.

Lemma 3.1: Multivector potentials.

For Maxwell's equation with electric sources, the total field F expressed in terms of the potentials of eq. (3.251) can be expressed in multivector potential form

$$F = \left\langle \left(\nabla - \frac{1}{c} \frac{\partial}{\partial t} \right)(-\phi + c\mathbf{A}) \right\rangle_{1,2}. \qquad (3.254)$$

For Maxwell's equation with only fictitious magnetic sources, the total field F expressed in terms of the potentials of eq. (3.253) can be expressed in multivector form

$$F = \left\langle \left(\nabla - \frac{1}{c} \frac{\partial}{\partial t} \right) I\eta \left(-\phi_m + c\mathbf{F} \right) \right\rangle_{1,2}. \qquad (3.255)$$

The reader should try to verify this themselves (exercise 3.12.)

Using superposition, we can form a multivector potential that includes all grades.

Definition 3.15: Multivector potential.

We call A, a multivector with all grades, the multivector potential, defining the total field as

$$\begin{aligned}
F &= \left\langle \left(\nabla - \frac{1}{c} \frac{\partial}{\partial t} \right) A \right\rangle_{1,2} \\
&= \left(\nabla - \frac{1}{c} \frac{\partial}{\partial t} \right) A - \left\langle \left(\nabla - \frac{1}{c} \frac{\partial}{\partial t} \right) A \right\rangle_{0,3}.
\end{aligned} \qquad (3.256)$$

Imposition of the constraint

$$\left\langle \left(\nabla - \frac{1}{c}\frac{\partial}{\partial t}\right)A\right\rangle_{0,3} = 0, \qquad (3.257)$$

is called the Lorentz gauge condition, and allows us to express F in terms of the potential without any grade selection filters.

Lemma 3.2: Conventional multivector potential.

Let

$$A = -\phi + c\mathbf{A} + I\eta\left(-\phi_m + c\mathbf{F}\right). \qquad (3.258)$$

With definition 3.15, this results in the conventional potential representation of the electric and magnetic fields

$$\begin{aligned}\mathbf{E} &= -\nabla\phi - \frac{\partial \mathbf{A}}{\partial t} - \frac{1}{\epsilon}\nabla \times \mathbf{F} \\ \mathbf{H} &= -\nabla\phi_m - \frac{\partial \mathbf{F}}{\partial t} + \frac{1}{\mu}\nabla \times \mathbf{A}.\end{aligned} \qquad (3.259)$$

In terms of potentials, the Lorentz gauge condition eq. (3.257) takes the form

$$\begin{aligned}0 &= \frac{1}{c}\frac{\partial \phi}{\partial t} + \nabla \cdot (c\mathbf{A}) \\ 0 &= \frac{1}{c}\frac{\partial \phi_m}{\partial t} + \nabla \cdot (c\mathbf{F}).\end{aligned} \qquad (3.260)$$

Proof. See exercise 3.13. □

Theorem 3.22: The potential wave equations.

3.11 MULTIVECTOR POTENTIAL.

The potentials are related to the sources by

$$\Box \phi = -\frac{\rho}{\epsilon} - \frac{\partial}{\partial t}\left(\nabla \cdot \mathbf{A} + \frac{1}{c^2}\frac{\partial \phi}{\partial t}\right)$$

$$\Box \mathbf{A} = -\mu \mathbf{J} + \nabla\left(\nabla \cdot \mathbf{A} + \frac{1}{c^2}\frac{\partial \phi}{\partial t}\right)$$

$$\Box \mathbf{F} = -\epsilon \mathbf{M} + \nabla\left(\nabla \cdot \mathbf{F} + \frac{1}{c^2}\frac{\partial \phi_m}{\partial t}\right)$$

$$\Box \phi_m = -\frac{\rho_m}{\mu} - \frac{\partial}{\partial t}\left(\nabla \cdot \mathbf{F} + \frac{1}{c^2}\frac{\partial \phi_m}{\partial t}\right).$$

Reminder: ($\Box$: see definition 2.11)

Proof. See exercise 3.14. $\square$

3.11.2 Gauge transformations.

Clearly it is desirable if potentials can be found for which $\nabla \cdot \mathbf{A} + (1/c^2)\partial_t \phi = \nabla \cdot \mathbf{F} + (1/c^2)\partial_t \phi_m = 0$. Finding such potentials relies on the fact that the potential representation is not unique. In particular, we have the freedom to add any spacetime gradient of any scalar or pseudoscalar potential without changing the field.

Theorem 3.23: Gauge invariance.

The spacetime gradient of a grade $(0, 3)$-multivector Ψ may be added to a multivector potential

$$A' = A + \left(\nabla + \frac{1}{c}\frac{\partial}{\partial t}\right)\Psi,$$

without changing the field. That is

$$F = \left\langle\left(\nabla - \frac{1}{c}\frac{\partial}{\partial t}\right)A\right\rangle_{1,2} = \left\langle\left(\nabla - \frac{1}{c}\frac{\partial}{\partial t}\right)A'\right\rangle_{1,2}.$$

Proof. To prove theorem 3.23 let

$$A' = A + \left(\nabla + \frac{1}{c}\frac{\partial}{\partial t}\right)(\psi + I\phi), \tag{3.261}$$

where ψ and ϕ are scalar functions. The field for potential A' is

$$\begin{aligned} F' &= \left\langle \left(\nabla - \frac{1}{c}\frac{\partial}{\partial t}\right) A' \right\rangle_{1,2} \\ &= \left\langle \left(\nabla - \frac{1}{c}\frac{\partial}{\partial t}\right) \left(A + \left(\nabla + \frac{1}{c}\frac{\partial}{\partial t}\right)(\psi + I\phi)\right) \right\rangle_{1,2} \\ &= \left\langle \left(\nabla - \frac{1}{c}\frac{\partial}{\partial t}\right) A \right\rangle_{1,2} + \left\langle \left(\nabla - \frac{1}{c}\frac{\partial}{\partial t}\right)\left(\nabla + \frac{1}{c}\frac{\partial}{\partial t}\right)(\psi + I\phi) \right\rangle_{1,2} \\ &= F + \langle \Box(\psi + I\phi) \rangle_{1,2}, \end{aligned} \qquad (3.262)$$

which is just F since the d'Alembertian operator $\Box$ is a scalar operator and $\psi + I\phi$ has no vector nor bivector grades. $\qquad\square$

We say that we are working in the Lorenz gauge, if the 0,3 grades of $\left(\nabla - \frac{1}{c}\frac{\partial}{\partial t}\right) A$ are zero, or a transformation that kills those grades is made.

Theorem 3.24: Lorentz gauge transformation.

Given any multivector potential A solution of Maxwell's equation, the transformation

$$A' = A - \left(\nabla + \frac{1}{c}\frac{\partial}{\partial t}\right)\Psi,$$

where

$$\Box\Psi = \left\langle \left(\nabla - \frac{1}{c}\frac{\partial}{\partial t}\right) A \right\rangle_{0,3},$$

allows Maxwell's equation to be written in wave equation form

$$\Box A' = J.$$

A potential satisfying this wave equation is said to be in the *Lorentz gauge*.

Proof. To prove theorem 3.24, let

$$A = A' + \left(\nabla + \frac{1}{c}\frac{\partial}{\partial t}\right)\Psi, \qquad (3.263)$$

so Maxwell's equation becomes

$$\begin{aligned} J &= \left(\nabla + \frac{1}{c}\frac{\partial}{\partial t}\right)\left\langle\left(\nabla - \frac{1}{c}\frac{\partial}{\partial t}\right)A\right\rangle_{1,2} \\ &= \Box A - \left(\nabla + \frac{1}{c}\frac{\partial}{\partial t}\right)\left\langle\left(\nabla - \frac{1}{c}\frac{\partial}{\partial t}\right)A\right\rangle_{0,3} \\ &= \Box A' + \Box\left(\nabla + \frac{1}{c}\frac{\partial}{\partial t}\right)\Psi - \left(\nabla + \frac{1}{c}\frac{\partial}{\partial t}\right)\left\langle\left(\nabla - \frac{1}{c}\frac{\partial}{\partial t}\right)A\right\rangle_{0,3} \\ &= \Box A' + \left(\nabla + \frac{1}{c}\frac{\partial}{\partial t}\right)\left(\Box\Psi - \left\langle\left(\nabla - \frac{1}{c}\frac{\partial}{\partial t}\right)A\right\rangle_{0,3}\right). \end{aligned} \qquad (3.264)$$

Requiring

$$\Box\Psi = \left\langle\left(\nabla - \frac{1}{c}\frac{\partial}{\partial t}\right)A\right\rangle_{0,3}, \qquad (3.265)$$

completes the proof. $\square$

Observe that Ψ has only grades 0,3 as required of a gauge function.

Such a transformation completely decouples Maxwell's equation, providing one scalar wave equation for each grade of $\Box A' = J$, relating each grade of the potential A' to exactly one grade of the source multivector current J. We are free to immediately solve for A' using the (causal) Green's function for the d'Alembertian

$$\begin{aligned} A'(\mathbf{x}, t) &= -\int dV' dt' \frac{\delta(|\mathbf{x} - \mathbf{x}'| - c(t - t'))}{4\pi \|\mathbf{x} - \mathbf{x}'\|} J(\mathbf{x}', t') \\ &= -\frac{1}{4\pi}\int dV' \frac{J(\mathbf{x}', t - \frac{1}{c}\|\mathbf{x} - \mathbf{x}'\|)}{\|\mathbf{x} - \mathbf{x}'\|}, \end{aligned} \qquad (3.266)$$

which is the sum of all the current contributions relative to the point $\mathbf{x}$ at the retarded time $t_r = t - (1/c)\|\mathbf{x} - \mathbf{x}'\|$. The field follows immediately by differentiation and grade selection

$$F = \left\langle\left(\nabla - \frac{1}{c}\frac{\partial}{\partial t}\right)A'\right\rangle_{1,2}. \qquad (3.267)$$

Again, using the Green's function for the d'Alembertian, the explicit form of the gauge function Ψ is

$$\Psi = -\frac{1}{4\pi}\int dV' \frac{\left\langle\left(\nabla - \frac{1}{c}\frac{\partial}{\partial t}\right)A(\mathbf{x}', t_r)\right\rangle_{0,3}}{\|\mathbf{x} - \mathbf{x}'\|}, \qquad (3.268)$$

however, we don't actually need to compute this. Instead, we only have to know we are free to construct a field from any solution A' of $\Box A' = J$ using eq. (3.267).

3.11.3 Far field.

Theorem 3.25: Far field magnetic vector potential.

Given a vector potential with a radial spherical wave representation

$$\mathbf{A} = \frac{e^{-jkr}}{r}\mathcal{A}(\theta,\phi),$$

the far field ($r \gg 1$) electromagnetic field is

$$F = -j\omega\left(1 + \hat{\mathbf{r}}\right)\left(\hat{\mathbf{r}} \wedge \mathbf{A}\right).$$

If $\mathbf{A}_\perp = \hat{\mathbf{r}}\left(\hat{\mathbf{r}} \wedge \mathbf{A}\right)$ represents the non-radial component of the potential, the respective electric and magnetic field components are

$$\mathbf{E} = -j\omega\mathbf{A}_\perp$$

$$\mathbf{H} = \frac{1}{\eta}\hat{\mathbf{r}} \times \mathbf{E}.$$

Proof. To prove theorem 3.25, we will utilize a spherical representation of the gradient

$$\nabla = \hat{\mathbf{r}}\partial_r + \nabla_\perp$$

$$\nabla_\perp = \frac{\hat{\boldsymbol{\theta}}}{r}\partial_\theta + \frac{\hat{\boldsymbol{\phi}}}{r\sin\theta}\partial_\phi. \tag{3.269}$$

The gradient of the vector potential is

$$\nabla\mathbf{A} = (\hat{\mathbf{r}}\partial_r + \nabla_\perp)\frac{e^{-jkr}}{r}\mathcal{A}$$

$$= \hat{\mathbf{r}}\left(-jk - \frac{1}{r}\right)\frac{e^{-jkr}}{r}\mathcal{A} + \frac{e^{-jkr}}{r}\nabla_\perp\mathcal{A}$$

$$= -\left(jk + \frac{1}{r}\right)\hat{\mathbf{r}}\mathbf{A} + O(1/r^2) \tag{3.270}$$

$$\approx -jk\hat{\mathbf{r}}\mathbf{A}.$$

Here, all the $O(1/r^2)$ terms, including the action of the non-radial component of the gradient on the $1/r$ potential, have been neglected. From eq. (3.270) the far field divergence and the (bivector) curl of $\mathbf{A}$ are

$$\nabla \cdot \mathbf{A} = -jk\hat{\mathbf{r}} \cdot \mathbf{A}$$

$$\nabla \wedge \mathbf{A} = -jk\hat{\mathbf{r}} \wedge \mathbf{A}. \tag{3.271}$$

Finally, the far field gradient of the divergence of $\mathbf{A}$ is

$$
\begin{aligned}
\nabla (\nabla \cdot \mathbf{A}) &= (\hat{\mathbf{r}} \partial_r + \nabla_\perp) (-jk \hat{\mathbf{r}} \cdot \mathbf{A}) \\
&\approx -jk \hat{\mathbf{r}} \partial_r (\hat{\mathbf{r}} \cdot \mathbf{A}) \\
&= -jk \hat{\mathbf{r}} \left(-jk - \frac{1}{r} \right) (\hat{\mathbf{r}} \cdot \mathbf{A}) \\
&\approx -k^2 \hat{\mathbf{r}} (\hat{\mathbf{r}} \cdot \mathbf{A}),
\end{aligned}
\qquad (3.272)
$$

again neglecting any $O(1/r^2)$ terms. The field is

$$
\begin{aligned}
F &= -j\omega \mathbf{A} - j\frac{c^2}{\omega} \nabla (\nabla \cdot \mathbf{A}) + c \nabla \wedge \mathbf{A} \\
&= -j\omega \mathbf{A} + j\omega \hat{\mathbf{r}} (\hat{\mathbf{r}} \cdot \mathbf{A}) - jkc \hat{\mathbf{r}} \wedge \mathbf{A} \\
&= -j\omega (\mathbf{A} - \hat{\mathbf{r}} (\hat{\mathbf{r}} \cdot \mathbf{A})) - j\omega \hat{\mathbf{r}} \wedge \mathbf{A} \\
&= -j\omega \hat{\mathbf{r}} (\hat{\mathbf{r}} \wedge \mathbf{A}) - j\omega \hat{\mathbf{r}} \wedge \mathbf{A} \\
&= -j\omega (\hat{\mathbf{r}} + 1) (\hat{\mathbf{r}} \wedge \mathbf{A}),
\end{aligned}
\qquad (3.273)
$$

which completes the first part of the proof. Extraction of the electric and magnetic fields can be done by inspection and is left to the reader to prove. □

One interpretation of this is that the (bivector) magnetic field is represented by the plane perpendicular to the direction of propagation, and the electric field by a vector in that plane.

Theorem 3.26: Far field electric vector potential.

Given a vector potential with a radial spherical wave representation

$$\mathbf{F} = \frac{e^{-jkr}}{r} \mathcal{F}(\theta, \phi),$$

the far field ($r \gg 1$) electromagnetic field is

$$F = -j\omega \eta I (\hat{\mathbf{r}} + 1) (\hat{\mathbf{r}} \wedge \mathbf{F}).$$

If $\mathbf{F}_\perp = \hat{\mathbf{r}}(\hat{\mathbf{r}} \wedge \mathbf{F})$ represents the non-radial component of the potential, the respective electric and magnetic field components are

$$\mathbf{E} = j\omega\eta\hat{\mathbf{r}} \times \mathbf{F}$$
$$\mathbf{H} = -j\omega\mathbf{F}_\perp.$$

The proof of theorem 3.26 is left to the reader.

Example 3.1: Vertical dipole potential.

We will calculate the far field along the propagation direction vector $\hat{\mathbf{k}}$ in the z-y plane

$$\hat{\mathbf{k}} = \mathbf{e}_3 e^{i\theta}$$
$$i = \mathbf{e}_{32},$$
(3.274)

for the infinitesimal dipole potential

$$\mathbf{A} = \frac{\mathbf{e}_3 \mu I_0 l}{4\pi r} e^{-jkr},$$
(3.275)

as illustrated in fig. 3.12.

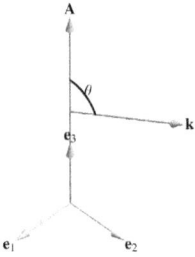

Figure 3.12: Vertical infinitesimal dipole and selected propagation direction.

The wedge of $\hat{\mathbf{k}}$ with $\mathbf{A}$ is proportional to

$$\begin{aligned}\hat{\mathbf{k}} \wedge \mathbf{e}_3 &= \left\langle \hat{\mathbf{k}} \mathbf{e}_3 \right\rangle_2 \\ &= \left\langle \mathbf{e}_3 e^{i\theta} \mathbf{e}_3 \right\rangle_2 \\ &= \left\langle \mathbf{e}_3^2 e^{-i\theta} \right\rangle_2 \\ &= -i \sin \theta,\end{aligned} \quad (3.276)$$

so from theorem 3.26 the field is

$$F = j\omega \left(1 + \mathbf{e}_3 e^{i\theta}\right) i \sin\theta \frac{\mu I_0 l}{4\pi r} e^{-jkr}. \quad (3.277)$$

The electric and magnetic fields can be found from the respective vector and bivector grades of eq. (3.277)

$$\begin{aligned}\mathbf{E} &= \frac{j\omega\mu I_0 l}{4\pi r} e^{-jkr} \mathbf{e}_3 e^{i\theta} i \sin\theta \\ &= \frac{j\omega\mu I_0 l}{4\pi r} e^{-jkr} \mathbf{e}_2 e^{i\theta} \sin\theta \\ &= \frac{jk\eta I_0 l \sin\theta}{4\pi r} e^{-jkr} \left(\mathbf{e}_2 \cos\theta - \mathbf{e}_3 \sin\theta \right),\end{aligned} \quad (3.278)$$

and

$$\begin{aligned}\mathbf{H} &= \frac{1}{I\eta} j\omega i \sin\theta_0 \frac{\mu I_0 l}{4\pi r} e^{-jkr} \\ &= \frac{1}{\eta} \mathbf{e}_{321} \mathbf{e}_{32} j\omega \sin\theta_0 \frac{\mu I_0 l}{4\pi r} e^{-jkr} \\ &= -\mathbf{e}_1 \frac{jk \sin\theta_0 I_0 l}{4\pi r} e^{-jkr}.\end{aligned} \quad (3.279)$$

The multivector electrodynamic field expression eq. (3.277) for F is more algebraically compact than the separate electric and magnetic field expressions, but this comes with the complexity of dealing with different types of imaginaries. There are two explicit unit imaginaries in eq. (3.277), the scalar imaginary j used to encode the time harmonic nature of the field, and $i = \mathbf{e}_{32}$ used to represent the plane that the far field propagation direction vector lay in. Additionally, when the magnetic field component was extracted, the pseudoscalar $I = \mathbf{e}_{123}$ entered into the mix. Care is required to keep these all separate, especially since I, j commute with all grades, but i does not.

3.11.4 Problems.

Exercise 3.10 Potentials for no-fictitious sources.

Starting with Maxwell's equation with only conventional electric sources

$$\left(\nabla + \frac{1}{c}\frac{\partial}{\partial t}\right)F = \langle J \rangle_{0,1}. \tag{3.280}$$

Show that this may be split by grade into three equations

$$\begin{aligned}\left\langle\left(\nabla + \frac{1}{c}\frac{\partial}{\partial t}\right)F\right\rangle_{0,1} &= \langle J \rangle_{0,1} \\ \nabla \wedge \mathbf{E} + \frac{1}{c}\frac{\partial}{\partial t}(I\eta\mathbf{H}) &= 0 \\ \nabla \wedge (I\eta\mathbf{H}) &= 0.\end{aligned} \tag{3.281}$$

Then use the identities $\nabla \wedge \nabla \wedge \mathbf{A} = 0$, for vector $\mathbf{A}$ and $\nabla \wedge \nabla \phi = 0$, for scalar ϕ to find the potential representation eq. (3.251).

Exercise 3.11 Potentials for fictitious sources.

Starting with Maxwell's equation with only fictitious magnetic sources

$$\left(\nabla + \frac{1}{c}\frac{\partial}{\partial t}\right)F = \langle J \rangle_{2,3}, \tag{3.291}$$

show that this may be split by grade into three equations

$$\begin{aligned}\left\langle\left(\nabla + \frac{1}{c}\frac{\partial}{\partial t}\right)IF\right\rangle_{0,1} &= I\langle J \rangle_{2,3} \\ -\eta\nabla \wedge \mathbf{H} + \frac{1}{c}\frac{\partial(I\mathbf{E})}{\partial t} &= 0 \\ \nabla \wedge (I\mathbf{E}) &= 0.\end{aligned} \tag{3.292}$$

Then use the identities $\nabla \wedge \nabla \wedge \mathbf{F} = 0$, for vector $\mathbf{F}$ and $\nabla \wedge \nabla \phi_m = 0$, for scalar ϕ_m to find the potential representation eq. (3.253).

Exercise 3.12 Total field in terms of potentials.

Prove lemma 3.1, either by direct expansion, or by trying to discover the multivector form of the field by construction.

Exercise 3.13   Fields in terms of potentials.

Prove lemma 3.2.

Exercise 3.14   Potential wave equations.

Prove theorem 3.22.

3.12 DIELECTRIC AND MAGNETIC MEDIA.

3.12.1 Statement.

Without imposing the constitutive relationships eq. (3.2) the geometric algebra form of Maxwell's equations requires a pair of equations, multivector fields, and multivector sources, instead of one of each.

Theorem 3.27: Maxwell's equations in media.

Maxwell's equations in media are

$$\left\langle \left(\nabla + \frac{1}{c}\frac{\partial}{\partial t}\right)F\right\rangle_{0,1} = J_e$$

$$\left\langle \left(\nabla + \frac{1}{c}\frac{\partial}{\partial t}\right)G\right\rangle_{2,3} = IJ_m,$$

where c is the group velocity of F, G in the medium, the fields are grade $(1, 2)$-multivectors

$$F = \mathbf{D} + \frac{I}{c}\mathbf{H}$$

$$G = \mathbf{E} + Ic\mathbf{B},$$

and the sources are grade $(0, 1)$-multivectors

$$J_e = \rho - \frac{1}{c}\mathbf{J}$$

$$J_m = c\rho_m - \mathbf{M}.$$

Proof. To prove theorem 3.27 we may simply expand the spacetime gradients and grade selection operations, and compare to eq. (3.1), the conventional representation of Maxwell's equations. For F we have

$$\begin{aligned}
\rho - \frac{\mathbf{J}}{c} &= \left\langle \left(\nabla + \frac{1}{c}\frac{\partial}{\partial t} \right) F \right\rangle_{0,1} \\
&= \left\langle \left(\nabla + \frac{1}{c}\frac{\partial}{\partial t} \right) \left(\mathbf{D} + \frac{I}{c}\mathbf{H} \right) \right\rangle_{0,1} \\
&= \left\langle \nabla \cdot \mathbf{D} + \nabla \wedge \mathbf{D} + \frac{I}{c}\nabla \cdot \mathbf{H} + \frac{I}{c}\nabla \wedge \mathbf{H} + \frac{1}{c}\frac{\partial \mathbf{D}}{\partial t} + I\frac{1}{c^2}\frac{\partial \mathbf{H}}{\partial t} \right\rangle_{0,1} \\
&= \nabla \cdot \mathbf{D} + \frac{1}{c}\frac{\partial \mathbf{D}}{\partial t} - \frac{1}{c}\nabla \times \mathbf{H},
\end{aligned}$$

(3.307)

and for G

$$\begin{aligned}
I(c\rho_m - \mathbf{M}) &= \left\langle \left(\nabla + \frac{1}{c}\frac{\partial}{\partial t} \right) G \right\rangle_{2,3} \\
&= \left\langle \left(\nabla + \frac{1}{c}\frac{\partial}{\partial t} \right) (\mathbf{E} + Ic\mathbf{B}) \right\rangle_{2,3} \\
&= \left\langle \nabla \cdot \mathbf{E} + \nabla \wedge \mathbf{E} + Ic\nabla \cdot \mathbf{B} + Ic\nabla \wedge \mathbf{B} + \frac{1}{c}\frac{\partial \mathbf{E}}{\partial t} + I\frac{\partial \mathbf{B}}{\partial t} \right\rangle_{2,3} \\
&= \nabla \wedge \mathbf{E} + Ic\nabla \cdot \mathbf{B} + I\frac{\partial \mathbf{B}}{\partial t} \\
&= I\left(\nabla \times \mathbf{E} + c\nabla \cdot \mathbf{B} + \frac{\partial \mathbf{B}}{\partial t} \right).
\end{aligned}$$

(3.308)

Applying further grade selection operations, rescaling (cancelling all factors of c and I), and a bit of rearranging, gives

$$\begin{aligned}
\nabla \cdot \mathbf{D} &= \rho \\
\nabla \times \mathbf{H} &= \mathbf{J} + \frac{\partial \mathbf{D}}{\partial t} \\
\nabla \cdot \mathbf{B} &= \rho_m \\
\nabla \times \mathbf{E} &= -\mathbf{M} - \frac{\partial \mathbf{B}}{\partial t},
\end{aligned}$$

(3.309)

which are Maxwell's equations. $\square$

3.12 DIELECTRIC AND MAGNETIC MEDIA.

Exercise 3.15   Maxwell's equations in media.

The proof above is somewhat unfriendly, as it works backwards from the answer. No motivation was given for why the particular multivector fields were chosen, nor why grade selection operations were required. To obtain some insight on why this works, prove theorem 3.27 from eq. (3.2) directly as follows:

1. Eliminate cross products using $\boldsymbol{\nabla} \times \mathbf{f} = I(\boldsymbol{\nabla} \wedge \mathbf{f})$.

2. Introduce a scalar constant c with dimensions of velocity and redimensionalize any time derivatives $\partial \mathbf{f}/\partial t = (1/c)\partial(c\mathbf{f})/\partial t$, so that $[(1/c)\partial/\partial t] = [\boldsymbol{\nabla}]$.

3. If required, multiply each of Maxwell's equations by a factor of I, to obtain a scalar and vector equation for $\mathbf{D}, \mathbf{H}$, and a bivector and pseudoscalar equation for $\mathbf{E}, \mathbf{B}$.

4. Sum the pairs of equations to form a multivector equation for each of $\mathbf{D}, \mathbf{H}$ and $\mathbf{E}, \mathbf{B}$.

5. Factor the terms in each equation into a product of the spacetime gradient and the respective fields F, G, and show the result may be simplified by grade selection.

3.12.2 Alternative form.

Theorem 3.28: Grade selection free equations.

Given multivector solutions F', G' to

$$J_e = \left(\boldsymbol{\nabla} + \frac{1}{c}\frac{\partial}{\partial t}\right) F'$$
$$I J_m = \left(\boldsymbol{\nabla} + \frac{1}{c}\frac{\partial}{\partial t}\right) G',$$

these can be related to solutions F, G of Maxwell's equations given by theorem 3.27 by

$$F = \langle F' \rangle_{1,2}$$
$$G = \langle G' \rangle_{1,2},$$

if
$$\left\langle\left(\nabla + \frac{1}{c}\frac{\partial}{\partial t}\right)\langle F'\rangle\right\rangle_{0,1} = 0$$
$$\left\langle\left(\nabla + \frac{1}{c}\frac{\partial}{\partial t}\right)\langle G'\rangle_3\right\rangle_{2,3} = 0.$$

Proof. To prove we select the grade 0,1 and grade 2,3 components from space time gradient equations of theorem 3.28. For the electric sources, this gives

$$J_e = \left\langle\left(\nabla + \frac{1}{c}\frac{\partial}{\partial t}\right)F'\right\rangle_{0,1}$$
$$= \left\langle\left(\nabla + \frac{1}{c}\frac{\partial}{\partial t}\right)\langle F'\rangle_{1,2}\right\rangle_{0,1} + \left\langle\left(\nabla + \frac{1}{c}\frac{\partial}{\partial t}\right)\langle F'\rangle\right\rangle_{0,1} + \left\langle\left(\nabla + \frac{1}{c}\frac{\partial}{\partial t}\right)\langle F'\rangle_3\right\rangle_{0,1},$$
(3.310)

however $\left(\nabla + \frac{1}{c}\frac{\partial}{\partial t}\right)\langle F'\rangle_3$ has only grade 2,3 components, leaving just

$$J_e = \left\langle\left(\nabla + \frac{1}{c}\frac{\partial}{\partial t}\right)\langle F'\rangle_{1,2}\right\rangle_{0,1} + \left\langle\left(\nabla + \frac{1}{c}\frac{\partial}{\partial t}\right)\langle F'\rangle\right\rangle_{0,1}, \quad (3.311)$$

as claimed. For the magnetic sources, we have

$$IJ_m = \left\langle\left(\nabla + \frac{1}{c}\frac{\partial}{\partial t}\right)G'\right\rangle_{2,3}$$
$$= \left\langle\left(\nabla + \frac{1}{c}\frac{\partial}{\partial t}\right)\langle G'\rangle_{1,2}\right\rangle_{2,3} + \left\langle\left(\nabla + \frac{1}{c}\frac{\partial}{\partial t}\right)\langle G'\rangle\right\rangle_{2,3} \quad (3.312)$$
$$+ \left\langle\left(\nabla + \frac{1}{c}\frac{\partial}{\partial t}\right)\langle G'\rangle_3\right\rangle_{2,3},$$

however $\left(\nabla + \frac{1}{c}\frac{\partial}{\partial t}\right)\langle G'\rangle_0$ has only grade 0,1 components, leaving just

$$IJ_m = \left\langle\left(\nabla + \frac{1}{c}\frac{\partial}{\partial t}\right)\langle G'\rangle_{1,2}\right\rangle_{2,3} + \left\langle\left(\nabla + \frac{1}{c}\frac{\partial}{\partial t}\right)\langle G'\rangle\right\rangle_{2,3}. \quad (3.313)$$

□

Theorem 3.28 is probably a more effect geometric algebra form for solution of Maxwell's equations in matter, as the grade selection free space-time gradients can be solved for F', G' directly using Green's function convolution. However, we have an open question of how to impose a zero scalar grade constraint on F' and a zero pseudoscalar grade constraint on G'.

3.12 DIELECTRIC AND MAGNETIC MEDIA.

Question: Is the solution as simple as grade selection of the convolution?

$$F = \int dt' dV' \langle G(\mathbf{x} - \mathbf{x}', t - t') J_e \rangle_{1,2}$$
$$G = \int dt' dV' \langle G(\mathbf{x} - \mathbf{x}', t - t') I J_m \rangle_{1,2},$$
(3.314)

where $G(\mathbf{x} - \mathbf{x}', t - t')$, is the Green's function for the space time gradient theorem 2.19, not to be confused with $G = \mathbf{E} + Ic\mathbf{B}$,

3.12.3 Gauge like transformations.

Because of the grade selection operations in theorem 3.27, we cannot simply solve for F, G using the Green's function for the spacetime gradient. However, we may make a gauge-like transformation of the fields. Additional exploration is required to determine if such transformations can be utilized to solve theorem 3.27.

Theorem 3.29: Multivector transformation of the fields.

If F, G are solutions to theorem 3.27, then so are

$$F' = F + \left\langle \left(\nabla - \frac{1}{c}\frac{\partial}{\partial t} \right) \Psi_{2,3} \right\rangle_{1,2}$$
$$G' = G + \left\langle \left(\nabla - \frac{1}{c}\frac{\partial}{\partial t} \right) \Psi_{0,1} \right\rangle_{1,2},$$

where $\Psi_{2,3}$ is any multivector with grades 2,3 and $\Psi_{0,1}$ is any multivector with grades 0,1.

Proof. To prove theorem 3.29 we need to show that

$$\left\langle \left(\nabla + \frac{1}{c}\frac{\partial}{\partial t} \right) F' \right\rangle_{0,1} = \left\langle \left(\nabla + \frac{1}{c}\frac{\partial}{\partial t} \right) F \right\rangle_{0,1}$$
$$\left\langle \left(\nabla + \frac{1}{c}\frac{\partial}{\partial t} \right) G' \right\rangle_{2,3} = \left\langle \left(\nabla + \frac{1}{c}\frac{\partial}{\partial t} \right) G \right\rangle_{2,3}.$$
(3.315)

Let's start with F

$$\left\langle \left(\nabla + \frac{1}{c}\frac{\partial}{\partial t}\right) F' \right\rangle_{0,1}$$
$$= \left\langle \left(\nabla + \frac{1}{c}\frac{\partial}{\partial t}\right) F \right\rangle_{0,1} + \left\langle \left(\nabla + \frac{1}{c}\frac{\partial}{\partial t}\right)\left\langle \left(\nabla - \frac{1}{c}\frac{\partial}{\partial t}\right) \Psi_{2,3} \right\rangle_{1,2} \right\rangle_{0,1}$$
$$= \left\langle \left(\nabla + \frac{1}{c}\frac{\partial}{\partial t}\right) F \right\rangle_{0,1} + \langle \Box \Psi_{2,3} \rangle_{0,1} - \left\langle \left(\nabla + \frac{1}{c}\frac{\partial}{\partial t}\right)\left\langle \left(\nabla - \frac{1}{c}\frac{\partial}{\partial t}\right) \Psi_{2,3} \right\rangle_{0,3} \right\rangle_{0,1}.$$
(3.316)

The second term is killed since $\Psi_{2,3}$ has no grade 0,1 components by definition, so neither does $\Box \Psi_{2,3}$. To see that the last term is zero, note that $\left(\nabla - \frac{1}{c}\frac{\partial}{\partial t}\right)\Psi_{2,3}$ can have only grades 1,2,3, so $\left\langle \left(\nabla - \frac{1}{c}\frac{\partial}{\partial t}\right)\Psi_{2,3}\right\rangle_{0,3}$ is a trivector. This means that $\left(\nabla + \frac{1}{c}\frac{\partial}{\partial t}\right)\left\langle \left(\nabla - \frac{1}{c}\frac{\partial}{\partial t}\right)\Psi_{2,3}\right\rangle_{0,3}$ has only grades 2,3, which are obliterated by the final grade 0,1 selection operation, leaving just $\left\langle \left(\nabla + \frac{1}{c}\frac{\partial}{\partial t}\right) F \right\rangle_{0,1}$. For G we have

$$\left\langle \left(\nabla + \frac{1}{c}\frac{\partial}{\partial t}\right) G' \right\rangle_{2,3}$$
$$= \left\langle \left(\nabla + \frac{1}{c}\frac{\partial}{\partial t}\right) G \right\rangle_{2,3} + \left\langle \left(\nabla + \frac{1}{c}\frac{\partial}{\partial t}\right)\left\langle \left(\nabla - \frac{1}{c}\frac{\partial}{\partial t}\right) \Psi_{0,1} \right\rangle_{1,2} \right\rangle_{2,3}$$
$$= \left\langle \left(\nabla + \frac{1}{c}\frac{\partial}{\partial t}\right) G \right\rangle_{2,3} + \langle \Box \Psi_{0,1} \rangle_{2,3}$$
$$- \left\langle \left(\nabla + \frac{1}{c}\frac{\partial}{\partial t}\right)\left\langle \left(\nabla - \frac{1}{c}\frac{\partial}{\partial t}\right) \Psi_{0,1} \right\rangle_{0,3} \right\rangle_{2,3}.$$
(3.317)

As before the d'Alembertian term is killed as it has no grades 2,3. To see that the last term is zero, note that $\left(\nabla - \frac{1}{c}\frac{\partial}{\partial t}\right)\Psi_{0,1}$ can have only grades 0,1,2, so $\left\langle \left(\nabla - \frac{1}{c}\frac{\partial}{\partial t}\right)\Psi_{0,1}\right\rangle_{0,3}$ is a scalar. This means that

$$\left(\nabla + \frac{1}{c}\frac{\partial}{\partial t}\right)\left\langle \left(\nabla - \frac{1}{c}\frac{\partial}{\partial t}\right)\Psi_{0,1}\right\rangle_{0,3},$$
(3.318)

has only grades 0,1, which are obliterated by the final grade 2,3 selection operation, leaving $\left\langle \left(\nabla + \frac{1}{c}\frac{\partial}{\partial t}\right) G \right\rangle_{2,3}$, completing the proof. $\square$

An additional variation of theorem 3.29 is also possible.

3.12 DIELECTRIC AND MAGNETIC MEDIA.

Theorem 3.30: Multivector transformation of the fields.

If F, G are solutions to theorem 3.27, then so are

$$F' = F + \left(\nabla - \frac{1}{c}\frac{\partial}{\partial t}\right)\Psi_{2,3}$$

$$G' = G + \left(\nabla - \frac{1}{c}\frac{\partial}{\partial t}\right)\Psi_{0,1}$$

where $\Psi_{2,3}$ is any multivector with grades 2,3 and $\Psi_{0,1}$ is any multivector with grades 0,1.

Proof. Theorem 3.30 can be proven by direct substitution. For F

$$\left\langle\left(\nabla + \frac{1}{c}\frac{\partial}{\partial t}\right)\left(F + \left(\nabla - \frac{1}{c}\frac{\partial}{\partial t}\right)\Psi_{2,3}\right)\right\rangle_{0,1} = \left\langle\left(\nabla + \frac{1}{c}\frac{\partial}{\partial t}\right)F + \Box\Psi_{2,3}\right\rangle_{0,1}$$

$$= \left\langle\left(\nabla + \frac{1}{c}\frac{\partial}{\partial t}\right)F\right\rangle,$$

and for G

$$\left\langle\left(\nabla + \frac{1}{c}\frac{\partial}{\partial t}\right)\left(G + \left(\nabla - \frac{1}{c}\frac{\partial}{\partial t}\right)\Psi_{0,1}\right)\right\rangle_{2,3} = \left\langle\left(\nabla + \frac{1}{c}\frac{\partial}{\partial t}\right)G + \Box\Psi_{0,1}\right\rangle_{2,3}$$

$$= \left\langle\left(\nabla + \frac{1}{c}\frac{\partial}{\partial t}\right)G\right\rangle.$$

which completes the proof. $\square$

3.12.4 Boundary value conditions.

Theorem 3.31: Boundary value relations.

The difference in the normal and tangential components of the electromagnetic field spanning a surface on which there are a surface current or surface charge or current densities $J_e = J_{es}\delta(n), J_m = J_{ms}\delta(n)$ can be related to those surface sources as follows

$$\langle\hat{n}(F_2 - F_1)\rangle_{0,1} = J_{es}$$
$$\langle\hat{n}(G_2 - G_1)\rangle_{2,3} = IJ_{ms},$$

where $F_k = \mathbf{D}_k + I\mathbf{H}_k/c$, $G_k = \mathbf{E}_k + Ic\mathbf{B}_k$, $k = 1, 2$ are the fields in the where $\hat{\mathbf{n}} = \hat{\mathbf{n}}_2 = -\hat{\mathbf{n}}_1$ is the outwards facing normal in the second medium. In terms of the conventional constituent fields, these may be written

$$\hat{\mathbf{n}} \cdot (\mathbf{D}_2 - \mathbf{D}_1) = \rho_s$$
$$\hat{\mathbf{n}} \times (\mathbf{H}_2 - \mathbf{H}_1) = \mathbf{J}_s$$
$$\hat{\mathbf{n}} \cdot (\mathbf{B}_2 - \mathbf{B}_1) = \rho_{ms}$$
$$\hat{\mathbf{n}} \times (\mathbf{E}_2 - \mathbf{E}_1) = -\mathbf{M}_s.$$

Figure 3.13 illustrates a surface where we seek to find the fields above the surface (region 2), and below the surface (region 1). These fields will be determined by integrating Maxwell's equation over the pillbox configuration, allowing the height n of that pillbox above or below the surface to tend to zero, and the area of the pillbox top to also tend to zero.

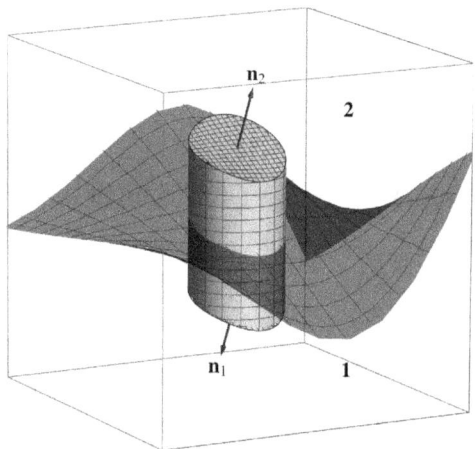

Figure 3.13: Pillbox integration volume.

Proof. We will work with theorem 3.27, Maxwell's equations in media, in their frequency domain form

$$\langle \boldsymbol{\nabla} F \rangle_{0,1} + jk\mathbf{D} = J_{es}\delta(n)$$
$$\langle \boldsymbol{\nabla} G \rangle_{2,3} + jkIc\mathbf{B} = IJ_{ms}\delta(n),$$
(3.319)

and integrate these over the pillbox volume in the figure. That is

$$\int dV \, \langle \nabla F \rangle_{0,1} + jk \int dV \, \mathbf{D} = \int dn dA \, J_{es} \delta(n)$$
$$\int dV \, \langle \nabla G \rangle_{2,3} + jkIc \int dV \, \mathbf{B} = I \int dn dA \, J_{ms} \delta(n).$$
(3.320)

The gradient integrals can be evaluated with theorem 2.11. Evaluating the delta functions picks leaves an area integral on the surface. Additionally, we assume that we are making the pillbox volume small enough that we can employ the mean value theorem for the $\mathbf{D}, \mathbf{B}$ integrals

$$\int_{\partial V} dA \, \langle \hat{\mathbf{n}} F \rangle_{0,1} + jk\Delta A \left(n_1 \tilde{\mathbf{D}}_1 + n_2 \tilde{\mathbf{D}}_2 \right) = \Delta A J_{es}$$
$$\int_{\partial V} dA \, \langle \hat{\mathbf{n}} G \rangle_{2,3} + jkIc\Delta A \left(n_1 \tilde{\mathbf{B}}_1 + n_2 \tilde{\mathbf{B}}_2 \right) = I\Delta A J_{ms}.$$
(3.321)

We now let n_1, n_2 tend to zero, which kills off the $\mathbf{D}, \mathbf{B}$ contributions, and also kills off the side wall contributions in the first pillbox surface integral. This leaves

$$\langle \hat{\mathbf{n}}_2 F_2 \rangle_{0,1} + \langle \hat{\mathbf{n}}_1 F_1 \rangle_{0,1} = J_{es}$$
$$\langle \hat{\mathbf{n}}_2 G_2 \rangle_{2,3} + \langle \hat{\mathbf{n}}_1 G_1 \rangle_{2,3} = J_{ms}.$$
(3.322)

Inserting $\hat{\mathbf{n}} = \hat{\mathbf{n}}_2 = -\hat{\mathbf{n}}_1$ completes the first part of the proof.

Expanding the grade selection operations, we find

$$\hat{\mathbf{n}} \cdot (\mathbf{D}_2 - \mathbf{D}_1) = \rho_s$$
$$I\hat{\mathbf{n}} \wedge (\mathbf{H}_2/c - \mathbf{H}_1/c) = -\mathbf{J}_s/c$$
$$\hat{\mathbf{n}} \wedge (\mathbf{E}_2 - \mathbf{E}_1) = -I\mathbf{M}_s$$
$$Ic\hat{\mathbf{n}} \cdot (\mathbf{B}_2 - \mathbf{B}_1) = Ic\rho_{ms},$$
(3.323)

and expansion of the wedge's as cross's using eq. (1.81) completes the proof. $\square$

In the special case where there are surface charge and current densities along the interface surface, but the media is uniform ($\epsilon_1 = \epsilon_2, \mu_1 = \mu_2$), then the field and current relationship has a particularly simple form [5]

$$\hat{\mathbf{n}}(F_2 - F_1) = J_s.$$
(3.324)

Exercise 3.16   Uniform media with currents and densities.

Prove that eq. (3.324) holds when $\epsilon_1 = \epsilon_2, \mu_1 = \mu_2$.

3.13 PROBLEM SOLUTIONS

Answer for Exercise 3.2

$$\nabla \mathbf{b} = \sum_{ij=1}^{3} \left(\mathbf{e}_i \frac{\partial}{\partial x_i} \right) (\mathbf{e}_j b_j)$$

$$= \sum_{ij=1}^{3} \mathbf{e}_i \mathbf{e}_j \frac{\partial b_j}{\partial x_i} \qquad (3.11)$$

$$= \sum_{i=1}^{3} \mathbf{e}_i \mathbf{e}_i \frac{\partial b_i}{\partial x_i} + \sum_{i \ne j} \mathbf{e}_i \mathbf{e}_j \frac{\partial b_j}{\partial x_i}.$$

Here we've decomposed the sum into symmetric and antisymmetric contributions. The symmetric part reduces easily to the divergence

$$\sum_{i=1}^{3} \mathbf{e}_i \mathbf{e}_i \frac{\partial b_i}{\partial x_i} = \sum_{i=1}^{3} \frac{\partial b_i}{\partial x_i} = \nabla \cdot \mathbf{b}. \qquad (3.12)$$

Because, for $i \ne j$, $\mathbf{e}_i \mathbf{e}_j = \mathbf{e}_i \wedge \mathbf{e}_j = I (\mathbf{e}_i \times \mathbf{e}_j)$, and both the wedge and dot products are zero for $i = j$, we can reintroduce the sum over all i, j indexes

$$\sum_{i \ne j} \mathbf{e}_i \mathbf{e}_j \frac{\partial b_j}{\partial x_i} = I \sum_{i \ne j} \mathbf{e}_i \times \mathbf{e}_j \frac{\partial b_j}{\partial x_i}$$

$$= I \sum_{ij=1}^{3} \mathbf{e}_i \times \mathbf{e}_j \frac{\partial b_j}{\partial x_i} \qquad (3.13)$$

$$= I \sum_{ij=1}^{3} \left(\mathbf{e}_i \frac{\partial}{\partial x_i} \right) \times (\mathbf{e}_j b_j)$$

$$= I (\nabla \times \mathbf{b}).$$

We've demonstrated the desired result, showing that our Laissez-faire substitution $\mathbf{a} = \nabla$ in $\mathbf{ab} = \mathbf{a} \cdot \mathbf{b} + I (\mathbf{a} \times \mathbf{b})$ was justified, despite the operator nature of the gradient.

Answer for Exercise 3.3

Our grade selection operators yield the following four equations

$$\langle \nabla \mathbf{E} \rangle = \eta c \rho$$
$$\left\langle \frac{1}{c}\frac{\partial \mathbf{E}}{\partial t} + \nabla I \eta \mathbf{H} \right\rangle_1 = -\eta \mathbf{J}$$
$$\left\langle \nabla \mathbf{E} + \frac{1}{c}\frac{\partial I \eta \mathbf{H}}{\partial t} \right\rangle_2 = -I\mathbf{M} \quad (3.14)$$
$$\langle \nabla I \eta \mathbf{H} \rangle_3 = I c \rho_m$$

Observe that $\eta c = 1/\epsilon$, so the first equation recovers Gauss's law

$$\nabla \cdot \mathbf{E} = \frac{\rho}{\epsilon}. \quad (3.15)$$

Dividing the vector equation through by $-\eta$, we have

$$\frac{-1}{c\eta}\frac{\partial \mathbf{E}}{\partial t} - I(\nabla \wedge \mathbf{H}) = \mathbf{J}, \quad (3.16)$$

or

$$-\frac{\partial \epsilon \mathbf{E}}{\partial t} + \nabla \times \mathbf{H} = \mathbf{J}, \quad (3.17)$$

the Ampére-Maxwell equation (with $\mathbf{D} = \epsilon \mathbf{E}$, and $\mathbf{H} = \mathbf{B}/\mu$.) Multiplying the bivector equation through by $-I$, and noting that $\eta/c = \mu$, we convert it to a vector equation

$$-I^2(\nabla \times \mathbf{E}) - I^2 \frac{\partial \mu \mathbf{H}}{\partial t} = I^2 \mathbf{M}, \quad (3.18)$$

which is the Maxwell-Faraday equation (augmented with the fictious magnetic current density.) Finally, dividing the pseudoscalar equation through by Ic, we find

$$\rho_m = \frac{\eta}{c}\nabla \cdot \mathbf{H} = \nabla \cdot (\mu \mathbf{H}), \quad (3.19)$$

which is Gauss's law for magnetism (with the fictious "engineering" magnetic charge density term.)

Answer for Exercise 3.4

With only magnetic sources, the current density multivector is

$$J = I(c\rho_m - \mathbf{M})$$
$$= Ic\left(1 - \frac{v}{c}\mathbf{e}_3\right)\lambda_m \delta(x)\delta(y). \qquad (3.83)$$

As in eq. (3.79), let the field observation point be $\mathbf{x} = \mathbf{x}_\perp + z\mathbf{e}_3$, so

$$F(\mathbf{x}) = \frac{1}{4\pi}\int_V dV' \frac{\left\langle (\mathbf{x} - \mathbf{x}')Ic\left(1 - \frac{v}{c}\mathbf{e}_3\right)\lambda_m \delta(x)\delta(y)\right\rangle_{1,2}}{\|\mathbf{x} - \mathbf{x}'\|^3}$$
$$= \frac{Ic\lambda_m}{4\pi}\int_{-\infty}^{\infty} dz' \frac{\left\langle (\mathbf{x}_\perp - z'\mathbf{e}_3)\left(1 - \frac{v}{c}\mathbf{e}_3\right)\right\rangle_{1,2}}{\left(\mathbf{x}_\perp^2 - (z-z')^2\right)^{3/2}} \qquad (3.84)$$

The grade selection reduces to three terms

$$\left\langle (\mathbf{x}_\perp - z'\mathbf{e}_3)\left(1 - \frac{v}{c}\mathbf{e}_3\right)\right\rangle_{1,2} = \mathbf{x}_\perp - z'\mathbf{e}_3 - \frac{v}{c}\mathbf{x}_\perp \mathbf{e}_3$$
$$= \mathbf{x}_\perp\left(1 - \frac{v}{c}\mathbf{e}_3\right) - z'\mathbf{e}_3. \qquad (3.85)$$

As before the integral with the z' dependence vanishes, and we are left with

$$F(\mathbf{x}) = \frac{Ic\lambda_m \mathbf{x}_\perp \left(1 - \frac{v}{c}\mathbf{e}_3\right)}{2\pi \mathbf{x}_\perp^2}. \qquad (3.86)$$

Substitution of $\mathbf{x}_\perp = R\hat{\rho}$ yields the desired result.

Finally, observe that

$$I\hat{\rho}\mathbf{e}_3 = I(\hat{\rho} \wedge \mathbf{e}_3)$$
$$= I^2(\hat{\rho} \times \mathbf{e}_3), \qquad (3.87)$$

so

$$F(\mathbf{x}) = Ic\mathbf{B} + \mathbf{B} \times \mathbf{v}, \qquad (3.88)$$

as claimed.

Answer for Exercise 3.7

The component of $\mathbf{v}(0)$ parallel to the plane Ω is

$$\begin{aligned}
\mathbf{v}(0)_{\parallel} &= \left(\mathbf{v}(0) \cdot \hat{\Omega}\right) \hat{\Omega}^{-1} \\
&= \left(\mathbf{v}(0) \cdot (I\hat{\mathbf{B}})\right)(-I\hat{\mathbf{B}}) \\
&= -\langle \mathbf{v}(0) I\hat{\mathbf{B}} \rangle_1 I\hat{\mathbf{B}} \\
&= -\langle I\left(\mathbf{v}(0) \cdot \hat{\mathbf{B}} + \mathbf{v}(0) \wedge \hat{\mathbf{B}}\right)\rangle_1 I\hat{\mathbf{B}} \\
&= -I\left(\mathbf{v}(0) \wedge \hat{\mathbf{B}}\right) I\hat{\mathbf{B}} \\
&= \left(\mathbf{v}(0) \wedge \hat{\mathbf{B}}\right) \hat{\mathbf{B}},
\end{aligned} \quad (3.193)$$

as claimed. This component of the velocity vector is entirely perpendicular to the normal $\hat{\mathbf{B}}$ for the plane Ω.

The component of this velocity vector that is perpendicular to the plane is

$$\begin{aligned}
\mathbf{v}(0)_{\perp} &= \left(\mathbf{v}(0) \wedge \hat{\Omega}\right) \hat{\Omega}^{-1} \\
&= -\langle \mathbf{v}(0) I\hat{\mathbf{B}} \rangle_3 I\hat{\mathbf{B}} \\
&= -\left(\mathbf{v}(0) \cdot \hat{\mathbf{B}}\right) I^2 \hat{\mathbf{B}} \\
&= \left(\mathbf{v}(0) \cdot \hat{\mathbf{B}}\right) \hat{\mathbf{B}},
\end{aligned} \quad (3.194)$$

also as claimed. Observe that the component of the initial velocity that is perpendicular to the plane Ω, must be, by construction, parallel to the normal ($\hat{\mathbf{B}}$) of the plane.

Finally, for the expansion of the vector and complex exponential product, first note that

$$\begin{aligned}
\Omega t &= -qIc\mathbf{B}t/(mc) \\
&= -qI\mathbf{B}t/m \\
&= -\frac{qBt}{m} I\hat{\mathbf{B}},
\end{aligned} \quad (3.195)$$

so we have

$$\mathbf{v}(0)_{\parallel} e^{\Omega t} = \mathbf{v}(0)_{\parallel} \left(\cos\left(\frac{qBt}{m}\right) - I\hat{\mathbf{B}} \sin\left(\frac{qBt}{m}\right)\right). \quad (3.196)$$

We can reduce the sine factor

$$\begin{aligned}
-\mathbf{v}(0)_{\parallel} I\hat{\mathbf{B}} &= \left(\mathbf{v}(0) \wedge \hat{\mathbf{B}}\right) \hat{\mathbf{B}} (-I\hat{\mathbf{B}}) = -I\left(\mathbf{v}(0) \wedge \hat{\mathbf{B}}\right) = \mathbf{v}(0) \times \hat{\mathbf{B}} \\
&= \left(\mathbf{v}(0)_{\parallel} + \mathbf{v}(0)_{\perp}\right) \times \hat{\mathbf{B}} \\
&= \mathbf{v}(0)_{\parallel} \times \hat{\mathbf{B}},
\end{aligned} \quad (3.197)$$

since $\mathbf{v}(0)_\perp$ is parallel to $\hat{\mathbf{B}}$.

Answer for Exercise 3.8

Using a difference of squares to leave a scalar only in the denominator of eq. (3.234), we find for F_t

$$F_t = j \frac{\frac{\omega}{c} + k\mathbf{e}_3}{\left(\frac{\omega}{c}\right)^2 - k^2} \boldsymbol{\nabla}_t F_z. \tag{3.235}$$

We may re-express F_z in terms of scalar field variables by noting First note that

$$\begin{aligned}\boldsymbol{\nabla}_t F_z &= \boldsymbol{\nabla}_t \mathbf{e}_3 \left(E_z + I\eta H_z\right) \\ &= -\mathbf{e}_3 \boldsymbol{\nabla}_t \left(E_z + I\eta H_z\right).\end{aligned} \tag{3.236}$$

so

$$F_t = -j \frac{\frac{\omega}{c}\mathbf{e}_3 + k}{\omega^2 \mu\epsilon - k^2} \boldsymbol{\nabla}_t \left(E_z + I\eta H_z\right). \tag{3.237}$$

This may now be split into electric and magnetic fields, but first note that the multivector operator

$$\begin{aligned}\mathbf{e}_3 \boldsymbol{\nabla}_t &= \mathbf{e}_3 \cdot \boldsymbol{\nabla}_t + \mathbf{e}_3 \wedge \boldsymbol{\nabla}_t \\ &= \mathbf{e}_3 \wedge \boldsymbol{\nabla}_t,\end{aligned} \tag{3.238}$$

has only a bivector component.

For the transverse electric field component, we have

$$\begin{aligned}\left\langle \left(\frac{\omega}{c}\mathbf{e}_3 + k\right) \boldsymbol{\nabla}_t \left(E_z + I\eta H_z\right) \right\rangle_1 &= k\boldsymbol{\nabla}_t E_z + \frac{\omega}{c}\mathbf{e}_3 \wedge \boldsymbol{\nabla}_t \left(I\eta H_z\right) \\ &= k\boldsymbol{\nabla}_t E_z - \frac{\eta\omega}{c}\mathbf{e}_3 \times \boldsymbol{\nabla}_t H_z.\end{aligned} \tag{3.239}$$

and for the magnetic field component

$$\left\langle \left(\frac{\omega}{c}\mathbf{e}_3 + k\right) \boldsymbol{\nabla}_t \left(E_z + I\eta H_z\right) \right\rangle_2 = \frac{\omega}{c}\mathbf{e}_3 \wedge \boldsymbol{\nabla}_t E_z + I\eta k \boldsymbol{\nabla}_t H_z \tag{3.240}$$

This means that

$$\begin{aligned}\mathbf{E}_t &= \frac{j}{\omega^2 \mu\epsilon - k^2} \left(-k\boldsymbol{\nabla}_t E_z + \frac{\eta\omega}{c}\mathbf{e}_3 \times \boldsymbol{\nabla}_t H_z\right) \\ \eta I \mathbf{H}_t &= -\frac{j}{\omega^2 \mu\epsilon - k^2} \left(\frac{\omega}{c}\mathbf{e}_3 \wedge \boldsymbol{\nabla}_t E_z + I\eta k \boldsymbol{\nabla}_t H_z\right)\end{aligned} \tag{3.241}$$

Cancelling out the ηI factors in the magnetic field component, and substituting $\eta/c = \mu$, $1/(c\eta) = \epsilon$, leaves us with

$$\begin{aligned} \mathbf{E}_t &= \frac{j}{\omega^2 \mu \epsilon - k^2} \left(-k \boldsymbol{\nabla}_t E_z + \mu \omega \mathbf{e}_3 \times \boldsymbol{\nabla}_t H_z \right) \\ \mathbf{H}_t &= -\frac{j}{\omega^2 \mu \epsilon - k^2} \left(\epsilon \omega \mathbf{e}_3 \times \boldsymbol{\nabla}_t E_z + k \boldsymbol{\nabla}_t H_z \right). \end{aligned} \qquad (3.242)$$

Answer for Exercise 3.9

Using a difference of squares to leave a scalar only in the denominator of eq. (3.234), we find for F_z

$$F_z = j \frac{\frac{\omega}{c} + k \mathbf{e}_3}{\left(\frac{\omega}{c}\right)^2 - k^2} \boldsymbol{\nabla}_t F_t. \qquad (3.243)$$

We seek the grade selections

$$\left\langle \left(\frac{\omega}{c} + k \mathbf{e}_3 \right) \boldsymbol{\nabla}_t F_t \right\rangle_{1,2} \qquad (3.244)$$

Performing each of these four grade selections in turn, for the $\boldsymbol{\nabla}_t F_t$ products we have

$$\begin{aligned} \langle \boldsymbol{\nabla}_t F_t \rangle_1 &= \langle \boldsymbol{\nabla}_t \left(\mathbf{E}_t + I \eta \mathbf{H}_t \right) \rangle_1 \\ &= \eta \langle I \boldsymbol{\nabla}_t \mathbf{H}_t \rangle_1 \\ &= \eta I \left(\boldsymbol{\nabla}_t \wedge \mathbf{H}_t \right) \\ &= -\eta \left(\boldsymbol{\nabla}_t \times \mathbf{H}_t \right). \end{aligned} \qquad (3.245)$$

The $\boldsymbol{\nabla}_t \mathbf{E}_t$ product has only grades 0,2, so it's grade-one selection is zero, leaving us with only $\mathbf{H}_t$ dependence.

For the grade two selection of the same, we have

$$\begin{aligned} \langle \boldsymbol{\nabla}_t F_t \rangle_2 &= \langle \boldsymbol{\nabla}_t \left(\mathbf{E}_t + I \eta \mathbf{H}_t \right) \rangle_2 \\ &= \boldsymbol{\nabla}_t \wedge \mathbf{E}_t \\ &= I \left(\boldsymbol{\nabla}_t \times \mathbf{E}_t \right). \end{aligned} \qquad (3.246)$$

This time we note that the vector-bivector product $\boldsymbol{\nabla}_t (I \mathbf{H}_t)$ has only 1,3 grades, and is killed by the grade-2 selection.

For the $\mathbf{e}_3 \boldsymbol{\nabla}_t F_t$ products, we have

$$\begin{aligned}
\langle \mathbf{e}_3 \boldsymbol{\nabla}_t F_t \rangle_1 &= \langle \mathbf{e}_3 \boldsymbol{\nabla}_t \left(\mathbf{E}_t + I\eta \mathbf{H}_t \right) \rangle_1 \\
&= \langle (\mathbf{e}_3 \cdot \boldsymbol{\nabla}_t + \mathbf{e}_3 \wedge \boldsymbol{\nabla}_t) \mathbf{E}_t \rangle_1 + \eta \langle I \mathbf{e}_3 \left(\boldsymbol{\nabla}_t \cdot \mathbf{H}_t + \boldsymbol{\nabla}_t \wedge \mathbf{H}_t \right) \rangle_1 \\
&= \langle I (\mathbf{e}_3 \times \boldsymbol{\nabla}_t) \mathbf{E}_t \rangle_1 \\
&= -(\mathbf{e}_3 \times \boldsymbol{\nabla}_t) \times \mathbf{E}_t.
\end{aligned}$$
(3.247)

Observe that we've made use of $\mathbf{e}_3 \cdot \boldsymbol{\nabla}_t = 0$, regardless of what it operates on. For the $\mathbf{H}_t$ dependence, we had a bivector-scalar product $(I\mathbf{e}_3)(\boldsymbol{\nabla}_t \cdot \mathbf{H}_t)$, and a bivector-bivector product $(I\mathbf{e}_3)(\boldsymbol{\nabla}_t \wedge \mathbf{H}_t)$, neither of which have any vector grades.

Finally

$$\begin{aligned}
\langle \mathbf{e}_3 \boldsymbol{\nabla}_t F_t \rangle_2 &= \eta \langle I \mathbf{e}_3 \boldsymbol{\nabla}_t \mathbf{H}_t \rangle_2 \\
&= -\eta \langle (\mathbf{e}_3 \times \boldsymbol{\nabla}_t) \mathbf{H}_t \rangle_2 \\
&= -\eta I (\mathbf{e}_3 \times \boldsymbol{\nabla}_t) \times \mathbf{H}_t.
\end{aligned}$$
(3.248)

Here we've discarded the $\mathbf{E}_t$ dependent terms, since the bivector-vector product $(\mathbf{e}_3 \wedge \boldsymbol{\nabla}_t) \mathbf{E}_t$ has only grades 1,3, and we seek grade 2.

Putting all the pieces together, noting that $\eta/c = \mu$ and $1/(c\eta) = \epsilon$, we have we have

$$\mathbf{E}_z = -\frac{j}{\omega^2 \mu \epsilon - k^2} \left(\omega \mu \left(\boldsymbol{\nabla}_t \times \mathbf{H}_t \right) + k \left(\mathbf{e}_3 \times \boldsymbol{\nabla}_t \right) \times \mathbf{E}_t \right),$$
(3.249)

and

$$\mathbf{H}_z = \frac{j}{\omega^2 \mu \epsilon - k^2} \left(\omega \epsilon \left(\boldsymbol{\nabla}_t \times \mathbf{E}_t \right) - k \left(\mathbf{e}_3 \times \boldsymbol{\nabla}_t \right) \times \mathbf{H}_t \right).$$
(3.250)

Answer for Exercise 3.10

Taking grade(0,1) and (2,3) selections of Maxwell's equation, we split our equations into source dependent and source free equations

$$\left\langle \left(\boldsymbol{\nabla} + \frac{1}{c} \frac{\partial}{\partial t} \right) F \right\rangle_{0,1} = \langle J \rangle_{0,1},$$
(3.282)

$$\left\langle \left(\boldsymbol{\nabla} + \frac{1}{c} \frac{\partial}{\partial t} \right) F \right\rangle_{2,3} = 0.$$
(3.283)

In terms of $F = \mathbf{E} + I\eta\mathbf{H}$, the source free equation expands to

$$\begin{aligned} 0 &= \left\langle \left(\boldsymbol{\nabla} + \frac{1}{c}\frac{\partial}{\partial t} \right) (\mathbf{E} + I\eta\mathbf{H}) \right\rangle_{2,3} \\ &= \langle \boldsymbol{\nabla}\mathbf{E} \rangle_2 + \langle I\eta\boldsymbol{\nabla}\mathbf{H} \rangle_3 + I\eta\frac{1}{c}\frac{\partial \mathbf{H}}{\partial t} \\ &= \boldsymbol{\nabla} \wedge \mathbf{E} + \boldsymbol{\nabla} \wedge (I\eta\mathbf{H}) + I\eta\frac{1}{c}\frac{\partial \mathbf{H}}{\partial t}, \end{aligned} \quad (3.284)$$

which can be further split into a bivector and trivector equation

$$0 = \boldsymbol{\nabla} \wedge \mathbf{E} + I\eta\frac{1}{c}\frac{\partial \mathbf{H}}{\partial t} \quad (3.285)$$

$$0 = \boldsymbol{\nabla} \wedge (I\eta\mathbf{H}). \quad (3.286)$$

It's clear that we want to write the magnetic field as a (bivector) curl, so we let

$$I\eta\mathbf{H} = Ic\mathbf{B} = c\boldsymbol{\nabla} \wedge \mathbf{A}, \quad (3.287)$$

or

$$\mathbf{H} = \frac{1}{\mu}\boldsymbol{\nabla} \times \mathbf{A}. \quad (3.288)$$

Equation (3.285) is reduced to

$$\begin{aligned} 0 &= \boldsymbol{\nabla} \wedge \mathbf{E} + I\eta\frac{1}{c}\frac{\partial \mathbf{H}}{\partial t} \\ &= \boldsymbol{\nabla} \wedge \mathbf{E} + \frac{1}{c}\frac{\partial}{\partial t}\boldsymbol{\nabla} \wedge (c\mathbf{A}) \\ &= \boldsymbol{\nabla} \wedge \left(\mathbf{E} + \frac{\partial \mathbf{A}}{\partial t} \right). \end{aligned} \quad (3.289)$$

We can now let

$$\mathbf{E} + \frac{\partial \mathbf{A}}{\partial t} = -\boldsymbol{\nabla}\phi. \quad (3.290)$$

We sneakily adjust the sign of the gradient so that the result matches the conventional representation.

Answer for Exercise 3.11

We multiply eq. (3.291) by I to find

$$\left(\boldsymbol{\nabla} + \frac{1}{c}\frac{\partial}{\partial t}\right) IF = I\langle J\rangle_{2,3}, \tag{3.293}$$

which can be split into

$$\left\langle\left(\boldsymbol{\nabla} + \frac{1}{c}\frac{\partial}{\partial t}\right) IF\right\rangle_{1,2} = I\langle J\rangle_{2,3}$$
$$\left\langle\left(\boldsymbol{\nabla} + \frac{1}{c}\frac{\partial}{\partial t}\right) IF\right\rangle_{0,3} = 0. \tag{3.294}$$

We expand the source free equation in terms of $IF = I\mathbf{E} - \eta\mathbf{H}$, to find

$$\begin{aligned}0 &= \left\langle\left(\boldsymbol{\nabla} + \frac{1}{c}\frac{\partial}{\partial t}\right)(I\mathbf{E} - \eta\mathbf{H})\right\rangle_{0,3} \\ &= \boldsymbol{\nabla}\wedge(I\mathbf{E}) + \frac{1}{c}\frac{\partial(I\mathbf{E})}{\partial t} - \eta\boldsymbol{\nabla}\wedge\mathbf{H},\end{aligned} \tag{3.295}$$

which has the respective bivector and trivector grades

$$0 = \boldsymbol{\nabla}\wedge(I\mathbf{E}) \tag{3.296}$$

$$0 = \frac{1}{c}\frac{\partial(I\mathbf{E})}{\partial t} - \eta\boldsymbol{\nabla}\wedge\mathbf{H}. \tag{3.297}$$

We can clearly satisfy eq. (3.296) by setting

$$I\mathbf{E} = -\frac{1}{\epsilon}\boldsymbol{\nabla}\wedge\mathbf{F}, \tag{3.298}$$

or

$$\mathbf{E} = -\frac{1}{\epsilon}\boldsymbol{\nabla}\times\mathbf{F}. \tag{3.299}$$

Here, once again, the sneaky inclusion of a constant factor $-1/\epsilon$ is to make the result match the conventional. Inserting this value for $I\mathbf{E}$ into our bivector equation yields

$$\begin{aligned}0 &= -\frac{1}{\epsilon}\frac{1}{c}\frac{\partial}{\partial t}(\boldsymbol{\nabla}\wedge\mathbf{F}) - \eta\boldsymbol{\nabla}\wedge\mathbf{H} \\ &= -\eta\boldsymbol{\nabla}\wedge\left(\frac{\partial\mathbf{F}}{\partial t} + \mathbf{H}\right),\end{aligned} \tag{3.300}$$

so we set

$$\frac{\partial \mathbf{F}}{\partial t} + \mathbf{H} = -\boldsymbol{\nabla}\phi_m, \tag{3.301}$$

and have a field representation that automatically satisfies the source free equations.

Answer for Exercise 3.12

Proof by expansion is straightforward, and left to the reader. Here we will start with eq. (3.251), and eq. (3.253), and form the respective total electromagnetic field $F = \mathbf{E} + I\eta\mathbf{H}$ for each case.

Starting with eq. (3.251), we find

$$\begin{aligned} F &= \mathbf{E} + I\eta\mathbf{H} \\ &= -\boldsymbol{\nabla}\phi - \frac{\partial \mathbf{A}}{\partial t} + I\frac{\eta}{\mu}\boldsymbol{\nabla}\times\mathbf{A} \\ &= -\boldsymbol{\nabla}\phi - \frac{1}{c}\frac{\partial(c\mathbf{A})}{\partial t} + \boldsymbol{\nabla}\wedge(c\mathbf{A}) \\ &= \left\langle -\boldsymbol{\nabla}\phi - \frac{1}{c}\frac{\partial(c\mathbf{A})}{\partial t} + \boldsymbol{\nabla}\wedge(c\mathbf{A}) \right\rangle_{1,2} \\ &= \left\langle -\boldsymbol{\nabla}\phi - \frac{1}{c}\frac{\partial(c\mathbf{A})}{\partial t} + \boldsymbol{\nabla}(c\mathbf{A}) \right\rangle_{1,2} \\ &= \left\langle \boldsymbol{\nabla}(-\phi + c\mathbf{A}) - \frac{1}{c}\frac{\partial(c\mathbf{A})}{\partial t} \right\rangle_{1,2} \\ &= \left\langle \left(\boldsymbol{\nabla} - \frac{1}{c}\frac{\partial}{\partial t}\right)(-\phi + c\mathbf{A}) \right\rangle_{1,2}. \end{aligned} \tag{3.302}$$

For the field for the fictitious source case, we start with eq. (3.253), and compute the result in the same way, inserting a no-op grade selection to allow us to simplify. We find

$$F = \mathbf{E} + I\eta\mathbf{H}$$
$$= -\frac{1}{\epsilon}\nabla \times \mathbf{F} + I\eta\left(-\nabla\phi_m - \frac{\partial \mathbf{F}}{\partial t}\right)$$
$$= \frac{1}{\epsilon c}I\left(\nabla \wedge (c\mathbf{F})\right) + I\eta\left(-\nabla\phi_m - \frac{1}{c}\frac{\partial(c\mathbf{F})}{\partial t}\right)$$
$$= I\eta\left(\nabla \wedge (c\mathbf{F}) + \left(-\nabla\phi_m - \frac{1}{c}\frac{\partial(c\mathbf{F})}{\partial t}\right)\right)$$
$$= I\eta\left\langle\nabla \wedge (c\mathbf{F}) + \left(-\nabla\phi_m - \frac{1}{c}\frac{\partial(c\mathbf{F})}{\partial t}\right)\right\rangle_{1,2} \qquad (3.303)$$
$$= I\eta\left\langle\nabla(c\mathbf{F}) - \nabla\phi_m - \frac{1}{c}\frac{\partial(c\mathbf{F})}{\partial t}\right\rangle_{1,2}$$
$$= I\eta\left\langle\nabla(-\phi_m + c\mathbf{F}) - \frac{1}{c}\frac{\partial(c\mathbf{F})}{\partial t}\right\rangle_{1,2}$$
$$= I\eta\left\langle\left(\nabla - \frac{1}{c}\frac{\partial}{\partial t}\right)(-\phi_m + c\mathbf{F})\right\rangle_{1,2}.$$

Answer for Exercise 3.13

We start by expanding $(\nabla - (1/c)\partial_t)A$ and then group by grade to find

$$\left(\nabla - \frac{1}{c}\frac{\partial}{\partial t}\right)A = \left(\nabla - \frac{1}{c}\frac{\partial}{\partial t}\right)(-\phi + c\mathbf{A} + \eta I(-\phi_m + c\mathbf{F}))$$
$$= -\nabla\phi + c\nabla \cdot \mathbf{A} + c\nabla \wedge \mathbf{A} + \frac{1}{c}\frac{\partial\phi}{\partial t} - \frac{\partial \mathbf{A}}{\partial t}$$
$$\quad + I\eta\left(-\nabla\phi_m + c\nabla \cdot \mathbf{F} + c\nabla \wedge \mathbf{F} + \frac{1}{c}\frac{\partial\phi_m}{\partial t} - \frac{\partial \mathbf{F}}{\partial t}\right)$$
$$= c\nabla \cdot \mathbf{A} + \frac{1}{c}\frac{\partial\phi}{\partial t}$$
$$\quad + \boxed{-\nabla\phi - \frac{\partial \mathbf{A}}{\partial t} - \frac{1}{\epsilon}\nabla \times \mathbf{F}} + \boxed{I\eta\left(-\nabla\phi_m - \frac{\partial \mathbf{F}}{\partial t} + \frac{1}{\mu}\nabla \times \mathbf{A}\right)}$$
$$\qquad\qquad\mathbf{E} \qquad\qquad\qquad\qquad\qquad I\eta\mathbf{H}$$
$$\quad + I\eta\left(c\nabla \cdot \mathbf{F} + \frac{1}{c}\frac{\partial\phi_m}{\partial t}\right),$$

(3.304)

which shows the claimed field split.

Observing that $F = \langle (\nabla - (1/c)\partial/\partial t) A \rangle_{1,2} = \mathbf{E} + I\eta \mathbf{H}$, completes the problem. We may write just $F = (\nabla - (1/c)\partial/\partial t) A$, if the Lorentz gauge condition is assumed, as the scalar and pseudoscalar components above are obliterated.

Answer for Exercise 3.14

In terms of the potentials Maxwell's equation $\left(\nabla + \frac{1}{c}\frac{\partial}{\partial t}\right) F = J$ is

$$\left(\nabla + \frac{1}{c}\frac{\partial}{\partial t}\right)\left\langle\left(\nabla - \frac{1}{c}\frac{\partial}{\partial t}\right) A\right\rangle_{1,2} = J, \qquad (3.305)$$

or

$$\Box A = J + \left(\nabla + \frac{1}{c}\frac{\partial}{\partial t}\right)\left\langle\left(\nabla - \frac{1}{c}\frac{\partial}{\partial t}\right) A\right\rangle_{0,3}. \qquad (3.306)$$

This is almost a wave equation. Inserting eq. (3.304) into eq. (3.306) and selecting each grade gives four almost-wave equations

$$-\Box \phi = \frac{\rho}{\epsilon} + \frac{1}{c}\frac{\partial}{\partial t}\left(c\nabla \cdot \mathbf{A} + \frac{1}{c}\frac{\partial \phi}{\partial t}\right)$$

$$c\Box \mathbf{A} = -\eta \mathbf{J} + \nabla\left(c\nabla \cdot \mathbf{A} + \frac{1}{c}\frac{\partial \phi}{\partial t}\right)$$

$$\eta c I \Box F = -I\mathbf{M} + \nabla \cdot \left(I\eta\left(c\nabla \cdot \mathbf{F} + \frac{1}{c}\frac{\partial \phi_m}{\partial t}\right)\right)$$

$$-I\eta \Box \phi_m = Ic\rho_m + \frac{1}{c}\frac{\partial}{\partial t}I\eta\left(c\nabla \cdot \mathbf{F} + \frac{1}{c}\frac{\partial \phi_m}{\partial t}\right)$$

Using $\eta = \mu c$, $\eta c \epsilon = 1$, and $\nabla \cdot (I\psi) = I\nabla \psi$ for scalar ψ, a bit of rearrangement completes the proof.

DISTRIBUTION THEOREMS.

Theorem A.1: K-vector dot and wedge product relations.

Given a k-vector B and a vector $\mathbf{a}$, the dot and wedge products have the following commutation relationships

$$\begin{aligned} B \cdot \mathbf{a} &= (-1)^{k-1} \mathbf{a} \cdot B \\ B \wedge \mathbf{a} &= (-1)^{k} \mathbf{a} \wedge B, \end{aligned} \tag{A.1}$$

and can be expressed as symmetric and antisymmetric sums depending on the grade of the blade

$$\begin{aligned} \mathbf{a} \wedge B &= \frac{1}{2}\left(\mathbf{a}B + (-1)^{k} B\mathbf{a}\right) \\ \mathbf{a} \cdot B &= \frac{1}{2}\left(\mathbf{a}B - (-1)^{k} B\mathbf{a}\right). \end{aligned} \tag{A.2}$$

For example, if B and $\mathbf{a}$ are both vectors, we recover theorem 1.14. As an other example, if B is a 2-vector, then

$$\begin{aligned} 2(\mathbf{a} \wedge B) &= \mathbf{a}B + B\mathbf{a} \\ 2(\mathbf{a} \cdot B) &= \mathbf{a}B - B\mathbf{a}. \end{aligned} \tag{A.3}$$

Observe that the dot(wedge) of two vectors is a (anti)symmetric sum of products, whereas the wedge(dot) of a vector and a bivector is an (anti)symmetric sum.

Proof. To prove theorem A.1, split the blade into components that intersect with and are disjoint from $\mathbf{a}$ as follows

$$B = \frac{1}{\mathbf{a}} \mathbf{n}_1 \mathbf{n}_2 \cdots \mathbf{n}_{k-1} + \mathbf{m}_1 \mathbf{m}_2 \cdots \mathbf{m}_k, \tag{A.4}$$

where $\mathbf{n}_i$ orthogonal to $\mathbf{a}$ and each other, and where $\mathbf{m}_i$ are all orthogonal. The products of B with $\mathbf{a}$ are

$$\begin{aligned}\mathbf{a}B &= \mathbf{a}\frac{1}{\mathbf{a}}\mathbf{n}_1\mathbf{n}_2\cdots\mathbf{n}_{k-1} + a\mathbf{m}_1\mathbf{m}_2\cdots\mathbf{m}_k \\ &= \mathbf{n}_1\mathbf{n}_2\cdots\mathbf{n}_{k-1} + a\mathbf{m}_1\mathbf{m}_2\cdots\mathbf{m}_k,\end{aligned} \quad (A.5)$$

and

$$\begin{aligned}B\mathbf{a} &= \frac{1}{\mathbf{a}}\mathbf{n}_1\mathbf{n}_2\cdots\mathbf{n}_{k-1}\mathbf{a} + \mathbf{m}_1\mathbf{m}_2\cdots\mathbf{m}_k\mathbf{a} \\ &= (-1)^{k-1}\mathbf{n}_1\mathbf{n}_2\cdots\mathbf{n}_{k-1} + (-1)^k a\mathbf{m}_1\mathbf{m}_2\cdots\mathbf{m}_k \\ &= (-1)^k\left(-\mathbf{n}_1\mathbf{n}_2\cdots\mathbf{n}_{k-1} + a\mathbf{m}_1\mathbf{m}_2\cdots\mathbf{m}_k\right),\end{aligned} \quad (A.6)$$

or

$$(-1)^k B\mathbf{a} = -\mathbf{n}_1\mathbf{n}_2\cdots\mathbf{n}_{k-1} + a\mathbf{m}_1\mathbf{m}_2\cdots\mathbf{m}_k. \quad (A.7)$$

Respective addition and subtraction of eq. (A.5) and eq. (A.7) gives

$$\begin{aligned}\mathbf{a}B + (-1)^k B\mathbf{a} &= 2a\mathbf{m}_1\mathbf{m}_2\cdots\mathbf{m}_k \\ &= 2\mathbf{a}\wedge B,\end{aligned} \quad (A.8)$$

and

$$\begin{aligned}\mathbf{a}B - (-1)^k B\mathbf{a} &= 2\mathbf{n}_1\mathbf{n}_2\cdots\mathbf{n}_{k-1} \\ &= 2\mathbf{a}\cdot B,\end{aligned} \quad (A.9)$$

proving eq. (A.2). Grade selection from eq. (A.7) gives

$$\begin{aligned}(-1)^k B\cdot\mathbf{a} &= -\mathbf{n}_1\mathbf{n}_2\cdots\mathbf{n}_{k-1} \\ &= -\mathbf{a}\cdot B,\end{aligned} \quad (A.10)$$

and

$$\begin{aligned}(-1)^k B\wedge\mathbf{a} &= a\mathbf{m}_1\mathbf{m}_2\cdots\mathbf{m}_k \\ &= \mathbf{a}\wedge B,\end{aligned} \quad (A.11)$$

which proves eq. (A.1). □

Theorem A.2: Vector-trivector dot product.

DISTRIBUTION THEOREMS.

Given a vector **a** and a blade $\mathbf{b} \wedge \mathbf{c} \wedge \mathbf{d}$ formed by wedging three vectors, the dot product of the two can be expanded as bivectors like

$$\begin{aligned}\mathbf{a} \cdot (\mathbf{b} \wedge \mathbf{c} \wedge \mathbf{d}) \\ = (\mathbf{b} \wedge \mathbf{c} \wedge \mathbf{d}) \cdot \mathbf{a} \\ = (\mathbf{a} \cdot \mathbf{b})(\mathbf{c} \wedge \mathbf{d}) - (\mathbf{a} \cdot \mathbf{c})(\mathbf{b} \wedge \mathbf{d}) + (\mathbf{a} \cdot \mathbf{d})(\mathbf{b} \wedge \mathbf{c}).\end{aligned} \quad (A.12)$$

Proof. The proof follows by expansion in coordinates

$$\mathbf{a} \cdot (\mathbf{b} \wedge \mathbf{c} \wedge \mathbf{d}) = \sum_{j \neq k \neq l} a_i b_j c_k d_l \langle \mathbf{e}_i \mathbf{e}_j \mathbf{e}_k \mathbf{e}_l \rangle_2. \quad (A.13)$$

The products within the grade two selection operator can be of either grade two or grade four, so only the terms where one of $i = j$, $i = k$, or $i = l$ contributes. Repeated anticommutation of the orthogonal unit vectors can put each such pair adjacent, where they square to unity. Those are respectively

$$\begin{aligned}\langle \mathbf{e}_i \mathbf{e}_i \mathbf{e}_k \mathbf{e}_l \rangle_2 &= \mathbf{e}_k \mathbf{e}_l \\ \langle \mathbf{e}_i \mathbf{e}_j \mathbf{e}_i \mathbf{e}_l \rangle_2 &= -\langle \mathbf{e}_j \mathbf{e}_i \mathbf{e}_i \mathbf{e}_l \rangle_2 = -\mathbf{e}_j \mathbf{e}_l \\ \langle \mathbf{e}_i \mathbf{e}_j \mathbf{e}_k \mathbf{e}_i \rangle_2 &= -\langle \mathbf{e}_j \mathbf{e}_i \mathbf{e}_k \mathbf{e}_i \rangle_2 = +\langle \mathbf{e}_j \mathbf{e}_k \mathbf{e}_i \mathbf{e}_i \rangle_2 = \mathbf{e}_j \mathbf{e}_k.\end{aligned} \quad (A.14)$$

Substitution back into eq. (1.117) gives

$$\begin{aligned}\mathbf{a} \cdot (\mathbf{b} \wedge \mathbf{c} \wedge \mathbf{d}) \\ = \sum_{j \neq k \neq l} a_i b_j c_k d_l \left(\mathbf{e}_i \cdot \mathbf{e}_j (\mathbf{e}_k \mathbf{e}_l) - \mathbf{e}_i \cdot \mathbf{e}_k (\mathbf{e}_j \mathbf{e}_l) + \mathbf{e}_i \cdot \mathbf{e}_l (\mathbf{e}_j \mathbf{e}_k) \right) \\ = (\mathbf{a} \cdot \mathbf{b})(\mathbf{c} \wedge \mathbf{d}) - (\mathbf{a} \cdot \mathbf{c})(\mathbf{b} \wedge \mathbf{d}) + (\mathbf{a} \cdot \mathbf{d})(\mathbf{b} \wedge \mathbf{c}).\end{aligned} \quad (A.15)$$

$\square$

Theorem A.2 is a specific case of the more general identity

Theorem A.3: Vector blade dot product distribution.

A vector dotted with a $n - blade$ distributes as

$$\mathbf{x} \cdot (\mathbf{y}_1 \wedge \mathbf{y}_2 \wedge \cdots \wedge \mathbf{y}_n)$$
$$= \sum_{i=1}^{n} (-1)^i (\mathbf{x} \cdot \mathbf{y}_i) (\mathbf{y}_1 \wedge \cdots \wedge \mathbf{y}_{i-1} \wedge \mathbf{y}_{i+1} \wedge \cdots \wedge \mathbf{y}_n).$$

This dot product is symmetric(antisymmetric) when the grade of the blade the vector is dotted with is odd(even).

For a proof of theorem A.3 (valid for all metrics) see [6].

Theorem A.4: Distribution of inner products

Given two blades A_s, B_r with grades subject to $s > r > 0$, and a vector $\mathbf{b}$, the inner product distributes according to

$$A_s \cdot (\mathbf{b} \wedge B_r) = (A_s \cdot \mathbf{b}) \cdot B_r.$$

Proof. The proof is straightforward, relying primarily on grade selection, but also mechanical. Start by expanding the wedge and dot products within a grade selection operator

$$A_s \cdot (\mathbf{b} \wedge B_r) = \langle A_s(\mathbf{b} \wedge B_r)\rangle_{s-(r+1)}$$
$$= \frac{1}{2}\langle A_s (\mathbf{b}B_r + (-1)^r B_r \mathbf{b})\rangle_{s-(r+1)}. \tag{A.16}$$

Solving for $B_r\mathbf{b}$ in

$$2\mathbf{b} \cdot B_r = \mathbf{b}B_r - (-1)^r B_r\mathbf{b}, \tag{A.17}$$

we have

$$A_s \cdot (\mathbf{b} \wedge B_r) = \frac{1}{2}\langle A_s \mathbf{b}B_r + A_s (\mathbf{b}B_r - 2\mathbf{b} \cdot B_r)\rangle_{s-(r+1)}$$
$$= \langle A_s \mathbf{b}B_r\rangle_{s-(r+1)} - \cancel{\langle A_s (\mathbf{b} \cdot B_r)\rangle_{s-(r+1)}}. \tag{A.18}$$

The last term above is zero since we are selecting the $s-r-1$ grade element of a multivector with grades $s-r+1$ and $s+r-1$, which has no terms for $r > 0$. Now we can expand the $A_s\mathbf{b}$ multivector product, for

$$A_s \cdot (\mathbf{b} \wedge B_r) = \langle (A_s \cdot \mathbf{b} + A_s \wedge \mathbf{b}) B_r\rangle_{s-(r+1)}. \tag{A.19}$$

The latter multivector (with the wedge product factor) above has grades $s+1-r$ and $s+1+r$, so this selection operator finds nothing. This leaves

$$A_s \cdot (\mathbf{b} \wedge B_r) = \langle (A_s \cdot \mathbf{b}) \cdot B_r + (A_s \cdot \mathbf{b}) \wedge B_r\rangle_{s-(r+1)}. \tag{A.20}$$

The first dot products term has grade $s-1-r$ and is selected, whereas the wedge term has grade $s-1+r \neq s-r-1$ (for $r > 0$). □

B

PROOF SKETCH FOR THE FUNDAMENTAL THEOREM OF GEOMETRIC CALCULUS.

We start with expanding the hypervolume integral, by separating the geometric product of the volume element and vector derivative direction vectors into dot and wedge contributions

$$\int_V F d^k \mathbf{x} \stackrel{\leftrightarrow}{\partial} G = \sum_i \int_V d^k u F I_k \mathbf{x}^i \stackrel{\leftrightarrow}{\partial}_i G$$
$$= \sum_i \int_V d^k u F \left(I_k \cdot \mathbf{x}^i + I_k \wedge \mathbf{x}^i \right) \stackrel{\leftrightarrow}{\partial}_i G. \quad (B.1)$$

Because $\mathbf{x}^i$ lies in span $\{\mathbf{x}_j\}$, the wedge product above is zero, leaving

$$\int_V F d^k \mathbf{x} \stackrel{\leftrightarrow}{\partial} G = \sum_i \int_V d^k u F \left(I_k \cdot \mathbf{x}^i \right) \stackrel{\leftrightarrow}{\partial}_i G$$
$$= \sum_i \int_V d^k u (\partial_i F) I_k \cdot \mathbf{x}^i G + \sum_i \int_V d^k u F \left(I_k \cdot \mathbf{x}^i \right) (\partial_i G)$$
$$= \sum_i \int_V d^k u \partial_i \left(F \left(I_k \cdot \mathbf{x}^i \right) G \right) - \int_V d^k u F \left(\sum_i \partial_i \left(I_k \cdot \mathbf{x}^i \right) \right) G. \quad (B.2)$$

The sum in the second integral turns out to be zero, but is somewhat messy to show in general. The $k = 1$ is a special case, as it is trivial

$$\partial_1 (\mathbf{x}_1 \cdot \mathbf{x}^1) = \partial_1 1 = 0. \quad (B.3)$$

The $k = 2$ case is illustrative

$$\sum_{i=1}^2 \partial_i \left(I_3 \cdot \mathbf{x}^i \right) = \partial_1 ((\mathbf{x}_1 \wedge \mathbf{x}_2) \cdot \mathbf{x}^1) + \partial_2 ((\mathbf{x}_1 \wedge \mathbf{x}_2) \cdot \mathbf{x}^2)$$
$$= \partial_1 (-\mathbf{x}_2) + \partial_2 \mathbf{x}_1 \quad (B.4)$$
$$= -\frac{\partial^2 \mathbf{x}}{\partial u_1 \partial_2} + \frac{\partial^2 \mathbf{x}}{\partial u_2 \partial_1},$$

which is zero by equality of mixed partials. To show that this sums to zero in general observe that cyclic permutation of the wedge factors in the pseudoscalar only changes the sign

$$\mathbf{x}_1 \wedge \mathbf{x}_2 \wedge \cdots \wedge \mathbf{x}_k = \mathbf{x}_2 \wedge \mathbf{x}_3 \wedge \cdots \wedge \mathbf{x}_k \wedge \mathbf{x}_1 (-1)^{1(k-1)}$$
$$= \mathbf{x}_3 \wedge \mathbf{x}_4 \wedge \cdots \wedge \mathbf{x}_k \wedge \mathbf{x}_1 \wedge \mathbf{x}_2 (-1)^{2(k-1)}$$
$$= \mathbf{x}_{i+1} \wedge \mathbf{x}_{i+2} \wedge \cdots \wedge \mathbf{x}_k \wedge \mathbf{x}_1 \wedge \mathbf{x}_2 \wedge \cdots \wedge \mathbf{x}_i (-1)^{i(k-1)}. \quad \text{(B.5)}$$

The pseudoscalar dot product $I_k \cdot \mathbf{x}^i$ is therefore

$$I_k \cdot \mathbf{x}^i = (\mathbf{x}_1 \wedge \mathbf{x}_2 \wedge \cdots \wedge \mathbf{x}_k) \cdot \mathbf{x}^i$$
$$= \mathbf{x}_{i+1} \wedge \mathbf{x}_{i+2} \wedge \cdots \wedge \mathbf{x}_k \wedge \mathbf{x}_1 \wedge \mathbf{x}_2 \wedge \cdots \wedge \mathbf{x}_{i-1} (-1)^{i(k-1)}, \quad \text{(B.6)}$$

and the sum is

$$\sum_i \partial_i \left(I_k \cdot \mathbf{x}^i \right)$$
$$= (\partial_{i,i+1}\mathbf{x}) \wedge \mathbf{x}_{i+2} \wedge \cdots \wedge \mathbf{x}_k \wedge \mathbf{x}_1 \wedge \mathbf{x}_2 \wedge \cdots \wedge \mathbf{x}_{i-1} (-1)^{i(k-1)}$$
$$+ \mathbf{x}_{i+1} \wedge (\partial_{i,i+2}\mathbf{x}) \wedge \cdots \wedge \mathbf{x}_k \wedge \mathbf{x}_1 \wedge \mathbf{x}_2 \wedge \cdots \wedge \mathbf{x}_{i-1} (-1)^{i(k-1)}$$
$$+ \quad \text{(B.7)}$$
$$\vdots$$
$$+ \mathbf{x}_{i+1} \wedge \mathbf{x}_{i+2} \wedge \cdots \wedge \mathbf{x}_k \wedge \mathbf{x}_1 \wedge \mathbf{x}_2 \wedge \cdots \wedge (\partial_{i,i-1}\mathbf{x}) (-1)^{i(k-1)}.$$

For each $i \neq j$ there will be one partial $\partial_{i,j}\mathbf{x}$ and one partial $\partial_{j,i}\mathbf{x}$ in this sum. Consider, for example, the $1, 2$ case which come from the $i = 1, 2$ terms in the sum

$$\partial_1(\mathbf{x}_2 \wedge \mathbf{x}_3 \wedge \cdots \wedge \mathbf{x}_{k-1} \wedge \mathbf{x}_k)(-1)^{1(k-1)}$$
$$+ \partial_2(\mathbf{x}_3 \wedge \mathbf{x}_4 \wedge \cdots \wedge \mathbf{x}_k \wedge \mathbf{x}_1)(-1)^{2(k-1)}$$
$$= (\partial_{1,2}\mathbf{x}) \wedge \mathbf{x}_3 \wedge \cdots \wedge \mathbf{x}_{k-1} \wedge \mathbf{x}_k)(-1)^{1(k-1)}$$
$$+ \mathbf{x}_3 \wedge \mathbf{x}_4 \wedge \cdots \wedge \mathbf{x}_k \wedge (\partial_{2,1}\mathbf{x})(-1)^{2(k-1)} + \cdots$$
$$= (-1)^{k-1}(\mathbf{x}_3 \wedge \cdots \wedge \mathbf{x}_{k-1} \wedge \mathbf{x}_k) \wedge \left((-1)^{k-2}\partial_{1,2}\mathbf{x} + (-1)^{k-1}\partial_{2,1}\mathbf{x}\right) + \cdots$$
$$= (\mathbf{x}_3 \wedge \cdots \wedge \mathbf{x}_{k-1} \wedge \mathbf{x}_k) \wedge \left(-\frac{\partial^2 \mathbf{x}}{\partial u_1 \partial u_2} + \frac{\partial^2 \mathbf{x}}{\partial u_2 \partial u_1}\right) + \cdots \quad \text{(B.8)}$$

By equality of mixed partials this difference of $1, 2$ partials are killed. The same argument holds for all other indexes, proving that $\sum_i \partial_i \left(I_k \cdot \mathbf{x}^i \right) = 0$.

Equation (B.2) is left with a sum of perfect differentials, each separately integrable

$$\int_V F d^k \mathbf{x} \, \overset{\leftrightarrow}{\partial} \, G = \sum_i \int_{\partial V} d^{k-1} u_i \int_{\Delta u_i} du_i \frac{\partial}{\partial u_i} \left(F \left(I_k \cdot \mathbf{x}^i \right) G \right)$$
$$= \sum_i \int_{\partial V} d^{k-1} u_i \left(F \left(I_k \cdot \mathbf{x}^i \right) G \right) \Big|_{\Delta u_i}, \tag{B.9}$$

which completes the sketch of the proof.

While much of the theoretical heavy lifting was carried by the reciprocal frame vectors, the final result does not actually require computing those vectors. When k equals the dimension of the space, as in $\mathbb{R}^3$ volume integrals, the vector derivative ∂ is identical to the ∇, in which case we do not even require the reciprocal frame vectors to express the gradient.

For a full proof of theorem 2.3, additional mathematical subtleties must be considered. Issues of connectivity of the hypervolumes (and integration theory in general) are covered very nicely in [18]. For other general issues required for a complete proof, like the triangulation of the volume and its boundary, please see [14], [6], and [22].

C

GREEN'S FUNCTIONS.

C.1 HELMHOLTZ OPERATOR.

The goal. The Helmholtz equation to solve is

$$\left(\nabla^2 + k^2\right) f(\mathbf{x}) = u(\mathbf{x}). \tag{C.1}$$

To solve using the Green's function of theorem 2.17, we require

$$\left(\nabla^2 + k^2\right) G(\mathbf{x}, \mathbf{x}') = \delta^3(\mathbf{x} - \mathbf{x}'). \tag{C.2}$$

Verifying this requires two steps, first considering points $\mathbf{x} \neq \mathbf{x}'$, and then considering an infinitesimal neighborhood around $\mathbf{x}'$.

Case I. $\mathbf{x} \neq \mathbf{x}'$. We will absorb the sign associated with the causal and acausal Green's function variations by writing $i = \pm j$, so that for points $\mathbf{x} \neq \mathbf{x}'$, (i.e. $r = \|\mathbf{x} - \mathbf{x}'\| \neq 0$), working in spherical coordinates, we find

$$\begin{aligned}
-4\pi \left(\nabla^2 + k^2\right) G(\mathbf{x}, \mathbf{x}') &= \frac{1}{r^2} \left(r^2 G'\right)' - 4\pi k^2 G \\
&= \frac{1}{r^2} \frac{d}{dr} \left(r^2 \left(\frac{ikr}{r} - \frac{1}{r^2}\right) e^{ikr}\right) + \frac{k^2}{r} e^{ikr} \\
&= \frac{1}{r^2} \frac{d}{dr} \left((rik - 1) e^{ikr}\right) + \frac{k^2}{r} e^{ikr} \\
&= \frac{1}{r^2} \left(ik + (rik - 1) ik\right) e^{ikr} + \frac{k^2}{r} e^{ikr} \\
&= \frac{1}{r^2} \left(-rk^2\right) e^{ikr} + k^2 \frac{e^{ikr}}{r} \\
&= 0.
\end{aligned} \tag{C.3}$$

Case II. In the neighborhood of $\|\mathbf{x} - \mathbf{x}'\| < \epsilon$ Having shown that we end up with zero everywhere that $\mathbf{x} \neq \mathbf{x}'$ we are left to consider an infinitesimal neighborhood of the volume surrounding the point $\mathbf{x}$ in our integral. Following the Coulomb treatment in §2.2 of [20] we use a spherical volume element centered around $\mathbf{x}$ of radius ϵ, and then convert a divergence to a surface area to evaluate the integral away from the problematic point.

$$\int \left(\nabla^2 + k^2\right) G(\mathbf{x}, \mathbf{x}') f(\mathbf{x}') dV'$$
$$= -\frac{1}{4\pi} \int_{\|\mathbf{x}-\mathbf{x}'\|<\epsilon} \left(\nabla^2 + k^2\right) \frac{e^{ik\|\mathbf{x}-\mathbf{x}'\|}}{\|\mathbf{x}-\mathbf{x}'\|} f(\mathbf{x}') dV' \qquad (C.4)$$
$$= -\frac{1}{4\pi} \int_{\|\mathbf{x}''\|<\epsilon} f(\mathbf{x}+\mathbf{x}'') \left(\nabla^2 + k^2\right) \frac{e^{ik\|\mathbf{x}''\|}}{\|\mathbf{x}''\|} dV'',$$

where a change of variables $\mathbf{x}'' = \mathbf{x}' - \mathbf{x}$, as illustrated in fig. C.1, has been performed.

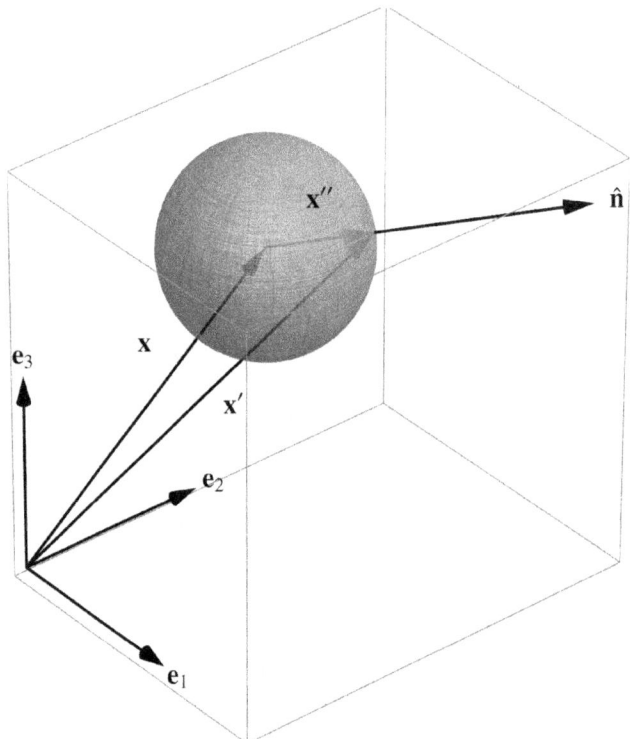

Figure C.1: Neighborhood $\|\mathbf{x} - \mathbf{x}'\| < \epsilon$.

We assume that $f(\mathbf{x})$ is sufficiently continuous and "well behaved" that it can be pulled it out of the integral, replaced with a mean value $f(\mathbf{x}^*)$ in the integration neighborhood around $\mathbf{x}'' = 0$.

$$\int \left(\nabla^2 + k^2\right) G(\mathbf{x}, \mathbf{x}') f(\mathbf{x}') dV' = \lim_{\epsilon \to 0} -\frac{f(\mathbf{x}^*)}{4\pi} \int_{\|\mathbf{x}''\|<\epsilon} \left(\nabla^2 + k^2\right) \frac{e^{ik\|\mathbf{x}''\|}}{\|\mathbf{x}''\|} dV''.$$

C.1 HELMHOLTZ OPERATOR. 299

$$\tag{C.5}$$

The k^2 term of eq. (C.5) can be evaluated with a spherical coordinate change of variables

$$\int_{\|\mathbf{x}''\|<\epsilon} k^2 \frac{e^{ik\|\mathbf{x}''\|}}{\|\mathbf{x}''\|} dV'' = \int_{r=0}^{\epsilon} \int_{\theta=0}^{\pi} \int_{\phi=0}^{2\pi} k^2 \frac{e^{ikr}}{r} r^2 dr \sin\theta d\theta d\phi$$
$$= 4\pi k^2 \int_{r=0}^{\epsilon} r e^{ikr} dr$$
$$= 4\pi \int_{u=0}^{k\epsilon} u e^{iu} du \tag{C.6}$$
$$= 4\pi (-iu+1) e^{iu} \Big|_0^{k\epsilon}$$
$$= 4\pi \left((-ik\epsilon+1) e^{ik\epsilon} - 1 \right).$$

To evaluate the Laplacian term of eq. (C.5), we can make a change of variables for the Laplacian

$$\nabla \frac{e^{ik\|\mathbf{x}''\|}}{\|\mathbf{x}''\|} = \nabla_{\mathbf{x}''}^2 \frac{e^{ik\|\mathbf{x}''\|}}{\|\mathbf{x}''\|} = \nabla_{\mathbf{x}''} \cdot \left(\nabla_{\mathbf{x}''} \frac{e^{ik\|\mathbf{x}''\|}}{\|\mathbf{x}''\|} \right), \tag{C.7}$$

and then employ the divergence theorem

$$\int_{\|\mathbf{x}''\|<\epsilon} \nabla^2 \frac{e^{ik\|\mathbf{x}''\|}}{\|\mathbf{x}''\|} dV'' = \int_{\|\mathbf{x}''\|<\epsilon} \nabla_{\mathbf{x}''} \cdot \left(\nabla_{\mathbf{x}''} \frac{e^{ik\|\mathbf{x}''\|}}{\|\mathbf{x}''\|} \right) dV''$$
$$= \int_{\partial V} \left(\nabla_{\mathbf{x}''} \frac{e^{ik\|\mathbf{x}''\|}}{\|\mathbf{x}''\|} \right) \cdot \hat{\mathbf{n}} dA'', \tag{C.8}$$

where ∂V represents the surface of the $\|\mathbf{x}''\| < \epsilon$ neighborhood, and $\hat{\mathbf{n}}$ is the unit vector directed along $\mathbf{x}'' = \mathbf{x}' - \mathbf{x}$. To evaluate this surface integral we will require only the radial portion of the gradient. With $r = \|\mathbf{x}''\|$, that is

$$\left(\nabla_{\mathbf{x}''} \frac{e^{ik\|\mathbf{x}''\|}}{\|\mathbf{x}''\|} \right) \cdot \hat{\mathbf{n}} = \left(\hat{\mathbf{n}} \frac{\partial}{\partial r} \frac{e^{ikr}}{r} \right) \cdot \hat{\mathbf{n}}$$
$$= \frac{\partial}{\partial r} \frac{e^{ikr}}{r}$$
$$= \left(ik\frac{1}{r} - \frac{1}{r^2} \right) e^{ikr} \tag{C.9}$$
$$= (ikr - 1) \frac{e^{ikr}}{r^2}.$$

Using a spherical area element $dA'' = r^2 \sin\theta d\theta d\phi$, we obtain

$$\int_{\|\mathbf{x}''\|<\epsilon} \nabla^2 \frac{e^{ik\|\mathbf{x}''\|}}{\|\mathbf{x}''\|} dV''$$
$$= \left. \int_{\theta=0}^{\pi} \int_{\phi=0}^{2\pi} (ikr-1) \frac{e^{ikr}}{r^2} r^2 \sin\theta d\theta d\phi \right|_{r=\epsilon} \quad \text{(C.10)}$$
$$= 4\pi (ik\epsilon - 1) e^{ik\epsilon}.$$

Putting everything back together we have

$$-\frac{1}{4\pi} \int (\nabla^2 + k^2) \frac{e^{ik\|\mathbf{x}-\mathbf{x}'\|}}{\|\mathbf{x}-\mathbf{x}'\|} f(\mathbf{x}') dV'$$
$$= \lim_{\epsilon \to 0} -f(\mathbf{x}^*) \left((-ik\epsilon + 1)e^{ik\epsilon} - 1 + (ik\epsilon - 1) e^{ik\epsilon} \right) \quad \text{(C.11)}$$
$$= \lim_{\epsilon \to 0} -f(\mathbf{x}^*) \left((-ik\epsilon + 1 + ik\epsilon - 1)e^{ik\epsilon} - 1 \right)$$
$$= \lim_{\epsilon \to 0} f(\mathbf{x}^*).$$

Observe the perfect cancellation of all the explicitly ϵ dependent terms. The mean value point $\mathbf{x}^*$ is also ϵ dependent, but tends to $\mathbf{x}$ in the limit, leaving

$$\boxed{f(\mathbf{x}) = -\frac{1}{4\pi} \int (\nabla^2 + k^2) \frac{e^{ik\|\mathbf{x}-\mathbf{x}'\|}}{\|\mathbf{x}-\mathbf{x}'\|} f(\mathbf{x}') dV'.} \quad \text{(C.12)}$$

This proves the delta function property that we claimed the Green's function had.

C.2 DELTA FUNCTION DERIVATIVES.

The Green's function for the spacetime gradient ends up with terms like

$$\frac{d}{dr} \delta(-r/c + t - t') \quad \text{(C.13)}$$
$$\frac{d}{dt} \delta(-r/c + t - t'),$$

where t' is the integration variable of the test function that the delta function will be applied to. If these were derivatives with respect to the integration variable, then we could use

$$\int_{-\infty}^{\infty} \left(\frac{d}{dt'} \delta(t') \right) \phi(t') = -\phi'(0), \quad \text{(C.14)}$$

C.2 DELTA FUNCTION DERIVATIVES.

which follows by chain rule, and an assumption that $\phi(t')$ is well behaved at the points at infinity. It is not clear that how, if at all, this could be applied to either of eq. (C.13).

Let's go back to square one, and figure out the meaning of these delta functions by their action on a test function. We wish to compute

$$\int_{-\infty}^{\infty} \frac{d}{du}\delta(au + b - t')f(t')dt'. \tag{C.15}$$

Let's start with a change of variables $t'' = au + b - t'$, for which we find

$$\begin{aligned} t' &= au + b - t'' \\ dt'' &= -dt' \\ \frac{d}{du} &= \frac{dt''}{du}\frac{d}{dt''} = a\frac{d}{dt''}. \end{aligned} \tag{C.16}$$

Substitution back into eq. (C.15) gives

$$\begin{aligned} a\int_{\infty}^{-\infty} \left(\frac{d}{dt''}\delta(t'')\right) f(au + b - t'')(-dt'') \\ = a\int_{-\infty}^{\infty} \left(\frac{d}{dt''}\delta(t'')\right) f(au + b - t'')dt'' \\ = a\delta(t'')f(au + b - t'')\Big|_{-\infty}^{\infty} \\ - a\int_{-\infty}^{\infty} \delta(t'')\frac{d}{dt''}f(au + b - t'')dt'' \\ = -a\frac{d}{dt''}f(au + b - t'')\Big|_{t''=0} \\ = a\frac{d}{ds}f(s)\Big|_{s=au+b}. \end{aligned} \tag{C.17}$$

This shows that the action of the derivative of the delta function (with respect to a non-integration variable parameter u) is

$$\boxed{\frac{d}{du}\delta(au + b - t') \sim a\frac{d}{ds}\Big|_{s=au+b}.} \tag{C.18}$$

D

ENERGY MOMENTUM TENSOR (VECTOR.)

Did you cry cheater because of the reliance on brute force computer assisted symbolic algebra to find the $T(\mathbf{a})$ relations of eq. (3.134)? Or did you try this as a problem, and need some assistance?

If so, here is an expansion of the energy momentum tensor for vector parameters. We start with

$$F\mathbf{e}_i = (\mathbf{E} + I\eta\mathbf{H})\,\mathbf{e}_i \qquad \text{(D.1)}$$
$$= E_i + \mathbf{E} \wedge \mathbf{e}_i + I\eta\,(H_i + \mathbf{H} \wedge \mathbf{e}_i).$$

To show that the scalar grades are related to the Poynting vector as $\langle T(\mathbf{e}_i)\rangle = -\mathbf{S}\cdot\mathbf{e}_i/c$, we need the scalar grades of the tensor for vector parameters

$$\begin{aligned}\frac{\epsilon}{2}\langle F\mathbf{e}_i F^\dagger\rangle &= \frac{\epsilon}{2}\langle \cancel{E_i\mathbf{E}} + I\eta\,(\mathbf{H}\wedge\mathbf{e}_i)\,\mathbf{E} + \cancel{(\mathbf{E}\wedge\mathbf{e}_i)\cdot\mathbf{E}} + \cancel{I\eta H_i\mathbf{E}}\rangle \\ &\quad + \frac{\epsilon}{2}\eta\langle -\cancel{E_i I\mathbf{H}} + \cancel{\eta\,(\mathbf{H}\wedge\mathbf{e}_i)\cdot\mathbf{H}} - (\mathbf{E}\wedge\mathbf{e}_i)\,I\mathbf{H} + \cancel{\eta H_i\mathbf{H}}\rangle \\ &= \frac{\epsilon}{2}\langle I^2\eta\,(\mathbf{H}\times\mathbf{e}_i)\,\mathbf{E} - \eta\,(\mathbf{E}\times\mathbf{e}_i)\,I^2\mathbf{H}\rangle \\ &= -(\mathbf{E}\times\mathbf{H})\cdot\mathbf{e}_i/c,\end{aligned} \qquad \text{(D.2)}$$

which is the Poynting relationship that was asserted. For the vector grades we have

$$\begin{aligned}&\langle F\mathbf{e}_i F^\dagger\rangle_1 \\ &= \langle (E_i + I\eta\,(\mathbf{H}\wedge\mathbf{e}_i) + \mathbf{E}\wedge\mathbf{e}_i + I\eta H_i)\,(\mathbf{E} - I\eta\mathbf{H})\rangle_1 \\ &= \langle E_i\mathbf{E} + \cancel{I\eta\,(\mathbf{H}\wedge\mathbf{e}_i)\mathbf{E}} + (\mathbf{E}\wedge\mathbf{e}_i)\cdot\mathbf{E} + \cancel{I\eta H_i\mathbf{E}}\rangle_1 \\ &\quad + \eta\langle -\cancel{E_i I\mathbf{H}} + \eta\,(\mathbf{H}\wedge\mathbf{e}_i)\cdot\mathbf{H} - \cancel{(\mathbf{E}\wedge\mathbf{e}_i)I\mathbf{H}} + \eta H_i\mathbf{H}\rangle_1 \\ &= E_i\mathbf{E} + \mathbf{E}E_i - \mathbf{E}^2\mathbf{e}_i + \eta^2 H\mathbf{H}_i - \eta^2\mathbf{H}^2\mathbf{e}_i + \eta^2 H_i\mathbf{H} \\ &= 2E_i\mathbf{E} - \mathbf{E}^2\mathbf{e}_i + \eta^2\left(2H_i\mathbf{H} - \mathbf{H}^2\mathbf{e}_i\right).\end{aligned} \qquad \text{(D.3)}$$

Assembling all the results, we have

$$\langle T(\mathbf{e}_i)\rangle_1\cdot\mathbf{e}_j = \frac{\epsilon}{2}\left(2E_iE_j - \mathbf{E}^2\delta_{ij} + \eta^2\left(2H_iH_j - \mathbf{H}^2\delta_{ij}\right)\right), \qquad \text{(D.4)}$$

which proves that $\langle T(\mathbf{e}_i)\rangle_1\cdot\mathbf{e}_j = -\Theta^{ij}$ as stated in definition 3.5.

E

DIFFERENTIAL FORMS COMPARISON.

It is likely that a student of electromagnetism will encounter differential forms in their studies. As with geometric algebra, Maxwell's equations also have a compact representation in differential forms. That formalism requires recasting the scalars or vectors of Maxwell's equations as 1-forms (differentials), 2-forms, or 3-forms

$$\begin{aligned}
\mathbf{E} &\to E_x\,dxcdt + E_y\,dycdt + E_z\,dzcdt, \\
\mathbf{B} &\to B_x\,dydz + B_y\,dzdx + B_z\,dxdy, \\
\mathbf{H} &\to -H_x\,dxcdt - H_y\,dycdt - H_z\,dzcdt, \\
\mathbf{J} &\to J_x\,dydzcdt + J_y\,dzdxcdt + J_z\,dxdycdt, \\
\rho &\to -\rho\,dxdydz.
\end{aligned} \qquad (\text{E.1})$$

This appendix is not intended to teach differential forms, nor electrodynamics using differential forms[1]. Instead, this appendix assumes some passing familiarity with differential forms, and provides an example that illustrates how differential forms and geometric calculus can be related.

The key to relating the two formalisms is the introduction of a parameterization. To consider these relations, consider a vector surface those span is controlled by two parameters

$$\mathbf{x} = \mathbf{x}(a, b). \qquad (\text{E.2})$$

In geometric calculus we introduce differentials that span the tangent plane at the point of evaluation

$$\begin{aligned}
dx_a &= \frac{\partial \mathbf{x}}{\partial a}\,da \\
dx_b &= \frac{\partial \mathbf{x}}{\partial b}\,db,
\end{aligned} \qquad (\text{E.3})$$

so the area element for this parameterization is

$$\begin{aligned}
d^2\mathbf{x} &= dx_a \wedge dx_b \\
&= \frac{\partial \mathbf{x}}{\partial a} \wedge \frac{\partial \mathbf{x}}{\partial b}\,dadb.
\end{aligned} \qquad (\text{E.4})$$

---

[1] The interested reader is referred to [9] for an introduction to both differential forms, and an introduction to their application to electrodynamics.

To relate this to differential forms, introduce an orthonormal basis $\mathbf{e}_k \cdot \mathbf{e}_j = 0$, $\mathbf{e}_k^2 = 1$. In this basis, the coordinate expansion (summation implied) of the vector $\mathbf{x}$ is

$$\mathbf{x} = \mathbf{e}_k x_k. \tag{E.5}$$

The coordinate expansion of the geometric area element is

$$\begin{aligned} d^2\mathbf{x} &= \frac{\partial x_k}{\partial a} \frac{\partial x_j}{\partial b} \mathbf{e}_k \wedge \mathbf{e}_j \, dadb \\ &= \sum_{\mu < \nu} \left(\frac{\partial x_k}{\partial a} \frac{\partial x_j}{\partial b} - \frac{\partial x_j}{\partial a} \frac{\partial x_k}{\partial b} \right) \mathbf{e}_k \wedge \mathbf{e}_j \, dadb \\ &= \sum_{\mu < \nu} \mathbf{e}_k \mathbf{e}_j \begin{vmatrix} \frac{\partial x_k}{\partial a} & \frac{\partial x_j}{\partial a} \\ \frac{\partial x_k}{\partial b} & \frac{\partial x_j}{\partial b} \end{vmatrix} dadb \\ &= \sum_{\mu < \nu} \mathbf{e}_k \mathbf{e}_j \frac{\partial(x_k, x_j)}{\partial(a, b)} dadb. \end{aligned} \tag{E.6}$$

Each element of this sum includes a product of a pseudoscalar, a Jacobian determinant, and a scalar two parameter differential.

Now consider a two parameter differential for the same vector. Recall that a differential (1-form) of a scalar function, again assuming two parameters, has the characteristic

$$df = \frac{\partial f}{\partial a} da + \frac{\partial f}{\partial b} db. \tag{E.7}$$

In particular, we may compute the differentials of the coordinate functions

$$\begin{aligned} dx_k &= \frac{\partial x_k}{\partial a} da + \frac{\partial x_k}{\partial b} db \\ dx_j &= \frac{\partial x_j}{\partial a} da + \frac{\partial x_j}{\partial b} db, \end{aligned} \tag{E.8}$$

from which we can compute a 2-form

$$\begin{aligned} dx_k \wedge dx_j &= \left(\frac{\partial x_k}{\partial a} da + \frac{\partial x_k}{\partial b} db \right) \wedge \left(\frac{\partial x_j}{\partial a} da + \frac{\partial x_j}{\partial b} db \right) \\ &= \frac{\partial x_k}{\partial a} \frac{\partial x_j}{\partial b} da \wedge db + \frac{\partial x_k}{\partial b} \frac{\partial x_j}{\partial a} db \wedge da \\ &= \begin{vmatrix} \frac{\partial x_k}{\partial a} & \frac{\partial x_j}{\partial a} \\ \frac{\partial x_k}{\partial b} & \frac{\partial x_j}{\partial b} \end{vmatrix} da \wedge db \\ &= \frac{\partial(x_k, x_j)}{\partial(a, b)} da \wedge db. \end{aligned} \tag{E.9}$$

We have almost the same structure as with geometric algebra, however, in differential forms, the antisymmetry of the surface area element is encoded in the 2-form $da \wedge db$ whereas in geometric calculus the required antisymmetry is encoded in a unit bivector.

Should we restrict our attention to a strictly planar subspace, the mapping between the two formalisms becomes more striking. We now have

$$d^2\mathbf{x} = \mathbf{e}_1\mathbf{e}_2 \frac{\partial(x_1, x_2)}{\partial(a, b)} dadb$$
$$dx_1 \wedge dx_2 = \frac{\partial(x_1, x_2)}{\partial(a, b)} da \wedge db. \tag{E.10}$$

That is, we can relate the formalisms by the mapping

$$\mathbf{e}_1\mathbf{e}_2\, dadb \leftrightarrow da \wedge db. \tag{E.11}$$

The 1-form has an intrinsic vectorial nature, the 2-form has a bivector nature, and a 3-form has a trivector nature.

F

HELPFUL FORMULAS.

F.1 VECTOR RELATIONS.

For vectors $\mathbf{a}, \mathbf{b}, \mathbf{c}, \mathbf{d}$

$$\begin{aligned} \mathbf{ab} &= \mathbf{a} \cdot \mathbf{b} + \mathbf{a} \wedge \mathbf{b} \\ &= \mathbf{a} \cdot \mathbf{b} + I\,(\mathbf{a} \times \mathbf{b}) \qquad (\mathbb{R}^3) \end{aligned} \tag{F.1}$$

$$\begin{aligned} \mathbf{a} \cdot \mathbf{b} &= \frac{1}{2}\,(\mathbf{ab} + \mathbf{ba}) \\ \mathbf{a} \wedge \mathbf{b} &= \frac{1}{2}\,(\mathbf{ab} - \mathbf{ba}) \end{aligned} \tag{F.2}$$

$$\mathbf{ba} = -\mathbf{ab} + 2\mathbf{a} \cdot \mathbf{b} \tag{F.3}$$

$$\mathbf{a} \wedge \mathbf{b} = \sum_{i<j} \begin{vmatrix} a_i & a_j \\ b_i & b_j \end{vmatrix} \mathbf{e}_i \mathbf{e}_j \tag{F.4}$$

$$\begin{aligned} \mathbf{a} \cdot (\mathbf{b} \wedge \mathbf{c}) &= (\mathbf{c} \wedge \mathbf{b}) \cdot \mathbf{a} = (\mathbf{a} \cdot \mathbf{b})\mathbf{c} - (\mathbf{a} \cdot \mathbf{c})\mathbf{b} \\ &= (\mathbf{b} \times \mathbf{c}) \times \mathbf{a} \qquad (\mathbb{R}^3) \end{aligned} \tag{F.5}$$

$$\begin{aligned} (\mathbf{a} \wedge \mathbf{b}) \cdot (\mathbf{c} \wedge \mathbf{d}) &= ((\mathbf{a} \wedge \mathbf{b}) \cdot \mathbf{c}) \cdot \mathbf{d} = (\mathbf{b} \cdot \mathbf{c})(\mathbf{a} \cdot \mathbf{d}) - (\mathbf{a} \cdot \mathbf{c})(\mathbf{b} \cdot \mathbf{d}) \\ &= -(\mathbf{a} \times \mathbf{b}) \cdot (\mathbf{c} \times \mathbf{d}) \qquad (\mathbb{R}^3) \end{aligned} \tag{F.6}$$

$$\begin{aligned} \mathbf{a} &= (\mathbf{a} \cdot \mathbf{b})\frac{1}{\mathbf{b}} + (\mathbf{a} \wedge \mathbf{b})\frac{1}{\mathbf{b}} = \text{Proj}_{\mathbf{b}}(\mathbf{a}) + \text{Rej}_{\mathbf{b}}(\mathbf{a}) \\ &= \left(\mathbf{a} \cdot \hat{\mathbf{b}}\right)\hat{\mathbf{b}} + \left(\mathbf{a} \wedge \hat{\mathbf{b}}\right)\hat{\mathbf{b}} \qquad (\mathbb{R}^N) \\ &= \left(\mathbf{a} \cdot \hat{\mathbf{b}}\right)\hat{\mathbf{b}} + \hat{\mathbf{b}} \times \left(\mathbf{a} \times \hat{\mathbf{b}}\right) \qquad (\mathbb{R}^3) \end{aligned} \tag{F.7}$$

$$\mathbf{a}^{-1} = \frac{\mathbf{a}}{\|\mathbf{a}\|^2} \tag{F.8}$$

$$\begin{aligned}
\mathbf{a} \cdot (\mathbf{b} \wedge \mathbf{c} \wedge \mathbf{d}) &\\
= (\mathbf{b} \wedge \mathbf{c} \wedge \mathbf{d}) \cdot \mathbf{a} &\\
= (\mathbf{a} \cdot \mathbf{b})(\mathbf{c} \wedge \mathbf{d}) - (\mathbf{a} \cdot \mathbf{c})(\mathbf{b} \wedge \mathbf{d}) + (\mathbf{a} \cdot \mathbf{d})(\mathbf{b} \wedge \mathbf{c}) &
\end{aligned} \tag{F.9}$$

$$\begin{aligned}
&\mathbf{a} \cdot (\mathbf{b}_1 \wedge \mathbf{b}_2 \wedge \cdots \wedge \mathbf{b}_n) \\
&= \sum_{i=1}^{n} (-1)^i (\mathbf{a} \cdot \mathbf{b}_i) (\mathbf{b}_1 \wedge \cdots \wedge \mathbf{b}_{i-1} \wedge \mathbf{b}_{i+1} \wedge \cdots \wedge \mathbf{b}_n)
\end{aligned} \tag{F.10}$$

$$\begin{aligned}
\mathbf{a} \wedge \mathbf{b} \wedge \mathbf{c} &= \sum_{i<j<k} \begin{vmatrix} a_i & a_j & a_k \\ b_i & b_j & b_k \\ c_i & c_j & c_k \end{vmatrix} \mathbf{e}_i \mathbf{e}_j \mathbf{e}_k \\
&= (\mathbf{a} \cdot (\mathbf{b} \times \mathbf{c})) I \qquad (\mathbb{R}^3)
\end{aligned} \tag{F.11}$$

F.2 BLADES.

For k-blade $A_k = \mathbf{a}_1 \mathbf{a}_2 \cdots \mathbf{a}_k$, j-blade $B_j = \mathbf{a}_1 \mathbf{a}_2 \cdots \mathbf{a}_j$, vector $\mathbf{a}$

$$A_k^\dagger = (-1)^{k(k-1)/2} A_k \tag{F.12}$$

$$\begin{aligned}
A_k \cdot \mathbf{a} &= (-1)^{k-1} \mathbf{a} \cdot A_k \\
A_k \wedge \mathbf{a} &= (-1)^k \mathbf{a} \wedge A_k
\end{aligned} \tag{F.13}$$

$$\begin{aligned}
\mathbf{a} \wedge A_k &= \frac{1}{2} \left(\mathbf{a} A_k + (-1)^k A_k \mathbf{a} \right) \\
\mathbf{a} \cdot A_k &= \frac{1}{2} \left(\mathbf{a} A_k - (-1)^k A_k \mathbf{a} \right)
\end{aligned} \tag{F.14}$$

$$A_k \cdot (\mathbf{a} \wedge B_j) = (A_k \cdot \mathbf{a}) \cdot B_j \qquad k > j > 0$$

F.3 MULTIVECTORS.

$$A \cdot B \equiv \sum_{i,j=0}^{N} \langle A_i B_j \rangle_{|i-j|}$$
$$A \wedge B \equiv \sum_{i,j=0}^{N} \langle A_i B_j \rangle_{i+j} \quad \text{(F.15)}$$

$$A \rfloor B = \sum_{i,j=0}^{N} \langle A_i B_j \rangle_{j-i}$$
$$A \lfloor B = \sum_{i,j=0}^{N} \langle A_i B_j \rangle_{i-j} \quad \text{(F.16)}$$
$$A * B = \sum_{i,j=0}^{N} \langle A_i B_j \rangle$$

F.4 VECTOR CALCULUS IDENTITIES.

Blade A,

$$\boldsymbol{\nabla} \wedge (\boldsymbol{\nabla} \wedge A) = 0 \quad \text{(F.17)}$$

$\mathbb{R}^3$, scalar f, vector $\mathbf{f}$

$$\boldsymbol{\nabla} \times (\boldsymbol{\nabla} f) = 0$$
$$\boldsymbol{\nabla} \cdot (\boldsymbol{\nabla} \times \mathbf{f}) = 0 \quad \text{(F.18)}$$

$$\begin{aligned}\boldsymbol{\nabla} (\mathbf{a} \cdot \mathbf{b}) &= (\mathbf{a} \cdot \boldsymbol{\nabla}) \mathbf{b} + (\mathbf{b} \cdot \boldsymbol{\nabla}) \mathbf{a} + (\boldsymbol{\nabla} \wedge \mathbf{b}) \cdot \mathbf{a} + (\boldsymbol{\nabla} \wedge \mathbf{a}) \cdot \mathbf{b} \\ &= (\mathbf{a} \cdot \boldsymbol{\nabla}) \mathbf{b} + (\mathbf{b} \cdot \boldsymbol{\nabla}) \mathbf{a} + \mathbf{a} \times (\boldsymbol{\nabla} \times \mathbf{b}) + \mathbf{b} \times (\boldsymbol{\nabla} \times \mathbf{a}) \quad (\mathbb{R}^3)\end{aligned} \quad \text{(F.19)}$$

$$\begin{aligned}\boldsymbol{\nabla} \cdot (\mathbf{a} \wedge \mathbf{b}) &= \mathbf{b} (\boldsymbol{\nabla} \cdot \mathbf{a}) - \mathbf{a} (\boldsymbol{\nabla} \cdot \mathbf{b}) - (\mathbf{b} \cdot \boldsymbol{\nabla}) \mathbf{a} + (\mathbf{a} \cdot \boldsymbol{\nabla}) \mathbf{b} \\ &= -\boldsymbol{\nabla} \times (\mathbf{a} \times \mathbf{b}) \quad (\mathbb{R}^3)\end{aligned} \quad \text{(F.20)}$$

$$\boldsymbol{\nabla} \wedge (f (\boldsymbol{\nabla} g \wedge \boldsymbol{\nabla} h)) = \boldsymbol{\nabla} f \wedge \boldsymbol{\nabla} g \wedge \boldsymbol{\nabla} h \quad \text{(F.21)}$$

For $\mathbb{R}^3$

$$\boldsymbol{\nabla} \cdot (f(\boldsymbol{\nabla} g \times \boldsymbol{\nabla} h)) = \boldsymbol{\nabla} f \cdot (\boldsymbol{\nabla} g \times \boldsymbol{\nabla} h) \tag{F.22}$$

$$\boldsymbol{\nabla} \wedge (\mathbf{a} \wedge \mathbf{b}) = \mathbf{b} \wedge (\boldsymbol{\nabla} \wedge \mathbf{a}) - \mathbf{a} \wedge (\boldsymbol{\nabla} \wedge \mathbf{b}) \tag{F.23}$$

For $\mathbb{R}^3$

$$\boldsymbol{\nabla} \cdot (\mathbf{a} \times \mathbf{b}) = \mathbf{b} \cdot (\boldsymbol{\nabla} \times \mathbf{a}) - \mathbf{a} \cdot (\boldsymbol{\nabla} \times \mathbf{b}) \tag{F.24}$$

INDEX

J_{es}, 273
J_{ms}, 273
$A \cdot B$, 43
$A * B$, 44
$A \lfloor B$, 44
$A \rfloor B$, 44
$A \wedge B$, 50
$A^{\dagger}$, 40
F, 186, 267
F_t, 252
F_z, 252
G, 267
I, 21
J, 187
J_{e}, 267
J_{m}, 267
M^*, 29
Q, 239
$T(\partial_t)$, 224
$T(\nabla)$, 224
$T(a)$, 221
T, 238
T_{ij}, 220
A, 256
B, 183
D, 183
E, 183
$\mathbf{E}_t, E_z$, 253
F, 256
H, 183
$\mathbf{H}_t, H_z$, 253
J, 183
$\mathbf{J}_{\text{s}}$, 273

M, 183
$\mathbf{M}_{\text{s}}$, 273
S, 219
T(a), 221
T(∇), 224
$\mathbf{a} \cdot \mathbf{b}$, 31
$\mathbf{a} \times \mathbf{b}$, 33
$\mathbf{a} \wedge \mathbf{b}$, 33
$\mathbf{e}_1, \mathbf{e}_2, \cdots$, 8
$\mathbf{e}_{ij}, \mathbf{e}_{ijk}, \cdots$, 20
k, 192
p, 185, 238
$\mathbf{p}_{\text{em}}$, 232
$\mathbf{p}_{\text{mech}}$, 232
$\mathbf{x}^{\phi}$
  polar, 112
  spherical, 116
$\mathbf{x}^{\rho}$
  polar, 112
$\mathbf{x}^{\theta}$
  spherical, 116
$\mathbf{x}^i$, 93
$\mathbf{x}^r$
  spherical, 116
$\mathbf{x}^{-1}$, 57
$\mathbf{x}_i$, 106
$\Theta^{\mu\nu}$, 220
$\alpha_{\text{R}}, \alpha_{\text{L}}$, 243, 250
$\mathbb{R}$, 1
$\mathscr{P}$, 219
$\bigwedge^k$, 40
∂, 109
$\mathscr{E}$, 185, 219, 238

□, 166
ϵ, 184
ϵ_0, 184
ϵ_r, 184
η, 186
$\langle M \rangle$, 19
$\langle M \rangle_k$, 19
$\hat{\mathbf{k}}$, 192
$\overleftarrow{\partial}$, 127
$\overleftrightarrow{\mathbf{T}}$, 220
$\overleftrightarrow{\partial}$, 127
$\overleftrightarrow{\nabla}'$, 162
$\overleftarrow{\nabla}'$, 161
μ, 184
μ_0, 184
μ_r, 184
ω, 192
$\overline{A}(x)$, 227
$\overline{T}(x)$, 228
∂_i, 109
ϕ, 243, 256
ϕ_m, 256
$\hat{\phi}$
    spherical, 116
$\propto$, 214
$\overrightarrow{\partial}$, 127
$\hat{\mathbf{r}}$
    spherical, 116
ρ, 183
ρ_m, 183
ρ_s, 273
ρ_{ms}, 273
$\overrightarrow{\nabla}'$, 161
∇', 161
∇, 109
∇_t, 253
$\hat{\theta}$
    spherical, 116
c, 186
c_1, c_2, 243, 250
$d^k\mathbf{x}$, 106
i, 21
q_e, 239
q_m, 239
$\delta_i{}^j$, 93
$\mathbb{R}^N$, 4
0-vector, 10
1-vector, 10
2-vector, 11
3-vector, 13

anticommutation, 17
antisymmetric sum, 65
area element, 124

basis, 6
bivector, 11
blade, 40
boundary values, 273

circular polarization, 245
colinear vectors
    wedge, 39
commutation, 17, 62
complex exponential, 24
complex imaginary, 21, 41
complex plane, 205
complex power, 236
conjugation, 62
convolution, 169
Cramer's rule, 71
cross product, 37
curvilinear bases, 102, 106, 110, 114

delta function, 169
determinant

wedge product, 36
differential form, 119, 121, 124, 139
dimension, 6
divergence theorem, 145
dot product, 7, 33, 49
dual, 29

electric charge density, 183
electric current density, 183
energy density, 219
energy flux, 219
energy momentum tensor, 220
Euler's formula, 24

far field, 262
Fourier transform, 154
frequency domain, 154
fundamental theorem of geometric calculus, 128

gauge transformation, 259
grade, 14, 40
grade selection, 19, 31
gradient
    spherical, 116
Gradient of the coordinates., 112
Green's function, 155, 159
    Helmholtz, 162
    Laplacian, 163

Helmholtz
    Green's function, 162
Helmholtz operator, 159
Helmholtz's theorem, 168

Jones vector, 243, 250

k-vector, 14

Laplacian, 170

Green's function, 163
left circular polarization, 245
length, 7
line charge, 202, 208
linear combination, 5
linear dependence, 5
linear independence, 6
linear system, 69
Lorenz gauge, 260

magnetic charge density, 183
magnetic current density, 183
Maxwell stress tensor, 220
Maxwell's equation, 186
momentum density, 219
multivector, 14
multivector dot product, 43
multivector space, 15
multivector wedge product, 50

oriented volume element, 106
orthogonal, 7
orthonormal, 8

parallelogram, 60
plane wave, 242
polar coordinates, 110
polar representation, 35
polarization, 242
Poynting vector, 219
projection, 53
pseudoscalar, 20, 22, 23, 41

quaternion, 42

reciprocal basis
    polar, 112
reciprocal frame, 93, 101
reflection, 67
rejection, 53

reverse, 40
right circular polarization, 245
rotation, 23, 64

scalar, 10
scalar selection, 49
span, 6
spherical coordinates, 114, 121, 208
standard basis, 8
subspace, 6
symmetric sum, 65

tangent space, 106
time harmonic, 154
toroid, 119
trivector, 13

unit pseudoscalar, 20
unit vector, 7

vector, 10
vector derivative, 109
vector inverse, 57
vector product, 31
vector space, 4
volume element, 106, 139
volume parameterization, 139

wedge factorization, 58
wedge product, 33
    linear solution, 69

BIBLIOGRAPHY

[1] Rafal Ablamowicz and Garret Sobczyk. *Lectures on Clifford (geometric) algebras and applications*, chapter Introduction to Clifford Algebras. Springer, 2004. (Cited on page 58.)

[2] Jose L. Aragón. *Geometric Algebra*, 2020. URL https://github.com/jlaragonvera/Geometric-Algebra. [Online; accessed 07-Sep-2020]. (Cited on page 72.)

[3] Constantine A Balanis. *Advanced engineering electromagnetics*. Wiley New York, 1989. (Cited on page 246.)

[4] Constantine A Balanis. *Antenna theory: analysis and design*. John Wiley & Sons, 3rd edition, 2005. (Cited on page 257.)

[5] James M Chappell, Samuel P Drake, Cameron L Seidel, Lachlan J Gunn, and Derek Abbott. Geometric algebra for electrical and electronic engineers. *Proceedings of the IEEE*, 102(9), 2014. (Cited on page 275.)

[6] C. Doran and A.N. Lasenby. *Geometric algebra for physicists*. Cambridge University Press New York, Cambridge, UK, 1st edition, 2003. (Cited on pages 159, 221, 242, 292, and 295.)

[7] L. Dorst, D. Fontijne, and S. Mann. *Geometric Algebra for Computer Science*. Morgan Kaufmann, San Francisco, 2007. (Cited on page 44.)

[8] R.P. Feynman, R.B. Leighton, and M.L. Sands. *Feynman lectures on physics, Volume II.[Lectures on physics]*, chapter The Maxwell Equations. Addison-Wesley Publishing Company. Reading, Massachusetts, 1963. URL https://www.feynmanlectures.caltech.edu/II_18.html. (Cited on page 256.)

[9] H. Flanders. *Differential Forms With Applications to the Physical Sciences*. Courier Dover Publications, 1989. (Cited on page 305.)

[10] David Jeffrey Griffiths and Reed College. *Introduction to electrodynamics*. Prentice hall Upper Saddle River, NJ, 3rd edition, 1999. (Cited on pages 218, 221, 224, and 256.)

[11] D. Hestenes. *New Foundations for Classical Mechanics*. Kluwer Academic Publishers, 1999. (Cited on pages 242 and 247.)

[12] David Hestenes. Proper dynamics of a rigid point particle. *Journal of Mathematical Physics*, 15(10):1778–1786, 1974. (Cited on page 242.)

[13] David Hestenes. Proper particle mechanics. *Journal of Mathematical Physics*, 15(10):1768–1777, 1974. (Cited on page 242.)

[14] David Hestenes, Garret Sobczyk, and James S Marsh. *Clifford Algebra to Geometric Calculus. A Unified Language for Mathematics and Physics*. AAPT, 1985. (Cited on page 295.)

[15] JD Jackson. *Classical Electrodynamics*. John Wiley and Sons, 2nd edition, 1975. (Cited on pages 47, 167, 221, 224, and 256.)

[16] Bernard Jancewicz. *Multivectors and Clifford algebra in electrodynamics*, chapter Appendix I. World Scientific, 1988. (Cited on page 197.)

[17] L.D. Landau and E.M. Lifshitz. *The classical theory of fields*. Butterworth-Heinemann, 1980. ISBN 0750627689. (Cited on pages 186 and 221.)

[18] A. Macdonald. *Vector and Geometric Calculus*. CreateSpace Independent Publishing Platform, 2012. (Cited on pages 110 and 295.)

[19] David M Pozar. *Microwave engineering*. John Wiley & Sons, 2009. (Cited on page 257.)

[20] M. Schwartz. *Principles of Electrodynamics*. Dover Publications, 1987. (Cited on page 297.)

[21] J Schwinger, LL DeRaad Jr, KA Milton, and W-Y Tsai. Classical electrodynamics, perseus. *Reading*, page 355, 1998. (Cited on pages 162 and 167.)

[22] Garret Sobczyk and Omar León Sánchez. Fundamental theorem of calculus. *Advances in Applied Clifford Algebras*, 21(1):221–231, 2011. URL https://arxiv.org/abs/0809.4526. (Cited on page 295.)

www.ingramcontent.com/pod-product-compliance
Lightning Source LLC
Chambersburg PA
CBHW052239220526
45471CB00001B/112